Springer Complexity

Springer Complexity is an interdisciplinary program publishing the best research and academic-level teaching on both fundamental and applied aspects of complex systems – cutting across all traditional disciplines of the natural and life sciences, engineering, economics, medicine, neuroscience, social and computer science.

Complex Systems are systems that comprise many interacting parts with the ability to generate a new quality of macroscopic collective behavior the manifestations of which are the spontaneous formation of distinctive temporal, spatial or functional structures. Models of such systems can be successfully mapped onto quite diverse "real-life" situations like the climate, the coherent emission of light from lasers, chemical reaction-diffusion systems, biological cellular networks, the dynamics of stock markets and of the internet, earthquake statistics and prediction, freeway traffic, the human brain, or the formation of opinions in social systems, to name just some of the popular applications.

Although their scope and methodologies overlap somewhat, one can distinguish the following main concepts and tools: self-organization, nonlinear dynamics, synergetics, turbulence, dynamical systems, catastrophes, instabilities, stochastic processes, chaos, graphs and networks, cellular automata, adaptive systems, genetic algorithms and computational intelligence.

The two major book publication platforms of the Springer Complexity program are the monograph series "Understanding Complex Systems" focusing on the various applications of complexity, and the "Springer Series in Synergetics", which is devoted to the quantitative theoretical and methodological foundations. In addition to the books in these two core series, the program also incorporates individual titles ranging from textbooks to major reference works.

Springer Series in Synergetics

Founding Editor: H. Haken

The Springer Series in Synergetics was founded by Herman Haken in 1977. Since then, the series has evolved into a substantial reference library for the quantitative, theoretical and methodological foundations of the science of complex systems.

Through many enduring classic texts, such as Haken's *Synergetics and Information and Self-Organization*, Gardiner's *Handbook of Stochastic Methods*, Risken's *The Fokker Planck-Equation* or Haake's *Quantum Signatures of Chaos*, the series has made, and continues to make, important contributions to shaping the foundations of the field.

The series publishes monographs and graduate-level textbooks of broad and general interest, with a pronounced emphasis on the physico-mathematical approach.

For further volumes:
http://www.springer.com/series/712

José María Amigó

Permutation Complexity in Dynamical Systems

Ordinal Patterns, Permutation Entropy and All That

 Springer

José María Amigó
Universidad Miguel Hernandez
Centro de Investigacion Operativa
Avda. de la Universidad, s/n
03202 Elche
Spain
jm.amigo@umh.es

ISSN 0172-7389
ISBN 978-3-642-04083-2   e-ISBN 978-3-642-04084-9
DOI 10.1007/978-3-642-04084-9
Springer Heidelberg Dordrecht London New York

Library of Congress Control Number: 2010920733

Cover design: Integra Software Services Pvt. Ltd., Pondicherry

Printed on acid-free paper

Springer is part of Springer Science+Business Media (www.springer.com)

To my parents

Preface

This is a research book on ordinal patterns, permutation entropy, and complexity, written at graduate level. The common denominator of the different topics presented in its pages is a hypothetical order structure of the state space, substantiated in form of ordinal patterns—permutations defined by the order relations among points in the orbits of dynamical systems. Here the state space is meant to be arbitrary (including discrete sets and n-dimensional intervals), as long as it is totally ordered, and the dynamical systems are meant to include stochastic processes (sometimes called random dynamical systems). Out of the order structure of the state space, a number of constructs will emerge to pave our way as we progress: admissible and forbidden patterns, order isomorphy, metric and topological permutation entropy, discrete entropy, regularity parameters, etc. The relation of these concepts to similar concepts in applied mathematics and computer science will be addressed as well, especially in the introductory part. The final result is a new approach to dynamical complexity characterized by conceptual simplicity, an algebraic flavor, and computational speed. The term *permutation complexity* in the title of this book intends to direct attention to this circle of ideas.

Complexity is a general concept that has different meanings in different contexts. For instance, complexity is related to "incompressibility" in information theory and computer science. In dynamical systems, complexity is usually measured by the topological entropy and reflects roughly speaking, the proliferation of periodic orbits with ever longer periods or the number of orbits that can be distinguished with increasing precision. In physics, the label "complex" is in principle attached to any nonlinear system whose numerical solutions exhibit a chaotic behavior. Neurologists claim that the human brain is the most complex system in the solar system, while entomologists teach us the baffling complexity of some insect societies. The list could be enlarged with examples from geometry, management science, communication and social networks, etc. In this book we will be mainly concerned with complexity from the viewpoint of discrete-time dynamical systems. In particular, permutation complexity refers to the dynamical features captured and quantified by tools based on order relations.

Permutation entropy was introduced in 2002 by C. Bandt and B. Pompe as a measure of complexity in time series. In a nutshell, permutation entropy replaces the probabilities of length-L symbol blocks in the definition of the Shannon entropy by the probabilities of length-L ordinal patterns. Since then this proposal has sparked new lines of research that capitalize on the order structure of the state space. Order is as well at the base of some classical results of combinatorial dynamics (notably, Sarkovskii's theorem), but the focus in these investigations is the periodicity structure of the map. Ordinal patterns provide an akin though different picture: akin because periodic points and ordinal patterns are closely related; different because ordinal patterns are amenable to numerical methods, while periodicity is not. A complete analysis of the relation between ordinal patterns and periodic points is still lacking.

As conventional entropy, permutation entropy comes in metric and topological versions, and these are limits of the corresponding rates of finite order. The metric and topological permutation entropies can be shown to coincide with their conventional counterparts under several assumptions. In applications, permutation entropy rates of finite order may be used to measure the complexity of a finite data sequence. Periodic or quasiperiodic sequences have vanishing or negligible complexity. At the opposite end, independent and identically distributed random sequences (white noise) have asymptotically divergent permutation entropies, owing to the fact that the number of allowed (or "admissible") ordinal patterns grows superexponentially with length. Between both ends lie the kind of sequences we are interested in; their permutation entropy rates of finite order can be calibrated by comparison with the corresponding rates of the white noise.

The study of permutation complexity, which we call *ordinal analysis*, can be envisioned as a new kind of symbolic dynamics whose basic blocks are ordinal patterns. Interesting enough, it turns out that under some mild mathematical assumptions, not all ordinal patterns can be materialized by the orbits of a given one- or multi-dimensional deterministic dynamics, not even if this dynamic is chaotic—contrarily to what happens with the symbol patterns. As a result, the existence of "forbidden" (i.e., not occurring) ordinal patterns is always a persistent dynamical feature, in opposition to properties such as proximity and correlation which die out with time in a chaotic dynamic. Moreover, if an ordinal pattern is forbidden, its absence pervades all longer patterns in form of more missing ordinal patterns, called outgrowth forbidden patterns. Admissible ordinal patterns grow exponentially with length, while forbidden patterns do superexponentially. Since random (unconstrained) dynamics has no forbidden patterns with probability 1, their existence can be used as a fingerprint of deterministic orbit generation.

This book is addressed to both researchers on dynamical systems and complexity and graduate students interested in these subjects. Some topics are already well established; others are asking for generalizations or more comprehensive analyses; still others, like the applications to space–time dynamics, are newcomers. The book consists of ten chapters, plus two technical annexes where the reader can find the mathematical background needed in the main text; overlaps between the main text and the annexes were unavoidable, but they have been kept at a minimum. The topics selected correspond to materials published by the author and collaborators in recent years, although they have been thoroughly revised and eventually reformulated for this occasion. The presentation is a compromise between mathematical rigor and getting the message across in a smooth way. Formal statements of results and their proofs allow knowing exactly which are the assumptions behind them, facilitating at the same time to refer to them from any place in the text. Examples illustrate the theory wherever convenient. Both the main text and the annexes contain also a sufficient number of exercises that invite the reader to explore beyond our exposition. Next we describe briefly the content of the different chapters.

Chapter 1 is an introduction to the main topics of this book, namely, patterns, complexity, and entropy. We show how these concepts are linked—sometimes in unexpected ways—in five different settings: information theory, symbolic dynamics, dynamical systems, computer science, and cellular automata. Ordinal patterns and permutation entropy make their first appearance in the second section, together with the forbidden patterns, one of the main characters of permutation complexity.

Once the stage has been set, Chap. 2 is a brief account on a few applications of ordinal analysis. We review four of them, to wit: entropy estimation, permutation complexity of time series, recovery of control parameters of unimodal maps from symbolic sequences, and characterization of the different kinds of synchronization between chaotic oscillators. This chapter should convey to the reader a first impression of the disparate possibilities of ordinal analysis, before going into technical details in Chaps. 3 through 7.

Chapter 3 is wholly devoted to the study of ordinal patterns and their main properties. Two of them are specially important in applications: existence of forbidden patterns in the orbits of dynamical systems (herein referred only to one-dimensional dynamics) and robustness of admissible and forbidden patterns against observational noise. Forbidden patterns are further classified into two groups: outgrowth and root forbidden patterns. The study of robustness is continued in Chap. 9.

In the relation between maps and the structure of their admissible and forbidden patterns there are far more questions than answers. It is therefore gratifying that this relation can be analyzed with great detail in the case of the shift and signed shift transformations. Due to its length, this topic has been divided into two parts: Chap. 4 and Chap. 5. Signed shifts include the standard ones but their handling is more difficult, and the results gotten till now are not so sharp. By order isomorphy, the results of these two chapters apply to perhaps more interesting cases, like the logistic map, baker map, sawtooth maps.

The next two chapters comprise an in-depth analysis of metric and topological permutation entropies. On defining the metric permutation entropy of maps in Chap. 6, we depart from the original approach to follow basically Kolmogorov's path, based on finite partitions. The pay-off is that the results are not limited to one-dimensional maps. For this reason we have to make a detour over symbolic dynamics (or, equivalently, finite-alphabet information sources), before getting ready to deal with maps. The main outcome is that the metric permutation entropy of ergodic maps coincides with the metric entropy (otherwise called measure-theoretical or Kolmogorov–Sinai entropy) of the map.

The same applies to the topological permutation entropy (Chap. 7), where now expansiveness is called in. An important consequence is the existence of forbidden patterns also in higher dimensional dynamics. Furthermore, numerical simulations provide ample evidence that forbidden patterns is a general feature of deterministic orbit generation.

Discrete entropy (Chap. 8) was proposed (together with the discrete Lyapunov exponent) as a tool of discrete chaos, a generalization of chaos to dynamical systems with discrete state spaces. Our approach follows the work of Bandt and Pompe on permutation entropy of time series. It is proved that discrete entropy converges to its "continuous" counterpart in an adequate sense.

Having shown in Chap. 7 that the existence of forbidden patterns is a landmark of determinism, Chap. 9 grapples with the implementation of this fact, the main obstacle being that real data are finite and noisy. The properties of ordinal patterns studied in Chap. 3 come here to the rescue, as well as the "dynamical robustness" discussed in the first section. Two methods are proposed, based on (i) the number of missing ordinal patterns and (ii) the distribution of visible ordinal patterns. The second resorts to a chi-square test, the null hypothesis being that the time series is white noise; its performance compares favorably to some widely used tests of statistical independence.

Cellular automata and coupled map lattices are, so to speak, toy models for real physics. And yet, what these dynamical systems lack in sophistication as compared to the usual space–time systems, they more than make up for in conceptual simplicity and modelization power. On applying some tools of ordinal analysis to cellular automata and coupled map lattices, as done in Chap. 10, we put to test the capabilities of this approach to discern different temporal structures in spatially extended systems. The task is formidable: trying to reduce the behavior of a space–time system to just a parameter seems to be more than what one could reasonably ask for. Nevertheless, the results reported in Chap. 10 are encouraging.

The book concludes with Chap. 11, where we remind the main messages of ordinal analysis and permutation complexity, gather some open problems scattered in the preceding chapters, and suggest future lines of research.

Much labor will be necessary to survey the full potential of ordinal analysis and the intricacies of permutation complexity at theoretical and practical levels. This book should be considered as a contribution to this task. One of the main challenges of complexity theory is to design conceptual and numerical tools to study, classify, and quantify the different degrees of complexity found in our mathematical models of the world around. Think, for example, of turbulence in fluid mechanics or the asymptotic behavior of cellular automata and coupled map lattices. Nonlinear physics has developed a battery of instruments that go by the name of power spectra, Lyapunov exponents, fractal dimensions of attractors, order parameters, etc. On the mathematical side, ergodic theory and topological dynamics study general properties of systems evolving in time. These disciplines have provided plenty of handles to understand complex dynamics, like deep concepts, invariants for classification purposes (most notably, the entropy), prototypes, and powerful theoretical and practical techniques. But order relations have been less exploited. One possible reason is that order relations are not invariant under metric and topological isomorphisms, which consistently only address measure-theoretical and topological properties. We hope that this book on permutation complexity convincingly shows that properties related to the temporal (and eventually also spatial) structure of a dynamics are useful and worth researching.

It is a great pleasure to thank all friends and colleagues who have collaborated with me on the topics of this book: Gonzalo Álvarez, David Arroyo, Rui Dilão, Sergi Elizalde, Matthew(Matt) B. Kennel, Ljupco Kocarev, Roberto Monetti, Ulrich Parlitz, Miguel A.F. Sanjuán, Janusz Szczepanski, Igor Tomovski, Elek Wajnryb, and Samuel Zambrano—without them this book had not been possible. In particular, Matt made the numerical simulations of Chaps. 6 and 7, Igor of Chap. 8, and Samuel of Chaps. 9 and 10; moreover Matt's ingenuity was decisive for the theoretical results of Chapter 6. For further assistance I am also indebted to Óscar Martínez Bonastre and Agustín Pérez Martín. Special thanks are due to Manfred Denker and Wolfgang Krieger for clarifying discussions on the generator problem. Most of the scientific articles this book is based on were written under the auspices of the Spanish Ministry for Education and Science (Project MTM2005-04948); this financial support is gratefully acknowledged. Furthermore, I want to express my gratitude to Ljupco Kocarev and Jürgen Kurths, Editorial and Programme Adviser of the Springer Series in Complexity, for encouraging me to write this book, as well as to Dr. Christian Caron, Executive Publishing Editor of Springer Verlag, for guiding me through the publication stages. Last but not least, I wish to highlight the enduring and stimulating collaboration of Samuel Zambrano; he has been much of a driving force in exploring new ideas, working out the applications and getting insights from the results.

Elche, Spain   José María Amigó

Contents

Chapter 1
What Is This All About?

This introductory chapter is meant as a tour of the main topics in this book: patterns, ordinal relations, complexity, and entropy. The approach is mostly informal; for the technicalities behind the different notions met on the way, the reader is referred to Annex A and Annex B.

1.1 Patterns, Complexity, and Entropy

Pattern is an abstract concept with different acceptations. In the context of dynamical systems, information theory, and computer science (the ones we are interested in), a pattern is a finite string of symbols, eventually chosen with some criterion. In the next sections we will meet some familiar instances of patterns in those contexts. Contrary to the concept of pattern, complexity does not lend itself to a short definition (would this be not a contradiction otherwise?) but, like poetry, it is very easy to recognize. For a panorama of complexity, see [77] or, at an introductory level, [158]. A third and also recurrent issue in the next pages will be entropy, one of the most important quantities when dealing with complexity in deterministic and random dynamical systems. Indeed, no matter how one counts the diversity of patterns generated by a data source, entropy enters the scene in some of its many disguises: Shannon entropy, metric entropy, topological entropy, etc.

1.1.1 Information Theory

Consider an information source outputting symbols or letters, one at a time, from a finite alphabet $S = \{s_1, \ldots, s_{|S|}\}$ (i.e., $|S|$ is the cardinality of S). Formally, an information source is a discrete-time, stationary stochastic process $\mathbf{X} = \{X_n\}_{n \in \mathbb{N}_0}$, where $\mathbb{N}_0 = \{0, 1, \ldots\}$ and X_n are random variables on a common probability space, taking on values in S. For the time being, we will dispense with the underlying probability space. A realization of $\mathbf{X}$ is a one-sided sequence, $x_0^\infty := (x_n)_{n \in \mathbb{N}_0}$, called[1] a

---

[1] The symbol ":=" means that the left side is defined by the right one; a corresponding meaning holds for "=:".

J.M. Amigó, *Permutation Complexity in Dynamical Systems*,
Springer Series in Synergetics, DOI 10.1007/978-3-642-04084-9_1,
© Springer-Verlag Berlin Heidelberg 2010

message. Correspondingly, the symbols $x_n \in S$ are sometimes called *letters*. A finite segment of a message, say, $x_k^{k+L-1} := x_k x_{k+1} \ldots x_{k+L-1}$ is called a *word* of length L. If $p(x_0^{L-1})$ denotes the probability of the word x_0^{L-1} to be output, then the (Shannon) entropy rate (or just *entropy*) of the data source $\mathbf{X}$ is defined as

$$h(\mathbf{X}) = -\lim_{L \to \infty} \frac{1}{L} \sum p(x_0^{L-1}) \log p(x_0^{L-1}), \tag{1.1}$$

where log usually stands for logarithm to base 2 ($h(\mathbf{X})$ is then measured in bits per symbol), and the sum is over all possible words of length L, numbering $|S|^L$, with the convention $0 \times \log 0 = \lim_{x \to 0+} x \log x = 0$. To indicate that a logarithm is to base e, we will write ln instead of log ($h(\mathbf{X})$ is then measured in nats per symbol). The convergence of limit (1.1) is proven in Sect. B.1.2.

In an information-theoretical setting, $\log p(x_0^{L-1})$ is the information conveyed by the output x_0^{L-1}, hence $h(\mathbf{X})$ is the average information per symbol conveyed by the messages of the information source $\mathbf{X}$ in the limit of arbitrarily long messages.

When the random variables X_n are independent, or (more often) intersymbol dependency is neglected for simplicity or limited influence, the information source is called *memoryless*. In this case $h(\mathbf{X})$ coincides with the entropy $H(X)$ of a random variable X with outcomes $x \in S$ and probabilities $p(x)$:

$$H(X) = -\sum_{x \in S} p(x) \log p(x).$$

Compression is any procedure that reduces the data requirements of a message without, in principle, losing information—although it can be acceptable as a trade-off between data reduction and information degradation. The idea of using codes or dictionaries for compression of information originates with the invention of the telegraph, since users were charged by the number of letters in the message. It is clear that data compression can be achieved by assigning short words to the most frequent outcomes of the information source. For example, in the Morse code, the most frequent symbol in English, namely the letter e, is represented by a single dot. This intuition is the guiding principle in the construction of the celebrated Huffman code for memoryless sources. Suppose that code words $w_1, \ldots, w_{|S|}$ of lengths $l_1, \ldots, l_{|S|}$, respectively, are assigned to the values $s_1, \ldots, s_{|S|}$ taken on by a random variable X with probabilities $p(s_1), \ldots, p(s_{|S|})$. The code words are combinations of characters taken from an alphabet $a_1, \ldots, a_D$, usually 0, 1 ($D = 2$) in modern communications. Then the Huffman code is a uniquely decipherable code that minimizes the average code-word length $\bar{l} = \sum_{n=1}^{|S|} p(s_n) l_n$, which according to the *noiseless coding theorem* is known to satisfy [22]

$$H(X) \le \bar{l} < H(X) + 1, \tag{1.2}$$

where the logarithm of $H(X)$ is taken to base D. But how to compress a message, say a digital picture to be sent by electronic mail or a text file written in a foreign language, if the probabilities of the corresponding symbols are not known? This feat requires a universal compressor.

Universal compressors are based on the fact that natural languages are not completely random but repeat patterns from time to time. In 1976 and 1978, A. Lempel and J. Ziv published two simple algorithms for universal data compression [137, 211], which work by parsing an input string of finite length into successive phrases. Some variants of the second (LZ78) are implemented in the most popular compressors currently used in electronic editing (like WinZip or pdf). For our purposes it is sufficient to consider the first scheme (LZ76); also, we will emphasize the interplay between complexity and entropy rather than the compression-related aspects.

In the LZ76, the message is sequentially parsed into strings that have not appeared so far in the initial segment ending at (and excluding) the current letter. For example, the binary word $x_0^{19} = 0101101000110111100010$ is parsed as

$$0, 1, 011, 0100, 011011, 1001, 0. \tag{1.3}$$

If, say, x_k is the first bit after a comma, then we check whether x_k appears in x_0^{k-1}. If it does not, then we write a comma after x_k and start a new block (this is the case for $k = 1$ in (1.3)). Otherwise, we check whether $x_k x_{k+1}$ appears in x_0^k; in negative case, we write a comma after x_{k+1}, otherwise the process continues till a pattern $x_k x_{k+1} \ldots x_{k+l}$ repeats (or the sequence finishes). The number of patterns found in the parsing of a word x_0^{L-1} is called its Lempel–Ziv (LZ) complexity, $C(x_0^{L-1})$. In example (1.3), $C(x_0^{19}) = 7$. Words x_0^{L-1} with a general alphabet S are parsed in an analogous way.

The formal definition of $C(x_0^{L-1})$ is recursive. A *block* of length l ($1 \leq l \leq L$) is just a segment of x_0^{L-1} of length l, i.e., a string of l consecutive letters, say $x_k^{k+l-1} = x_k x_{k+1} \ldots x_{k+l-1}$ ($0 \leq k \leq L - l$). In particular, letters are blocks of length 1. Set $B_0 = x_0$ and suppose that after $k \geq 1$ steps, we have parsed x_0^{L-1} as

$$B_0, B_1, \ldots, B_{k-1},$$

where $B_1 = x_1^{n_1}, \ldots, B_{k-1} = x_{n_{k-2}+1}^{n_{k-1}}$, and $n_{i-1} + 1 \leq n_i < L - 1$ for $i = 1, \ldots, k - 1$ (with $n_0 = 0$). Define

$$B_k := x_{n_{k-1}+1}^{n_k} \quad (n_{k-1} + 1 \leq n_k \leq L - 1),$$

to be the shortest block such that it does not occur in the sequence $x_0^{n_k - 1}$. (In the LZ78 algorithm, one checks instead whether the current block $x_{n_{k-1}+1}^{n_k}$ coincides with one of the previous blocks, $B_0, B_1, \ldots, B_{k-1}$.) Proceeding in this way, we obtain a (uniquely defined) decomposition of x_0^{L-1} in "minimal" blocks, say

$$x_0^{L-1} = B_0, B_1, \ldots, B_{p-1}, \tag{1.4}$$

in which only the last block can occasionally appear twice. Then,

$$C(x_0^{L-1}) := p.$$

For computational efficiency, one uses the well-known "suffix-tree" data structure and search algorithms for quickly finding substrings of the input string.

From the foregoing description, we may say that $C(x_0^{L-1})$ measures the complexity of the word x_0^{L-1}; words with a periodic or almost periodic structure have a small LZ complexity, while those displaying a random-looking structure have a high count of distinct patterns, hence a great LZ complexity. It can be proven [211] that if the source $\mathbf{X}$ is ergodic (i.e., the probability of any length-L word equals its frequency in a single, "typical" sequence), then

$$\limsup_{L \to \infty} \frac{C(x_0^{L-1})}{L/\log_{|S|} L} = h(\mathbf{X}) \tag{1.5}$$

with probability 1. The normalization factor in (1.5) is the LZ complexity of a memoryless, equidistributed source. Let us mention in passing that (1.5) shows that the ideal compression factor of the LZ76 algorithm, in the limit of long messages, is $h(\mathbf{X})$. The same is true for the LZ78 scheme.

Equations (1.2) and (1.5) provide examples in which the concepts of complexity (here related to "incompressibility") and entropy (here related to "uncertainty") are linked in a perhaps unexpected way. As a by-product, LZ complexity can be used as an estimator of the entropy. A principal advantage of this approach is that the LZ algorithm is entirely automatic with no free parameters (unlike naive plug-in methods or methods which estimate $h(\mathbf{X})$ via block entropies; see [167] and Sect. 2.1). Another practical issue is the convergence speed with L: the normalized LZ76 complexity converges to the entropy faster than the LZ78, what makes it a better choice in practice [6]. A variance estimator for the entropy estimation by means of the LZ76 complexity can be found in [9].

1.1.2 Symbolic Dynamics

Symbolic dynamics, first proposed by Morse and Hedlund [160], is an approach to complex dynamics that aims to capture the essential aspects of complexity by studying conceptually simple models. As it often happens in mathematics, symbolic dynamics has developed in short time from an auxiliary tool to an independent field [139, 123], with applications to the study of formal languages. As a result, dynamical systems connect through symbolic dynamics to computer science, information theory, and automata.

To motivate symbolic dynamics, consider the dynamics generated by a self-map f of a set Ω. Of course, the dynamics is introduced in the *state space* Ω via the repeated action of f on Ω. Given $x \in \Omega$, the *orbit* or *trajectory* of x under f is defined as $\mathcal{O}_f(x) = \{f^n(x) : n \in \mathbb{N}_0\}$, where $f^0(x) := x$ and $f^n(x) := f(f^{n-1}(x))$. If f is invertible, then one can distinguish between the *full orbit* $\mathcal{O}_f(x) = \{f^n(x) : n \in \mathbb{Z}\}$ and the *forward orbit* $\mathcal{O}_f^+(x) = \{f^n(x) : n \in \mathbb{N}_0\}$. The name "orbit" clearly hints to the interpretation of the iteration index n as discrete time: each application of f on the point $x_n = f(x_{n-1})$ updates the "movement" of the *initial condition* x in Ω. If the resulting dynamics is complicated, we might content ourselves with a "blurred" picture of the orbit behavior. This can be done as follows. Divide Ω into a finite number of disjoint pieces A_i, $i = 0, 1, \ldots, k - 1$, and keep track of the trajectory of $x \in \Omega$ with the precision set by the decomposition $\alpha = \{A_0, \ldots, A_{k-1}\}$. (We reserve the name partition for a measurable decomposition, provided Ω is endowed with a sigma algebra; see below.) Specifically, we assign to x a (one-sided) sequence[2] $\Phi(x) = (\xi_0, \xi_1, \ldots, \xi_n, \ldots)$, the nth entry $\xi_n \in \{0, 1, \ldots, k - 1\}$ telling us in which element of α the iterate $f^n(x)$ is to be found. When f is invertible, we can also assign a two-sided sequence $\Phi(x) = (\ldots, \xi_{-1}, \xi_0, \xi_1, \ldots, \xi_n, \ldots)$, the entries with negative indices corresponding to the locations of $f^{-n}(x)$, $n \geq 1$. For brevity we focus on the general case. We call Φ a *coding map*, and $\Phi(x)$ the *itinerary* of x with respect to the decomposition α. Formally,

$$\Phi_n(x) = i \text{ iff } f^n(x) \in A_i, \tag{1.6}$$

where $n \in \mathbb{N}_0$ and $\Phi_n(x)$ denotes the nth component of the sequence $\Phi(x)$.

Let us reformulate this simple idea in a more general way. Given the finite alphabet $S = \{0, 1, \ldots, k-1\}$, denote by $S^{\mathbb{N}_0}$ the space of one-sided sequences of symbols from S:

$$S^{\mathbb{N}_0} = \{(\xi_n)_{n \in \mathbb{N}_0} = (\xi_0, \xi_1, \ldots, \xi_n, \ldots): \xi_n \in S\}.$$

Hence, $\Phi(x) \in S^{\mathbb{N}_0}$. The space $S^{\mathbb{N}_0}$ (and also $S^{\mathbb{Z}}$) is generically referred to as a *sequence* or *symbolic space*. One can put on a sequence space different (non-equivalent) metrics d making it a compact space. For example,

$$d((\xi_n)_{n \in \mathbb{N}_0}, (\eta_n)_{n \in \mathbb{N}_0}) = \begin{cases} 0 & \text{if } \xi_n = \eta_n \text{ for all } n \in \mathbb{N}_0, \\ 2^{-N} & \text{if } \xi_n = \eta_n \text{ for } n < N \text{ and } \xi_N \neq \eta_N. \end{cases} \tag{1.7}$$

Thus, two one-sided sequences are apart 2^{-N} in this metric if their first N entries coincide (and the $(N + 1)$th ones do not). In $S^{\mathbb{Z}}$, two sequences $(\xi_n)_{n \in \mathbb{Z}}$ and $(\eta_n)_{n \in \mathbb{Z}}$ are at distance 2^{-N} if their entries coincide from $-(N - 1)$ to $N - 1$, i.e., if $\xi_n = \eta_n$ for $|n| < N$. In Annex A.2 we consider other metrics.

Having introduced the sequence spaces, observe now that the action of f on the orbit of $x \in \Omega$, namely, $f^n(x) \mapsto f(f^n(x)) = f^{n+1}(x)$, translates into the action $(\Phi(x))_n \mapsto (\Phi(x))_{n+1}$ on the components of the itineraries. For this reason one introduces the (one-sided) *shift transformation* (or just *shift*) $\Sigma: S^{\mathbb{N}_0} \to S^{\mathbb{N}_0}$ as follows:

$$\Sigma: (\xi_0, \xi_1, \ldots, \xi_n, \ldots) \mapsto (\xi_1, \xi_2, \ldots, \xi_{n+1}, \ldots). \tag{1.8}$$

---

[2] The dependence of $\Phi(x)$ on f and α is not made explicit in order to keep the notation simple.

In words, Σ deletes the first component of $(\xi_n)_{n\in\mathbb{N}_0}$ and shifts the other components one position to the left. It is easily shown that Σ is a continuous transformation. As observed above, the diagram

$$
\begin{array}{ccc}
\Omega & \xrightarrow{f} & \Omega \\
\Phi \downarrow & & \downarrow \Phi \\
S^{\mathbb{N}_0} & \xrightarrow{\Sigma} & S^{\mathbb{N}_0}
\end{array}
$$

commutes, i.e., $\Phi \circ f = \Sigma \circ \Phi$. Note that Σ is not invertible (indeed, it is a k-to-1 map), although f might be invertible—unless two-sided itineraries are used.

As a simple illustration (see Fig. 1.1), consider the *sawtooth* (also called *dyadic*, *shift*, etc.) *map* $E_2:[0, 1] \rightarrow [0, 1]$, defined as

$$E_2(x) = 2x \bmod 1,$$

and decompose $[0, 1]$ into the intervals $A_0 = [0, \frac{1}{2})$ and $A_1 = [\frac{1}{2}, 1]$, so the alphabet is $S = \{0, 1\}$. In this case, the orbit $E_2^n(x)$, $n \in \mathbb{N}_0$, is coded to an infinitely long 0–1 string $\Phi(x)$, where

$$
(\Phi(x))_n = \begin{cases} 0 & \text{if } E_2^n(x) \in A_0, \\ 1 & \text{if } E_2^n(x) \in A_1. \end{cases}
$$

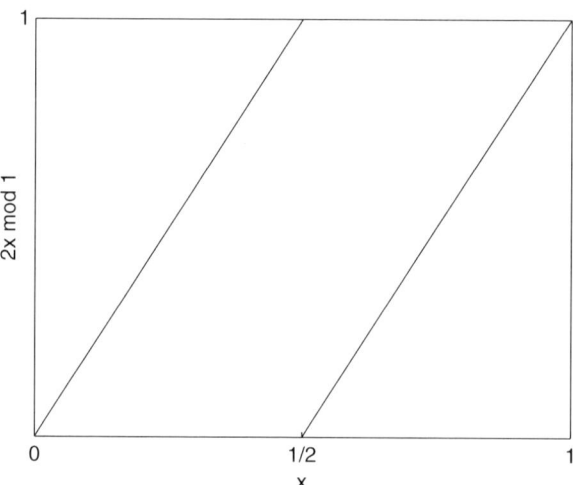

Fig. 1.1 The function $E_2(x) = 2x \bmod 1$

Let

$$x = \frac{b_0}{2} + \frac{b_1}{2^2} + \cdots + \frac{b_k}{2^{k+1}} + \cdots = \sum_{k=0}^{\infty} b_k 2^{-(k+1)} =: 0.b_0 b_1 \ldots b_k \ldots,$$

$b_n \in \{0, 1\}$, be a binary expansion of $x \in [0, 1]$. Then

$$E_2(0.b_0 b_1 \ldots b_k \ldots) = 0.b_1 b_2 \ldots b_{k+1} \ldots$$

for $x \in [0, 1)$ and $E_2(1) = E_2(0.1^\infty) = 0 = 0.0^\infty$, where here and throughout the upper label "∞" attached to a symbol means indefinite repetition of that symbol. The *dyadic rationals* in $(0, 1)$ (i.e., numbers of the form $m/2^n$, $m = 1, 2, \ldots, 2^n - 1$) are characterized by possessing two binary expansions: one terminating with 0^∞ and other terminating with 1^∞. Indeed, $0.10^\infty = 0.01^\infty$ and $0.b_0 \ldots b_{k-1} 10^\infty = 0.b_0 \ldots b_{k-1} 01^\infty$, $k \geq 1$, since

$$\sum_{n=k+1}^{\infty} 2^{-(n+1)} = 2^{-(k+2)} \sum_{n=0}^{\infty} 2^{-n} = 2^{-(k+2)} \cdot 2 = 2^{-(k+1)}.$$

If $x = 0.b_0 b_1 \ldots \in (0, 1)$ is not a dyadic rational, then

$$E_2^n(x) = 0.b_n b_{n+1} \ldots \in A_i \quad \text{iff } b_n = i \in \{0, 1\},$$

hence

$$\Phi(x) = (b_n)_{n \in \mathbb{N}_0} = (b_0, b_1, \ldots, b_n, \ldots). \tag{1.9}$$

Furthermore, $\Phi(0) = (0^\infty)$ and $\Phi(1) = (1, 0^\infty)$. If $x \in (1, 0)$ is a dyadic rational, then x is a preimage of 0 under E_2, thus (1.9) is fulfilled provided $(b_n)_{n \in \mathbb{N}_0}$ corresponds to the binary expansion of x ending with 0^∞. We conclude that given any binary sequence $(b_n)_{n \in \mathbb{N}_0}$ not terminating with 1^∞, there exists always $x \in [0, 1]$, namely $x = 0.b_0 b_1 \ldots$, such that its itinerary with respect to the decomposition $\alpha = \{A_0, A_1\}$ under E_2 is precisely that sequence. In particular, given a finite word b_0^n, there exist infinitely many points in $[0, 1]$, to wit:

$$x \in \left[\frac{b_0 2^n + b_1 2^{n-1} + \cdots + b_n}{2^{n+1}}, \frac{b_0 2^n + b_1 2^{n-1} + \cdots + b_n + 1}{2^{n+1}} \right)$$
$$= [0.b_0 \ldots b_n, 0.b_0 \ldots b_n + 2^{-(n+1)}), \tag{1.10}$$

whose itineraries $\Phi(x)$ "realize" the pattern b_0^{L-1} in the sense that $\Phi(x)_0^n = b_0^n$. The fact that all finite words of a symbolic space can be materialized as segments of itineraries for a wide class of maps (Sect. 3.1) contrasts with the situation we shall come upon when studying the so-called ordinal patterns in Sect. 1.2.

Shifts are a special instance of the so-called subshifts. If K is a closed and Σ-invariant (i.e., $\Sigma(K) \subset K$) subset of $S^{\mathbb{N}_0}$, the restriction of the shift transformation to K, written as $\Sigma|_K$, is called a *subshift*. Sometimes Σ is called a *full* shift to distinguish it from the subshifts proper ($K \neq S^{\mathbb{N}_0}$).

A special class of subshifts are of great interest in applications. Let $A = (a_{ij})_{0 \leq i, j \leq k-1}$ be a $k \times k$ matrix of 0's and 1's and define

$$S_A^{\mathbb{N}_0} = \left\{ (\xi_n)_{n \in \mathbb{N}_0} \in S^{\mathbb{N}_0} : a_{\xi_n \xi_{n+1}} = 1 \text{ for all } n \in \mathbb{N}_0 \right\}.$$

Put in simple terms, the matrix A determines which letters $\xi_{n+1} \in S = \{0, 1, \ldots, k - 1\}$ may follow the letter ξ_n in the word $(\xi_n)_{n \in \mathbb{N}_0}$. Thus $S_A^{\mathbb{N}_0}$ is a closed and Σ-invariant subset of the sequence space $S^{\mathbb{N}_0}$ that contains all well-formed or *admissible* sequences. Alternatively, one can also describe $S_A^{\mathbb{N}_0}$ by listing the forbidden words. This explains the connection between symbolic dynamics and the theory of formal languages we mentioned above. The restriction of Σ to $S_A^{\mathbb{N}_0}$, written as Σ_A, is called a *subshift of finite type, Markov subshift*, or a *topological Markov chain*. If $a_{ij} = 1$ for every $0 \leq i, j \leq k - 1$, we recover the full shift. At the opposite end, $S_A^{\mathbb{N}_0}$ may be empty. This happens if and only if the matrix A is nilpotent (i.e., $A^n = 0$ for some $n \in \mathbb{N}$).

As way of example, take $k = 2$ and

$$A = \begin{pmatrix} 1 & 1 \\ 1 & 0 \end{pmatrix}.$$

Since $a_{11} = 0$, this means that the binary sequence $(\xi_n)_{n \in \mathbb{N}_0}$ is admissible if and only if it does not contain two consecutive 1's. In this case, the only forbidden block of length 2 is 11.

Let $\mathbb{K} = \mathbb{N}_0$ or $\mathbb{Z}$, and $(S_A^{\mathbb{K}}, \Sigma_A)$, $(T_B^{\mathbb{K}}, \Sigma_B)$ be two subshifts of finite type possibly with different alphabets S and T, respectively. Suppose $F : S_A^{\mathbb{K}} \to T_B^{\mathbb{K}}$ is a shift-commuting map, that is, $F \circ \Sigma_A = \Sigma_B \circ F$. The continuous, shift-commuting maps from a subshift of finite type $S_A^{\mathbb{K}}$ to another $T_B^{\mathbb{K}}$ were characterized in [92] as those maps for which there exist integers $l \leq r$ and a "local rule" $f : S^{r-l+1} \to T$ such that for any $\xi = (\xi_n)_{n \in \mathbb{K}} \in S_A^{\mathbb{K}}$ and $i \in \mathbb{K}$,

$$F(\xi)_i = f(\xi_{i+l}, \ldots, \xi_{i+r}). \tag{1.11}$$

If F is not the constant map, then a maximal l and a minimal r with this property exist; they are called left and right radii of F, respectively. If $\mathbb{K} = \mathbb{N}_0$, then $l \geq 0$. When $\mathbb{K} = \mathbb{Z}$, $p = \max\{-l, r\}$ is called the *radius* of F. In this case,

$$F(\xi)_i = f(\xi_{i-p}, \ldots, \xi_i, \ldots \xi_{i+p}),$$

where $\xi = (\xi_n)_{n \in \mathbb{Z}}$. A map between two subshifts of finite type of the form (1.11) is called a *block map* [123]. Block maps provide the mathematical underpinnings of cellular automata (Sect. 1.5).

Markov subshifts not only do provide conceptually simple prototypes for important dynamical properties, but they are basic components of some physical systems (e.g., think of Smale's horseshoes in Hamiltonian dynamical systems). To be more specific, we point out next that Markov subshifts can exhibit all properties of low-dimensional chaos.

Let us recall some basic definitions first. A 0–1 matrix A is said to be *transitive* if A^m is positive (i.e., all its entries are positive) for some $m \in \mathbb{N}$. A continuous self-map f of a metric space M is *topologically transitive* if there exists $x \in M$ such that $\mathcal{O}_f(x) = (f^n)_{n \in \mathbb{N}_0}$ is dense in M; if f is invertible, then the requirement for topological transitivity is that $\mathcal{O}_f(x) = (f^n)_{n \in \mathbb{Z}}$ is dense in M for some $x \in M$. It holds [91] that if A is a transitive $k \times k$ matrix, then the topological Markov chain Σ_A is topologically transitive and its periodic orbits are dense in $S_A^{\mathbb{N}_0}$ ($S = \{0, 1, \ldots, k - 1\}$), therefore Σ_A is chaotic in the sense of Devaney [69]; in particular, Σ_A has sensitive dependence on initial conditions (see Sect. A.2). This result includes the full shifts. The corresponding statements for f invertible and $M = S_A^{\mathbb{Z}}$ hold true as well.

1.1.3 Dynamical Systems

We shall encounter two kinds of dynamical systems in this book. A *continuous* (or *topological*) *dynamical system* consists of a topological space (e.g., a metrical space) M and a continuous map $f: M \to M$. This being the case, these systems will be denoted by the pair (M, f). Subshifts are examples of continuous systems, (K, Σ_K). A *measure-theoretical dynamical system* is comprised of a *measurable space* $(\Omega, \mathcal{B})$, a measurable map $f: \Omega \to \Omega$, and a *non-singular measure* μ on $(\Omega, \mathcal{B})$. Thus, Ω is a non-empty set, $\mathcal{B}$ is a sigma-algebra of subsets of Ω, $f^{-1}B \in \mathcal{B}$ for all $B \in \mathcal{B}$, and $B \in \mathcal{B}$ is a μ-zero set iff $f^{-1}B$ is a μ-zero set. Only finite-measure spaces will be considered henceforth. Therefore, $(\Omega, \mathcal{B}, \mu)$ may be assumed without restriction to be a probability space, with μ being a probability on the space of "events" $(\Omega, \mathcal{B})$. Measure-theoretical systems will be denoted by $(\Omega, \mathcal{B}, \mu, f)$. To promote a continuous system (M, f) to a measure-theoretical one, it suffices to endow the topological space M with its Borel sigma-algebra (i.e., the sigma-algebra generated by the open sets), and the corresponding Lebesgue measure. In topological dynamics, the attention focuses on continuous systems. In ergodic theory, the framework is set by *measure-preserving* self-maps of (usually) probability spaces. We say that $f: \Omega \to \Omega$ preserves a measure μ on $(\Omega, \mathcal{B})$, if $\mu(f^{-1}B) = \mu(B)$ for all $B \in \mathcal{B}$. Alternatively, we say that the measure-theoretical system $(\Omega, \mathcal{B}, \mu, f)$ is μ-preserving, or that μ is f-invariant. Sometimes, measure-preserving, invertible maps are called *automorphisms*, while the name *endomorphisms* is reserved for the non-invertible ones.

The dynamical complexity of a measure-preserving system $(\Omega, \mathcal{B}, \mu, f)$ can be quantified by its metric entropy. So to speak, the metric entropy measures the uncertainty of the forward evolution of the system when the initial condition is not exactly known —the higher the uncertainty, the greater the complexity. The original proposal of A. Kolmogorov (later completed by Y. Sinai) amounts to the following recipe: coarse-grain the state space of the dynamical system and calculate the Shannon entropy of the resulting stochastic process. Let us follow this path.

A *partition* of a measure space $(\Omega, \mathcal{B}, \mu)$ (or just Ω for brevity) is a disjoint family of elements of $\mathcal{B}$, called atoms, whose union is Ω. Partitions will be denoted by small Greek letters. Two extreme examples of partitions of Ω are the *trivial partition* $\{\emptyset, \Omega\}$ and the *point partition* (or partition of Ω into separate points)

$$\epsilon = \{\{x\} : x \in \Omega\}. \tag{1.12}$$

Except for ϵ, we consider only finite partitions, i.e., partitions with a finite number of atoms. If, furthermore, Ω is a compact metric space with metric d, then the "size" or "coarseness" of a partition $\alpha = \{A_0, A_1, \ldots, A_{|\alpha|-1}\}$ is measured by its *norm* (sometimes also called *diameter*),

$$\|\alpha\| = \sup_{0 \leq k \leq |\alpha|-1} \{d(x, y) : x, y \in A_k\}. \tag{1.13}$$

We saw already in the last section that a discretization of the state space Ω may provide useful insights into a complicated dynamic. In measure-preserving systems this is even more certain since, as we are going to see presently, partitions allow establishing a connection with stochastic and information theory.

Given a finite partition $\alpha = \{A_0, A_1, \ldots, A_{|\alpha|-1}\}$ of $(\Omega, \mathcal{B}, \mu)$, the maps[3] $X_n : \Omega \to S = \{0, 1, \ldots, |\alpha| - 1\}$, $n \in \mathbb{N}_0$, defined as

$$X_n(x) = i \quad \text{iff } f^n(x) \in A_i$$

are random variables on the probability space $(\Omega, \mathcal{B}, \mu)$. Indeed,

$$X_n^{-1}(i) = f^{-n}(A_i) \in \mathcal{B}$$

because f is measurable. Observe that $X_n(x)$ is the nth component of the itinerary of x with respect to α. The difference now with respect to the itineraries of Sect. 1.1.2 is the existence of an invariant measure, which allows to promote $\mathbf{X} = \{X_n\}_{n \in \mathbb{N}_0}$ to a stationary stochastic process. In fact

(i) The probability (mass) function of X_n is given by

---

[3] The dependence of X_n on α is not made explicit here in order to keep the notation simple.

$$\Pr\{X_n = i\} = \mu\left\{x \in \Omega : f^n(x) \in A_i\right\} = \mu(f^{-n}A_i) = \mu(A_i),$$

because f is μ-preserving. As for the joint probability function of $X_0, \ldots,$ $X_n = X_0^n$,

$$\Pr\left\{X_0^n = i_0, \ldots, i_n\right\} = \mu\left\{x \in \Omega : x \in A_{i_0}, \ldots, f^n(x) \in A_{i_n}\right\}$$
$$= \mu\left(A_{i_0} \cap \ldots \cap f^{-n}A_{i_n}\right).$$

(ii) The stochastic process $\{X_n : n \in \mathbb{N}_0\}$ is stationary:

$$\Pr\left\{X_k^{k+n} = i_0, \ldots, i_n\right\} = \mu\left\{x \in \Omega : f^k(x) \in A_{i_0}, \ldots, f^{k+n}(x) \in A_{i_n}\right\}$$
$$= \mu\left(f^{-k}(A_{i_0} \cap \cdots \cap f^{-n}A_{i_n})\right)$$
$$= \mu\left(A_{i_0} \cap \cdots \cap f^{-n}A_{i_n}\right)$$

because f is μ-preserving. Therefore,

$$\Pr\{X_k = i_0, \ldots, X_{k+n} = i_n\} = \Pr\{X_0 = i_0, \ldots, X_n = i_n\}$$

for every $n, k \in \mathbb{N}_0$.

It follows that the stochastic process $\mathbf{X} = \{X_n\}_{n \in \mathbb{N}_0}$ is an information source with alphabet $S = \{0, 1, \ldots, |\alpha| - 1\}$. The *metric entropy of f with respect to the partition* α is defined to be the Shannon entropy (rate) of $\mathbf{X}$:

$$h_\mu(f, \alpha) = -\lim_{n \to \infty} \frac{1}{n} \sum \Pr\{X_0^{n-1} = i_0, \ldots, i_{n-1}\} \log \Pr\{X_0^{n-1} = i_0, \ldots, i_{n-1}\}$$
$$= -\lim_{n \to \infty} \frac{1}{n} \sum \mu(A_{i_0} \cap \cdots \cap f^{-n}A_{i_n}) \log \mu(A_{i_0} \cap \cdots \cap f^{-n}A_{i_n}),$$

where the summation is over all $i_0, \ldots, i_{n-1} \in S$. If we define the *refinement*

$$\bigvee_{i=0}^{n-1} f^{-i}\alpha = \{A_{j_0} \cap f^{-1}A_{j_1} \cap \cdots \cap f^{-(n-1)}A_{j_{n-1}} : 0 \le j_0, \ldots, j_{n-1} \le |\alpha| - 1\}$$

of the partition $\alpha = \{A_0, \ldots, A_{|\alpha|-1}\}$, and the function

$$H_\mu(\beta) = -\sum_{j=0}^{|\beta|-1} \mu(B_j) \log(B_j)$$

for any partition $\beta = \{B_0, \ldots, B_{|\beta|-1}\}$ of $(\Omega, \mathcal{B}, \mu)$, then we recover the usual expression of $h_\mu(f, \alpha)$:

$$h_\mu(f, \alpha) = \lim_{n \to \infty} \frac{1}{n} H_\mu \left(\bigvee_{i=0}^{n-1} f^{-i} \alpha \right). \tag{1.14}$$

The convergence of this limit is proven in Sect. B.2.

If an application of f is interpreted as a passage of one unit of time, then $\bigvee_{i=0}^{n-1} f^{-i} \alpha$ represents the combined experiment of performing n consecutive times the original experiment, represented by α. Then $h_\mu(f, \alpha)$ is the average information per unit of time that one gets from performing the original experiment every unit of time [202].

The metric (Kolmogorov–Sinai or measure-theoretical) entropy of f is then the supremum of $h_\mu(f, \alpha)$ over all finite partitions of $(\Omega, \mathcal{B}, \mu)$:

$$h_\mu(f) = \sup_\alpha h_\mu(f, \alpha). \tag{1.15}$$

Continuing with the previous information-theoretical interpretation, $h_\mu(f)$ provides the maximum average information per unit of time obtainable by performing the same experiment every unit of time.

In general there are several obstacles preventing an exact calculation of $h(f)$. First, except in simple cases limit (1.14) itself is not computable, so we must be content with an evaluation of $\frac{1}{n} H_\mu \left(\bigvee_{i=0}^{n-1} f^{-i} \alpha \right)$ for some large value of n. Second, considerable computation is necessary to identify the elements of the refined partitions $\bigvee_{i=0}^{n-1} f^{-i} \alpha$, the computational effort being exponential in n. Third, the measure μ is usually unknown to us in closed form. Fortunately, there are exceptions, for instance, when one can find a partition α for which $h_\mu(f, \alpha) = h_\mu(f)$. Such partitions are called generators or *generating partitions* with respect to f. A finite partition α is a *one-sided generator* for f if

$$\bigvee_{i=0}^{\infty} f^{-i} \alpha = \epsilon, \tag{1.16}$$

where ϵ is the point partition of Ω (see (1.12)). Moreover, if f is even an automorphism and $\bigvee_{i=-\infty}^{\infty} f^{-i} \alpha = \epsilon$, then α is called a *two-sided generator* or just a generator for f. Automorphisms may have not only generators but also one-sided generators. According to the Kolmogorov–Sinai theorem (Annex B.13), if α is a generator (one-sided or not) for f, then $h_\mu(f, \alpha) = h_\mu(f)$.

As way of illustration, consider the *symmetric tent map* $\Lambda : [0, 1] \to [0, 1]$ defined as (Fig. 1.2)

$$\Lambda(x) = 1 - |1 - 2x| = \begin{cases} 2x & \text{if } 0 \le x \le \frac{1}{2}, \\ 2(1-x) & \text{if } \frac{1}{2} \le x \le 1. \end{cases} \tag{1.17}$$

If we equip $[0, 1]$ with the Borel sigma-algebra (generated by the intersections of open intervals of $\mathbb{R}$ with $[0, 1)$), then Λ is easily seen to preserve the Lebesgue

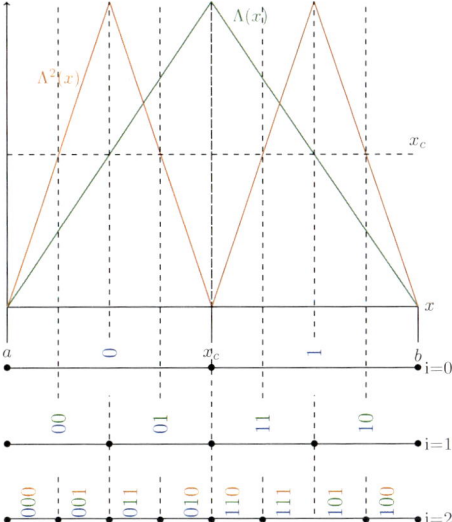

Fig. 1.2 Symbolic intervals generated by the symmetric tent map Λ and its second iterate $\Lambda^2 (x_c = \frac{1}{2})$

measure. As in the previous section, let $\alpha = \{A_0, A_1\}$, where

$$A_0 = [0, \tfrac{1}{2}), \ A_1 = [\tfrac{1}{2}, 1].$$

Then,

$$\Lambda^{-1}A_0 = [0, \tfrac{1}{4}) \cup (\tfrac{3}{4}, 1], \ \Lambda^{-1}A_1 = [\tfrac{1}{4}, \tfrac{3}{4}].$$

Hence

$$\alpha \cap \Lambda^{-1}\alpha = \{A_{00}, A_{01}, A_{11}, A_{10}\},$$

with

$$A_{00} = A_0 \cap \Lambda^{-1}A_0 = [0, \tfrac{1}{4}), \quad A_{01} = A_0 \cap \Lambda^{-1}A_1 = [\tfrac{1}{4}, \tfrac{1}{2}),$$

$$A_{11} = A_1 \cap \Lambda^{-1}A_1 = [\tfrac{1}{2}, \tfrac{3}{4}], \quad A_{10} = A_1 \cap \Lambda^{-1}A_0 = (\tfrac{3}{4}, 1].$$

The sets of $\alpha, \alpha \cap \Lambda^{-1}\alpha$, and $\alpha \cap \Lambda^{-1}\alpha \cap \Lambda^{-2}\alpha$ are shown in Fig. 1.2. In general,

$$\bigcap_{i=0}^{k} \Lambda^{-i}\alpha = \left\{A_{b_0 b_1 \dots b_k} : b_0, b_1, \dots, b_k \in \{0, 1\}\right\},$$

where the 2^{k+1} disjoint sets

$$A_{b_0 b_1 \ldots b_k} = A_{b_0} \cap \Lambda^{-1} A_{b_1} \cap \cdots \cap \Lambda^{-k} A_{b_k} \qquad (1.18)$$

build a family of ever-shorter intervals that covers uniformly the unit interval. As a matter of fact, the sets $A_{b_0 b_1 \ldots b_k}$ are a permutation of the dyadic intervals (1.10), except eventually for the endpoints. It follows that $\bigcap_{i=0}^{k} \Lambda^{-i} \alpha$ converges to the point partition of $[0,1]$, hence α is a one-sided generator for Λ. If λ denotes the Lebesgue measure, $\lambda(dx) = dx$, then

$$
\begin{aligned}
h_\lambda(\Lambda) &= -\lim_{n\to\infty} \frac{1}{n} \sum_{b_0 \ldots b_{n-1} \in \{0,1\}} \lambda(A_{b_0 \ldots b_{n-1}}) \log \lambda(A_{b_0 \ldots b_{n-1}}) \\
&= -\lim_{n\to\infty} \frac{1}{n} \sum_{b_0 \ldots b_{n-1} \in \{0,1\}} 2^{-n} \log 2^{-n} \\
&= \log 2.
\end{aligned}
$$

A similar argument can be applied to other maps, like the *logistic map* $g\colon[0,1] \to [0,1]$,

$$g(x) = 4x(1-x). \qquad (1.19)$$

In this case, the absolutely continuous measure[4]

$$\mu(dx) = \frac{dx}{\pi \sqrt{x(1-x)}} \qquad (1.20)$$

is g-invariant. This measure is called the *natural* or *physical invariant measure* of g because it is the one obtained in numerical experiments [72].

Since $(\Omega, \mathcal{B}, \mu)$ is a probability space, dynamical complexity can be given a probabilistic meaning. In this sense we can say that the entropy $h_\mu(f)$ (or other related concepts, like the Lyapunov exponents greater than 1) measures the randomness or, rather, the pseudo-randomness of the dynamic induced by the map f.

The complexity of continuous dynamical systems is usually measured by the topological entropy. As we shall presently see, this quantity is related to the periodic structure in some relevant systems. Rather than going into the definition of topological entropy, which is quite technical (see Sect. B.3), we only recall here its expression for a one-sided or two-sided Markov subshift Σ_A. It can be shown [91] that

$$h_{\text{top}}(\Sigma_A) = \limsup_{n\to\infty} \frac{1}{n} \log^+ P_n(\Sigma_A),$$

---

[4] Absolute continuity of measures will refer to the Lebesgue measure throughout this book.

where $h_{top}(\Sigma_A)$ is the topological entropy of Σ_A (in general, $h_{top}(f)$ stands for the topological entropy of a continuous self-map f), $P_n(\Sigma_A)$ is the number of periodic points of period n of Σ_A, and $\log^+ x = \log x$ if $x \geq 1$, and 0 otherwise. To explicitly calculate the right-hand side of this expression, we need the following two properties: (i) If B is a non-negative matrix, then there exists an eigenvalue $\lambda_{max} \geq 0$ such that no other eigenvalue of B has absolute value greater than λ_{max} (this is part of the Perron–Frobenius theorem [202]) and (ii) the number of periodic points of period $p \in \mathbb{N}$ of a Markov subshift Σ_A is the trace of A^p (i.e., the sum of the diagonal elements), denoted as $\operatorname{tr} A^p$. For the full shift on k symbols, $(A^n)_{ij} = k^{n-1}$, for all $0 \leq i, j \leq k - 1$, hence the trace of A^n is k^n. This yields

$$h_{top}(\Sigma) = \log k.$$

In general, $\operatorname{tr} A^p = \lambda_1^p + \cdots + \lambda_k^p$, where λ_i are the k eigenvalues (eventually repeated) of the matrix A. It follows that [91]

$$h_{top}(\Sigma_A) = \log^+ \lambda_{max}.$$

1.1.4 Computer Science

The origin of algorithmic complexity has to be sought in the efforts of R. Solomonoff, A. Kolmogorov, and G. Chaitin to define the elusive concept of "randomness" of finite-alphabet sequences [79, 133, 201]. The basic intuition is that random sequences are "patternless," hence there is no efficient way to describe them other than giving the sequence itself. The *algorithmic complexity* of a string $s_0^{n-1} = s_0 s_1 \ldots s_{n-1}$, written as $K(s_0^{n-1})$, can be consistently defined as the length of the shortest binary program that, run on a universal prefix-free Turing machine, outputs s_0^{n-1} and halts [59, 67, 138]. As in the case of information theory, this definition of complexity is linked to the general concept of compressibility, this time with respect to all possible algorithms that produce the sequence in question.

Somewhat paradoxically, algorithmic complexity is not a computable quantity. Then suppose that K_n is claimed to be the complexity of a length-n string s_0^{n-1}. In order to check this, we remove one bit from the hypothetically shortest program and let it run. There are two possibilities: either the $(K_n - 1)$-bit program outputs a string different from s_0^{n-1} and halts or else it runs longer than we have time to wait. In the second case, there is no way to know whether the program will halt (this is the famous Turing's halting problem), eventually revealing the actual complexity to be $K_n - 1$.

Any finite sequence s_0^{n-1} can be certainly output by the copy program: "PRINT $s_0, \ldots, s_n$." Without loss of generality, we may restrict to binary sequences for the time being. Since patternless n-bit sequences cannot be computed by any algorithm significantly shorter than the copy program, their complexity is given by $K_n \leq n + C$, where C is a constant that accounts for the computational overhead (like the operating system). At the opposite end stands the sequences consisting of a repeated bit, say 0. The complexity of the program "PRINT 0, n TIMES" can be bounded as $K_n \leq \log_2 n + C'$, where $\log_2 n$ is the number of bits needed to specify the length n and, again, C' is the computational overhead. Observe that if these programs are run on a computer other than a universal Turing machine, the constants C and C' may depend on the machine, but they are independent of the actual sequence being calculated. In the limit of very long sequences, the algorithmic complexity will practically range between $\log_2 n$ and n. This being the case, one may state that the binary sequence s_0^{n-1} is random if $K(s_0^{n-1}) \simeq n$. (In the non-binary case, $K(s_0^{n-1}) \simeq nb$ for random sequences, where b is the minimal number of bits needed to code the symbols s_i, $0 \leq i \leq n-1$.) Formally, a sequence $(s_n) \in S^{\mathbb{N}_0}$ is said to be *incompressible* when there exists a constant C such that

$$K(s_0^{n-1}) \geq n - C$$

for all $n \geq 1$.

Randomness can also be defined as *typicality*, meaning that typical sequences have no feature that makes them special in any sense. This was the path taken by Martin-Löf to come to grips with the concept of random sequence. Rather than addressing the technicalities of this approach, which are beyond the scope of this book, we will proceed directly to the conclusions: random sequences are realizations of stochastic processes.

Let $(\Omega, \mathcal{B}, \mu)$ be a probability space. The realizations of a stochastic process $\{X_n\}_{n \in \mathbb{N}_0}$ on $(\Omega, \mathcal{B}, \mu)$ with a finite number of possible outcomes can be identified with the elements of a (one-sided) sequence space. Specifically, if $X_n : \Omega \to S$ with $S = \{s_1, \ldots, s_{|S|}\}$ for every $n \in \mathbb{N}_0$, then $(X_n(\omega))_{n \in \mathbb{N}_0} \in S^{\mathbb{N}_0}$ for every $\omega \in \Omega$. The general method to place a probability m on $S^{\mathbb{N}_0}$ induced by the probability μ is explained in Sect. A.3. At present we only need to resort to the so-called (p, q)-Bernoulli shifts or systems on two symbols, which are measure-preserving systems $(S^{\mathbb{N}_0}, \mathcal{B}, m, \Sigma)$, where

(i) $S = \{0, 1\}$,
(ii) $\mathcal{B}$ is the sigma-algebra generated by the so-called *cylinder sets*,

$$C_{s_0 \ldots s_{n-1}} = \{\xi_0^\infty \in S^{\mathbb{N}_0} : \xi_0 = s_0, \ldots, \xi_{n-1} = s_{n-1}\},$$

(iii) the probability m of the binary string $s_0^{n-1} = s_0 s_1 \ldots s_{n-1}$ is defined as

$$m(s_0^{n-1}) = m(C_{s_0 \ldots s_{n-1}}) = p^k q^{n-k},$$

where $p + q = 1$, k is the number of 1's in s_0^{n-1}, and $n - k$ is the number of 0's, and
(iv) Σ is the shift transformation on $S^{\mathbb{N}_0}$.

In the language of probability theory, the cylinder sets correspond to the elementary events; in the language of computer science, $C_{s_0 \ldots s_{n-1}}$ comprises all sequences with the prefix $w = s_0, \ldots, s_{n-1}$. The (p, q)-Bernoulli system models an independent, dichotomous process, one outcome (say, "success") having probability p to occur and the other ("failure") probability $q = 1 - p$. Think, for example, of a random experiment consisting in tossing forever a coin with the odds p for head and q for tail. The shift Σ corresponds to the "time" translation $n \mapsto n + 1$. The fact that Σ preserves m (or, equivalently, that m is Σ-invariant) accounts for the probabilities being the same in every draw.

In particular, the $(\frac{1}{2}, \frac{1}{2})$-Bernoulli system is a model for the tossing of a fair coin. If $0.b_0 b_1 \ldots b_n \ldots$ is a binary expansion and $\Phi:[0, 1] \to \{0, 1\}^{\mathbb{N}_0}$ is the map

$$\Phi:0.b_0 b_1 \ldots b_n \ldots \mapsto (b_0, b_1, \ldots, b_n, \ldots)$$

we met already in (1.9), then

$$\Phi([0.b_0 b_1 \ldots b_n, \ 0.b_0 b_1 \ldots b_n + 2^{-(n+1)})) = C_{b_0 b_1 \ldots b_n}.$$

Thus, Φ allows to identify the cylinder set $C_{b_0 b_1 \ldots b_n}$ of $\{0, 1\}^{\mathbb{N}_0}$ with the interval $[0.b_0 b_1 \ldots b_{n-1}, 0.b_0 b_1 \ldots b_{n-1} + 2^{-n})$ of $[0, 1]$. But even more is true. If m denotes the measure of the $(\frac{1}{2}, \frac{1}{2})$-Bernoulli system and λ the Lebesgue measure of $[0, 1]$, then

$$m(C_{b_0 b_1 \ldots b_{n-1}}) = \frac{1}{2^n} = \lambda([0.b_0 b_1 \ldots b_{n-1}, \ 0.b_0 b_1 \ldots b_{n-1} + 2^{-n})).$$

Since the cylinder sets generate the sigma-algebra of the Bernoulli systems and the semi-open dyadic intervals do the same for the Borel sigma-algebra of $[0, 1]$, we conclude $m = \lambda \circ \Phi^{-1}$, i.e., m corresponds to the Lebesgue (or uniform) measure on $[0, 1]$.

Levin, Schnorr, and Chaitin proved that a binary sequence is typical with respect to the $(\frac{1}{2}, \frac{1}{2})$-Bernoulli measure (i.e., it can be considered the result of tossing a fair coin indefinitely) if and only if it is incompressible. In this way, two seemingly different concepts of randomness incompressibility and typicality are shown to coincide in a natural setting.

Remarkably enough, this result is not the only achievement connecting concepts related to complexity but stemming from different areas. Let us provide another one in which algorithmic complexity and metric entropy are brought together.

Given a measure-preserving dynamical system $(\Omega, \mathcal{B}, \mu, f)$, each $x \in \Omega$ generates an infinitely long sequence under the action of f, namely, its (forward) orbit $\mathcal{O}_f(x) = \{f^n(x) : n \in \mathbb{N}_0\}$. Let $s_0^\infty = s_0^\infty(x, \alpha)$ be the itinerary of x with respect to the partition $\alpha = \{A_0, \ldots, A_{|\alpha|-1}\}$ of Ω, that is, $s_k = i$ iff $f^k(x) \in A_i$, $i \in \{0, \ldots, |\alpha| - 1\}$. The *algorithmic complexity* of $\mathcal{O}_f(x)$, written as $k(f, x)$, is measured by the largest algorithmic complexity per symbol of $s_0^\infty(x, \alpha)$ over all possible finite partitions α:

$$k(f, x) = \sup_\alpha \limsup_{n \to \infty} \frac{1}{n} K(s_0^{n-1}(x, \alpha)).$$

Of course, one expects that random-like trajectories are computationally more difficult to reproduce than the regular ones. This expectation can be rigorously proved under the proviso that f is ergodic with respect to the invariant measure μ. In this case [39],

$$k(f, x) = h_\mu(f) \quad \mu\text{-almost everywhere.}$$

1.1.5 Cellular Automata

A cellular automaton is a discrete-time dynamical system with discrete space and discrete states. The state variables are defined on the sites of a D-dimensional regular lattice ($\mathbb{Z}^D$)—the cells of the D-dimensional automaton—taking on values in a finite alphabet $S = \{0, 1, \ldots, k - 1\}$. The set of all possible states (formally the set of all possible mappings $\mathbb{Z}^D \to S$) is called the *configuration space*. For numerical simulations it is convenient that the lattice of sites is finite or has a non-trivial topology, like a circle or a 2-torus; these requirements can be implemented with quiescent cells or with periodic conditions, respectively. In order to accommodate this disparity of possibilities, the configuration space will be denoted by a neutral Ω. The states of the cells evolve synchronously in discrete time steps according to identical rules. But what makes cellular automata special is the evolution rule: the state of a particular cell is determined by the previous states of a neighborhood of cells around it.

Cellular automata were introduced by Ulam [199] and von Neumann [161] as simple models of universal computation and machine self-reproduction, respectively. Indeed, a remarkable property of cellular automata is their ability to simulate other symbol processors. Another one is self-organization, even when started from disordered configurations. Two-dimensional cellular automata became quite popular in the 1970s thanks to the article that Martin Gardner devoted to John Conway's *Game of Life* in his section "Mathematical Games" of *Scientific American* [84]. A purely mathematical approach was initiated by Hedlund and collaborators, who studied the endomorphisms and automorphisms of the shift dynamical system [92]. Apart from the many subsequent papers on their dynamical and ergodic properties from this point of view, cellular automata have also been the object of intensive study in mathematical physics, computer science, biology, etc. [207]. Being at the crossroads of symbolic dynamical systems and computation, it is not surprising that the theory of cellular automata benefits from both areas, at the same time that cross-pollinate them, as we try to show in the next lines. For a readable account on cellular automata and their remarkable performance in physical modeling, see, e.g., [198].

For simplicity we will consider only one-dimensional cellular automata. In this case, the configuration space is the two-sided sequence space $S^{\mathbb{Z}}$. One-sided sequences or even finite sequences, corresponding to lattices adequately flanked by quiescent cells, may also be considered along the same lines. A *neighborhood* of size $l \geq 1$ of the cell $i \in \mathbb{Z}$, written as $\mathcal{U}_l(i)$, is the set of $2l + 1$ cells

$$i - l, i - l + 1, \ldots, i, \ldots, i + l.$$

The state of cell i at time $t \geq 0$ will be denoted as $s_t(i)$. At each time step $t + 1$, the previous state at each cell i, $s_t(i) \in S$, is updated according to the states of $\mathcal{U}_l(i)$ by a *local rule* $f : S^{2l+1} \to S$ of the form

$$s_{t+1}(i) = f(s_t(i - l), s_t(i - l + 1), \ldots, s_t(i + l)).$$

Note that f does not depend on i nor t, but only on the states of $\mathcal{U}_l(i)$; if f is allowed to depend on i, then one speaks of *hybrid* cellular automata.

The local rule f leads to a *global transition map* of the configuration space, $F : \Omega \to \Omega$, defined in the obvious way:

$$F(\ldots, s_t(i), \ldots) = (\ldots, f(s_t(i - l), s_t(i - l + 1), \ldots, s_t(i + l)), \ldots)$$
$$= (\ldots, s_{t+1}(i), \ldots).$$

Observe that F is a block map from a full shift to itself of radius l. As pointed out in Sect. 1.1.2, it follows that F is continuous and shift-commuting. (This characterization generalizes to D-dimensional cellular automata just by replacing the sequence space $S^{\mathbb{Z}}$ by $S^{\mathbb{Z}^D}$.)

As way of illustration, Fig. 1.3 depicts the time evolution of a one-dimensional, binary cellular automaton with periodic boundary conditions: $s_t(N + 1) = s_t(1)$ and $s_t(0) = s_t(N)$ for all $t \geq 0$. Here $N = 250$, the horizontal axis represents space (label i), and time (label t) elapses along the vertical direction, from top to bottom. Once the initial configuration has been fixed, the global map F determines the dynamics of the automaton on the configuration space.

The relation between the properties of the local rule f and the properties of the global transition map F is one of the most important and difficult problems in the

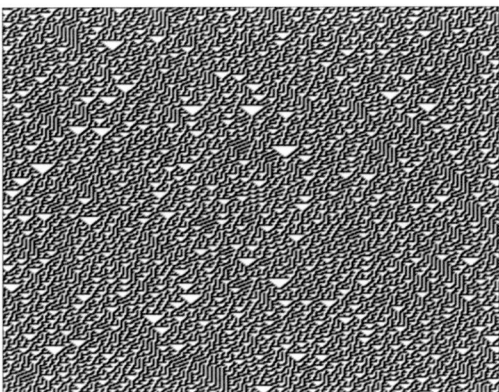

Fig. 1.3 A typical space–time evolution diagram of a one-dimensional cellular automaton with 250 sites and periodic boundary conditions. Time elapses from top to bottom

theory of cellular automata. This problem has been proved to be algorithmically unsolvable for some properties (surjectivity and injectivity for dimension $D > 1$, nilpotency for $D \geq 1$, etc.), and it is believed to be unsolvable for others (ergodicity, sensitivity, etc.).

On a more practical level, hybrid cellular automata with binary state variables and null boundaries (i.e., the cells delimiting the site lattice are permanently in the 0-state) have been explicitly shown to emulate linear feedback shift registers (LFSRs), which are widely used in cryptography as pseudo-random bit generators for stream ciphers. Specifically, given the primitive polynomial of an LFSR [151], then the algorithm given in [48] allows to "synthesize" a null-boundary, hybrid binary cellular automaton that emulates the said LFSR using only the local rules $f(p, q, r) = p + r \bmod 2 \equiv p \oplus q$ and $f(p, q, r) = p + q + r \bmod 2 \equiv p \oplus q \oplus r$. Most importantly, the same is true for the so-called self-shrunken LSFRs [149], which are nonlinear structures featured in some designs of stream ciphers. Since the previous local rules are linear, this fact allows to cryptanalyze such ciphers using cellular automata.

Suppose that the configuration space Ω is $S^{\mathbb{Z}}$. In the topology induced by the cylinder sets

$$C_{s_{-n}, \dots, s_0, \dots, s_n} = \{\xi_0^{\infty} \in S^{\mathbb{Z}} : \xi_k = s_k, |k| \leq n\},$$

the global transition map $F : \Omega \to \Omega$ that updates the states of the cellular automaton is continuous, which makes (Ω, F) a continuous dynamical system. Hence, we can measure the complexity of its time evolution with the topological entropy $h_{\text{top}}(F)$; see Sect. B.3 for different ways of calculating the topological entropy of a continuous dynamical system. Alternatively, let $R(w, t)$ be the number of distinct rectangles of width w and height (temporal extent) t occurring in a space–time evolution diagram of (Ω, F); see Fig. 1.4. Then [62]

$$h_{\text{top}}(F) = \lim_{w \to \infty} \lim_{t \to \infty} \frac{1}{t} \log R(w, t). \tag{1.21}$$

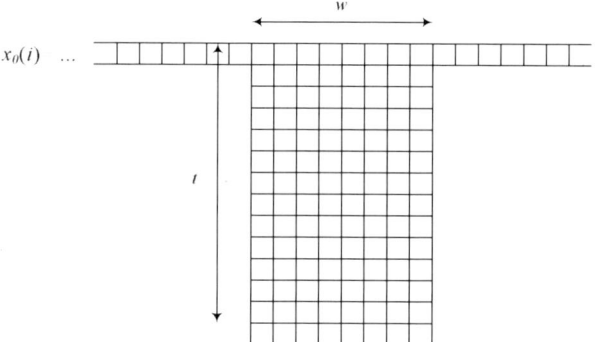

Fig. 1.4 Geometrical illustration of the rectangles $R(w, t)$ used in (1.21)

Therefore, the complexity of (Ω, F) can be measured by the number of distinct words or patterns per time unit generated by the global transition map F as time evolves. It follows that

$$h_{\text{top}}(F) \leq 2l \log k,$$

where l is the neighborhood size of the automaton and $k = |S|$.

Topological entropy belongs also to the dynamical properties that cannot be algorithmically computed for general cellular automata [101]. More generally, whether metric and/or topological entropy is effectively computable (i.e., can be approximated with an arbitrary small error) is an open question for most dynamical systems.

1.2 Admissible and Forbidden Ordinal Patterns

The concept of ordinal pattern of length L only demands a totally ordered set $(\Omega, \leq)$. Let us caution the reader that there are several definitions of ordinal patterns in the literature; the one used in this book follows Bandt et al. [28, 29]. In the simplest setting, the ordinal pattern defined by the elements $x_0, \ldots, x_{L-1} \in \Omega$ can be viewed as the permutation π of $\{0, 1, \ldots, L-1\}$ that arrange those elements according to their order in Ω: $x_{\pi_0} < x_{\pi_1} < \cdots < x_{\pi_{L-1}}$. In case $x_i = x_j$, we agree that $x_i < x_j$ if $i < j$. We write $\pi = \langle \pi_0, \pi_1, \ldots, \pi_{L-1} \rangle$ to summarize that x_{π_0} is the smallest element, x_{π_1} is the second smallest element, etc., in the length-L sequence $x_0, \ldots, x_{L-1}$. For example, if $\Omega = \mathbb{R}$ (endowed with the standard order), and $x_0 = \sqrt{3}, x_1 = e, x_2 = 2$, and $x_3 = -1.7$, then $\pi = \langle 3, 0, 2, 1 \rangle$. In an extended setting where we have a self-map f of Ω, the sets of points to be arranged by π are naturally provided by the initial segments of the f-orbits: $x_n = f^n(x), 0 \leq n \leq L-1$. In this case, one usually dispenses with periodic orbits of period smaller than L. The set of ordinal L-patterns will be denoted by $\mathcal{S}_L$ throughout this book. Ordinal patterns are sometimes called permutations.

As a minor technical point, let us mention that a permutation $\tau : i \mapsto \tau(i), i \in \{0, 1, \ldots, L-1\}$, is written in combinatorics as

$$\begin{pmatrix} 0 & 1 & \cdots & L-1 \\ \tau(0) & \tau(1) & \cdots & \tau(L-1) \end{pmatrix} =: [\tau(0), \tau(1), \ldots, \tau(L-1)]. \tag{1.22}$$

Observe that an ordinal pattern $\pi = \langle \pi_0, \ldots, \pi_{L-1} \rangle$ does not correspond—as one might think— to the permutation $[\pi_0, \ldots, \pi_{L-1}]$, but rather to its inverse: $\pi_0 \mapsto 0, \ldots, \pi_{L-1} \mapsto L-1$, i.e.,

$$\langle \pi_0, \ldots, \pi_{L-1} \rangle = \begin{pmatrix} \pi_0 & \pi_1 & \cdots & \pi_{L-1} \\ 0 & 1 & \cdots & L-1 \end{pmatrix} = [\pi_0, \ldots, \pi_{L-1}]^{-1}. \tag{1.23}$$

For example, the ordering $x_2 < x_0 < x_1$ defines the ordinal pattern $\langle 2, 0, 1 \rangle$ but the permutation $0 = \pi_1 \mapsto 1, 1 = \pi_2 \mapsto 2$, and $2 = \pi_0 \mapsto 0$, which in the conventional notation reads

$$[1, 2, 0] = [2, 0, 1]^{-1}.$$

In sum, an ordinal pattern $\pi \in \mathcal{S}_L$ corresponds actually to the permutation $\pi_i \mapsto i$, $0 \leq i \leq L - 1$, which will be denoted as $[\pi]^{-1}$ whenever needed:

$$[\pi]^{-1} = [\pi_0, \pi_1, \ldots, \pi_{L-1}]^{-1}. \tag{1.24}$$

Furthermore, if $\pi = \langle \pi_0, \ldots, \pi_{L-1} \rangle$ and $\pi' = \langle \pi'_0, \ldots, \pi'_{L-1} \rangle$, a (non-commutative) product $\pi \circ \pi'$ can be defined in $\mathcal{S}_L$ via composition

$$
\begin{aligned}
\pi \circ \pi' &= \begin{pmatrix} \pi_0 & \pi_1 & \cdots & \pi_{L-1} \\ 0 & 1 & \cdots & L-1 \end{pmatrix} \begin{pmatrix} \pi'_0 & \pi'_1 & \cdots & \pi'_{L-1} \\ 0 & 1 & \cdots & L-1 \end{pmatrix} \\
&= \begin{pmatrix} \pi'_{\pi_0} & \pi'_{\pi_1} & \cdots & \pi'_{\pi_{L-1}} \\ 0 & 1 & \cdots & L-1 \end{pmatrix} \\
&= \langle \pi'_{\pi_0}, \pi'_{\pi_1}, \ldots, \pi'_{\pi_{L-1}} \rangle.
\end{aligned} \tag{1.25}
$$

Endowed with this product, $\mathcal{S}_L$ becomes a non-Abelian group of order $L!$. The neutral element of the group $(\mathcal{S}_L, \circ)$ is the identity permutation $\langle 0, 1, \ldots, L-1 \rangle$. Ordinal patterns will be studied in detail in Chap. 3.

After these algebraic prolegomena, consider now a function $f : I \to I$, where I is a closed interval of $\mathbb{R}$. Given the finite orbit $\{f^n(x) : 0 \leq n \leq L - 1\}$ of $x \in I$, we say that x defines the ordinal pattern of length L (or ordinal L-pattern) $\pi = \pi(x) = \langle \pi_0, \pi_1, \ldots, \pi_{L-1} \rangle$ if

$$f^{\pi_0}(x) < f^{\pi_1}(x) < \cdots < f^{\pi_{L-1}}(x). \tag{1.26}$$

We say also that π is realized by x or that x is of type π.

If, for example, $I = [0, 1]$ and g is the *logistic map*, $g(x) = 4x(1 - x)$, then we find to four digit precision.

$$\mathcal{O}_g(0.6416) = 0.6416, 0.9198, 0.2951, 0.8320, 0.5590, 0.9861, \ldots$$

hence $x = 0.6416$ is of the types

$$\langle 0, 1 \rangle, \langle 2, 0, 1 \rangle, \langle 2, 0, 3, 1 \rangle, \langle 2, 4, 0, 3, 1 \rangle, \langle 2, 4, 0, 3, 1, 5 \rangle, \ldots$$

Instead of fixing x and varying L, we can do the opposite, as in the following illustration with $L = 3$:

$$\mathcal{O}_g(0.15) = 0.15, 0.51, 0.9996, \dots \quad \text{hence } 0.15 \text{ is of type } \langle 0, 1, 2 \rangle \,,$$
$$\mathcal{O}_g(0.30) = 0.30, 0.84, 0.5376, \dots \quad \text{hence } 0.30 \text{ is of type } \langle 0, 2, 1 \rangle \,,$$
$$\mathcal{O}_g(0.55) = 0.55, 0.99, 0.0396, \dots \quad \text{hence } 0.55 \text{ is of type } \langle 2, 0, 1 \rangle \,,$$
$$\mathcal{O}_g(0.80) = 0.80, 0.64, 0.9216, \dots \quad \text{hence } 0.80 \text{ is of type } \langle 1, 0, 2 \rangle \,,$$
$$\mathcal{O}_g(0.95) = 0.95, 0.19, 0.6156, \dots \quad \text{hence } 0.95 \text{ is of type } \langle 1, 2, 0 \rangle \,.$$

Points and ordinal patterns provide complementary perspectives of the same picture. Thus, as in the first instance, one can be more interested in the ordinal patterns defined by a given point or, as in the second instance, in the points that realize a given pattern. In order to introduce the second point of view, we define following [29] the sets

$$P_\pi = \{x \in I : x \text{ defines } \pi \in \mathcal{S}_L\}. \tag{1.27}$$

If $P_\pi \neq \emptyset$, then π is said to be an *allowed* or *admissible* (ordinal) *pattern* for f; otherwise π is called a *forbidden* (ordinal) *pattern* for f. In words, $\pi \in \mathcal{S}_L$ is allowed or admissible if there exists $x \in I$ such that x is of type π, whereas it is forbidden if no x is of type π. We will see shortly that maps have forbidden patterns (in fact, infinitely many of them) under quite general assumptions.

The properties of the sets $P_\pi \neq \emptyset$ are closely related to the properties of f. Thus, P_π is a union of open intervals if f is continuous or the union of intervals (including none, one, or both endpoints) if f is piecewise continuous. The endpoints of P_π are determined by the periodic points of f. All these facts can be easily exposed via the graphs of the map and their iterates. First of all, draw the graph of the identity (f^0) in the square $I \times I \subset \mathbb{R}^2$, which is the diagonal $y = x$, $x \in I$, on the Cartesian plane $\{(x, y) \in \mathbb{R} \times \mathbb{R}\}$. Then draw the graphs of the functions $y = f(x), \dots, y = f^{L-1}(x)$, $x \in I$. The components of the distinct P_π's, $\pi \in \mathcal{S}_L$, are separated by the intersection points of all those graphs. Indeed, if $x \in P_\pi$ "moves" leftward or rightward, it will leave the current component of P_π at the left or right endpoint, respectively, as soon as the condition

$$f^{\pi_i}(x) = f^{\pi_{i+1}}(x) \tag{1.28}$$

holds for some $i = 0, 1, \dots, L - 2$, unless it leaves the interval I before. Note that condition (1.28) implies that $f^{\min\{\pi_i, \pi_{i+1}\}}(x)$ is a periodic point of period $|\pi_i - \pi_{i+1}|$, thus x is a $\min\{\pi_i, \pi_{i+1}\}$th preimage of such a point. In this case, $\min\{\pi_i, \pi_{i+1}\} + |\pi_i - \pi_{i+1}| = \max\{\pi_i, \pi_{i+1}\} \leq L - 1$. In particular, if $\pi_i = 0$ or $\pi_{i+1} = 0$, then x is a periodic point.

In short, the endpoints of the intervals $P_\pi \neq \emptyset$, $\pi \in \mathcal{S}_L$, are given by the periodic points of f of periods $p \leq L - 1$, and their preimages up to the order $L - 2$. We conclude that the admissible ordinal patterns for f are determined by its periodic structure.

As a simple illustration, consider again the logistic map $g(x) = 4x(1 - x)$, $0 \leq x \leq 1$. For $L = 2$ we have, see Fig. 1.5,

$$P_{\langle 0,1\rangle} = \left(0, \tfrac{3}{4}\right), \quad P_{\langle 1,0\rangle} = \left(\tfrac{3}{4}, 1\right).$$

The separating point $x = \tfrac{3}{4}$ between $P_{\langle 0,1\rangle}$ and $P_{\langle 1,0\rangle}$ is given by the condition $g^{\pi_0}(x) = g^{\pi_1}(x)$, where $\pi_0, \pi_1 \in \{0,1\}$, i.e.,

$$g(x) = x.$$

For $L = 3$ ($g^2(x) = -64x^4 + 128x^3 - 80x^2 + 16x$), Fig. 1.6 shows that

$$
\begin{aligned}
P_{\langle 0,1,2\rangle} &= \left(0, \tfrac{1}{4}\right), & P_{\langle 0,2,1\rangle} &= \left(\tfrac{1}{4}, \tfrac{5-\sqrt{5}}{8}\right), & P_{\langle 2,0,1\rangle} &= \left(\tfrac{5-\sqrt{5}}{8}, \tfrac{3}{4}\right), \\
P_{\langle 1,0,2\rangle} &= \left(\tfrac{3}{4}, \tfrac{5+\sqrt{5}}{8}\right), & P_{\langle 1,2,0\rangle} &= \left(\tfrac{5+\sqrt{5}}{8}, 1\right).
\end{aligned}
\tag{1.29}
$$

The separating points of the intervals P_π, $\pi \in \mathcal{S}_3$, are given now by the conditions $g^{\pi_i}(x) = g^{\pi_{i+1}}(x)$, $\pi_i, \pi_{i+1} \in \{0,1,2\}$, i.e.,

$$g(x) = x, \ g^2(x) = x, \ g^2(x) = g(x).$$

We conclude that the common endpoints of the intervals P_π for $\pi \in \mathcal{S}_3$ are now the points of period 1 (fixed points), period 2, and first preimages of period-1 points. Moreover, when going from $L = 2$ to $L = 3$, we see that $P_{\langle 0,1\rangle}$ splits into the subintervals $P_{\langle 0,1,2\rangle}$, $P_{\langle 0,2,1\rangle}$, and $P_{\langle 2,0,1\rangle}$ at the eventually period-1 point $\tfrac{1}{4}$ (preimage of the fixed point $\tfrac{3}{4}$) and at the period-2 point $\tfrac{5-\sqrt{5}}{8}$. Likewise, $P_{\langle 1,0\rangle}$ splits into $P_{\langle 1,0,2\rangle}$ and $P_{\langle 1,2,0\rangle}$ at the period-2 point $\tfrac{5+\sqrt{5}}{8}$.

Ordinal patterns are the main ingredient of *permutation entropy* which, as the standard concept of entropy, comes also in metric and topological versions.

Suppose that μ is an f-invariant measure. Then the definition of the *metric permutation entropy* of f is formally similar to the definition of the Shannon entropy of an information source:

$$h_\mu^*(f) = -\lim_{L\to\infty} \frac{1}{L} \sum_{\pi \in \mathcal{S}_L} \mu(P_\pi) \log \mu(P_\pi), \tag{1.30}$$

provided the limit exists. Note that $\mu(P_\pi)$ is the probability for the ordinal L-pattern π to occur (while in the expression for the Shannon entropy, (1.1), the corresponding probabilities refer to length-L blocks x_0^{L-1}). Sometimes the factor $1/(L-1)$ is used instead of $1/L$ —of course, this is inconsequential in the limit $L \to \infty$.

As for the *topological permutation entropy* of f, one just counts distinct allowed patterns:

$$h_{\text{top}}^*(f) = -\lim_{L\to\infty} \frac{1}{L} \log |\{P_\pi \neq \emptyset : \pi \in \mathcal{S}_L\}|, \tag{1.31}$$

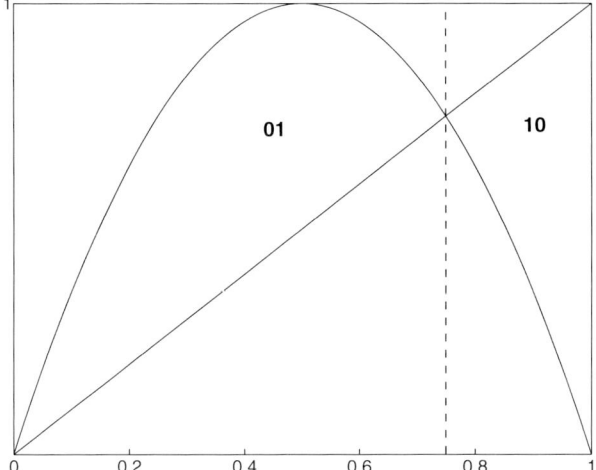

Fig. 1.5 Points in the interval $(0, \frac{3}{4})$ are of type $\langle 0, 1 \rangle$ (shorthanded 01), while points in the interval $(\frac{3}{4}, 0)$ are of type $\langle 1, 0 \rangle$ (shorthanded 10)

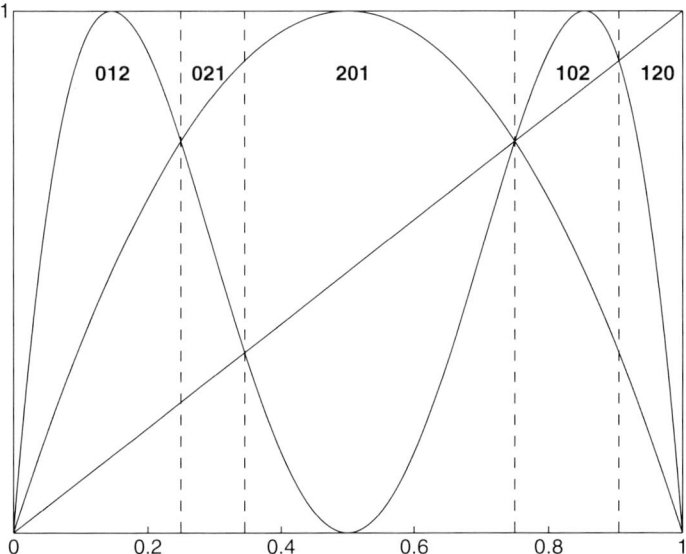

Fig. 1.6 The sets P_π, $\pi \in \mathcal{S}_3$, are graphically obtained by raising vertical lines at the crossing points of the curves $y = x$, $y = f(x)$, and $y = f^2(x)$. The three digits on the upper part of the figure are shorthand for ordinal patterns (e.g., 012 stands for $\langle 0, 1, 2 \rangle$). Observe that $P_{\langle 2,1,0 \rangle} = \emptyset$

where $|\cdot|$ denotes here cardinality. We are assuming again that this limit converges, otherwise $h^*_{\mathrm{top}}(f)$ is not defined.

An interval map $f:I \to I$ is called *piecewise monotone* if there is a finite partition of I into intervals, such that f is continuous and monotone on each of those intervals. A nice result of Bandt, Keller, and Pompe [29] states that if f is piecewise monotone, then (i) the metric permutation entropy of f coincides with its metric entropy and (ii) the topological permutation entropy of f coincides with its topological entropy. In mathematical notation:

$$\text{(i)}\ h^*_\mu(f) = h_\mu(f) \quad \text{and} \quad \text{(ii)}\ h^*_{\mathrm{top}}(f) = h_{\mathrm{top}}(f). \tag{1.32}$$

From (ii) and (1.31), it follows that if f is piecewise monotone and its topological entropy is finite, then

$$|\{P_\pi \neq \emptyset : \pi \in \mathcal{S}_L\}| \sim e^{L/h_{\mathrm{top}}(f)}, \tag{1.33}$$

where the symbol $\sim$ stands for "asymptotically as $L \to \infty$." Hence, the number of allowed L-patterns for f grows exponentially with L. On the other hand,

$$|\{P_\pi : \pi \in \mathcal{S}_L\}| = L! \sim e^{L(\ln L - 1) + 1/2 \ln 2\pi L}, \tag{1.34}$$

according to Stirling's formula for the factorial of a positive integer. Comparison of (1.33) and (1.34) not only does show that piecewise monotone maps have necessarily forbidden L-patterns for L sufficiently large but also that their number grows superexponentially with L.

From (1.29) we see that already for $L = 3$ there is one forbidden pattern for the logistic map, namely, $\langle 2, 1, 0 \rangle$. But this is not the end of the story. The absence of the ordinal pattern $\pi = \langle 2, 1, 0 \rangle$ triggers, in turn, an avalanche of longer missing patterns. To begin with, all the patterns $\langle *, 2, *, 1, *, 0, * \rangle$ (where the wildcard $*$ stands eventually for any other entries of the pattern) cannot be realized by any $x \in [0, 1]$ since the inequalities

$$\cdots < g^2(x) < \cdots < g(x) < \cdots < x < \cdots \tag{1.35}$$

cannot occur. By the same token, the patterns $\langle *, 3, *, 2, *, 1, * \rangle$, $\langle *, 4, *, 3, *, 2, * \rangle$, and, more generally,

$$\langle *, n + 2, *, n + 1, *, n, * \rangle \in \mathcal{S}_L, \ 0 \leq n \leq L - 3, \tag{1.36}$$

cannot be realized either for the same reason (replace x by $g^n(x)$ in (1.35)). We conclude that each forbidden pattern generates an infinite trail of ever-longer forbidden patterns. This issue will be revisited in full generality in Chap. 3.

Let us clarify this last point with the logistic map once more and $L = 4$. In Fig. 1.7, which is Fig. 1.6 with the curve

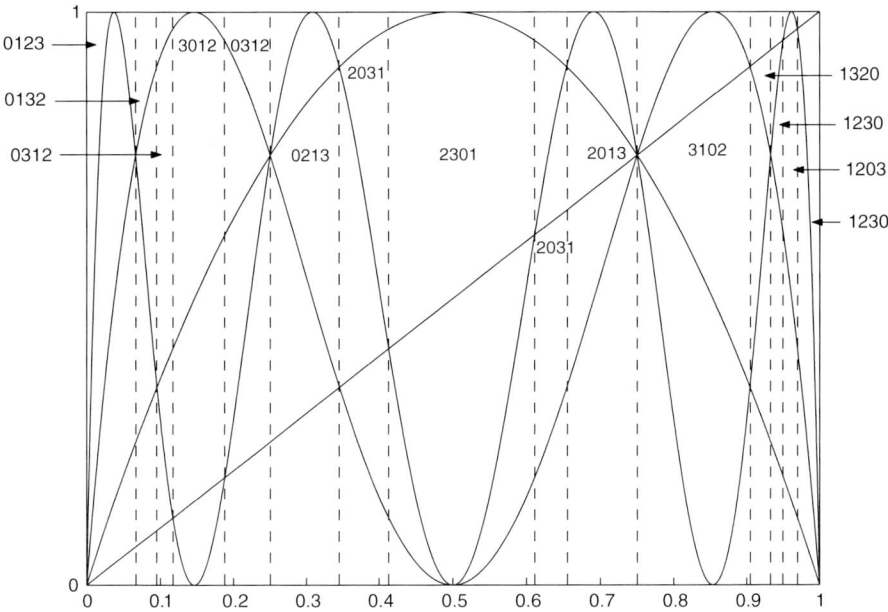

Fig. 1.7 The 12 allowed ordinal 4-patterns for the logistic map. Note the two components of $P_{\langle 0,3,1,2\rangle}$, $P_{\langle 2,0,3,1\rangle}$, and $P_{\langle 1,2,3,0\rangle}$

$$
\begin{aligned}
y &= g^3(x) \\
&= -16\,384x^8 + 65\,536x^7 - 106\,496x^6 + 90\,112x^5 \\
&\quad -42\,240x^4 + 10\,752x^3 - 1344x^2 + 64x
\end{aligned}
$$

superimposed, we can see the 12 allowed 4-patterns for the logistic map. Since there are 24 possible patterns of length 4, we conclude that 12 of them are forbidden. Seven forbidden 4-patterns belong to trail (1.36) of $\langle 2, 1, 0\rangle$ (observe that $\langle 3, 2, 1, 0\rangle$ is repeated):

$$
\begin{aligned}
(n = 0) \quad &\langle 3, 2, 1, 0\rangle, \langle 2, 3, 1, 0\rangle, \langle 2, 1, 3, 0\rangle, \langle 2, 1, 0, 3\rangle \\
(n = 1) \quad &\langle 0, 3, 2, 1\rangle, \langle 3, 0, 2, 1\rangle, \langle 3, 2, 0, 1\rangle, \langle 3, 2, 1, 0\rangle
\end{aligned}
\tag{1.37}
$$

Therefore, the remaining five forbidden 4-patterns,

$$
\langle 0, 2, 3, 1\rangle, \langle 1, 0, 2, 3\rangle, \langle 1, 0, 3, 2\rangle, \langle 1, 3, 0, 2\rangle, \langle 3, 1, 2, 0\rangle,
\tag{1.38}
$$

are seeds for new trails of forbidden patterns of lengths $L \geq 5$ that eventually can overlap.

In Fig. 1.7 one can also follow the first two splittings of the intervals P_π:

$$P_{\langle 0,1 \rangle} \rightarrow \begin{cases} P_{\langle 0,1,2 \rangle} \rightarrow P_{\langle 0,1,2,3 \rangle}, P_{\langle 0,1,3,2 \rangle}, P_{\langle 0,3,1,2 \rangle}, P_{\langle 3,0,1,2 \rangle}, \\ P_{\langle 0,2,1 \rangle} \rightarrow P_{\langle 0,2,1,3 \rangle}, \\ P_{\langle 2,0,1 \rangle} \rightarrow P_{\langle 2,0,1,3 \rangle}, P_{\langle 2,0,3,1 \rangle}, P_{\langle 2,3,0,1 \rangle}, \end{cases}$$

$$P_{\langle 1,0 \rangle} \rightarrow \begin{cases} P_{\langle 1,0,2 \rangle} \rightarrow P_{\langle 3,1,0,2 \rangle}, \\ P_{\langle 1,2,0 \rangle} \rightarrow P_{\langle 1,2,0,3 \rangle}, P_{\langle 1,2,3,0 \rangle}, P_{\langle 1,3,2,0 \rangle}. \end{cases}$$

The splitting of the intervals P_π can be understood in terms of periodic points and their preimages. Thus, the splitting of $P_{\langle 0,1 \rangle}$ is due to the points $\frac{1}{4}$ (first preimage of the period-1 point $\frac{3}{4}$) and $\frac{5-\sqrt{5}}{8}$ (a period-2 point); the second period-2 point, $\frac{5-\sqrt{5}}{8}$, is responsible for the splitting of $P_{\langle 1,0 \rangle}$. On the contrary, $P_{\langle 0,2,1 \rangle}$ and $P_{\langle 1,0,2 \rangle}$ do not split because they contain neither period-3 point nor first preimages of period-2 points nor second preimages of fixed points.

Chapter 2
First Applications

In this chapter we present four applications of permutation entropy and ordinal patterns: entropy estimation, complexity analysis, recovery of parameters from itineraries, and synchronization analysis of time series. The scope is to give the reader a multifaceted picture of ordinal analysis in action. Two more applications (to determinism detection and to space–time chaos) will be discussed at length in Chaps. 9 and 10, respectively.

2.1 Entropy Estimation

Real or numerical time series, say $(x_n)_{n \in \mathbb{N}_0}$ with $x_n \in \mathbb{R}$, can be produced in principle by discrete-time or continuous-time dynamical systems, which for convenience we think as including also the corresponding stochastic systems. In the continuous-time case, x_n can be thought as readouts of an analogue signal at discrete times, as it actually happens in practice. Formally, continuous-time dynamical systems are constructed from the solutions of ordinary differential equations and are called *flows* [98]. When solving differential equations numerically, the time variable is discretized anyway [173].

Permutation entropy made its first appearance in the analysis of univariate time series, i.e., sequences of real numbers—the only ones we will consider in this section. Given a finite time series[1] $x_0^{N-1} = x_0, x_1, \ldots, x_{N-1}$, take a sliding window of size $2 \le L \ll N$ along the time series (each window comprising a symbol block $x_n^{n+L-1} = x_n, \ldots, x_{n+L-1}, 0 \le n \le N-L$) and count the number of blocks realizing a particular ordinal pattern $\pi \in \mathcal{S}_L$. The relative frequency of each $\pi \in \mathcal{S}_L$ in the sequence x_0^{N-1} is then

$$\hat{p}(\pi) = \frac{\left| \{ n : 0 \le n \le N-L, \, x_n^{n+L-1} \text{ is of type } \pi \} \right|}{N-L+1}. \tag{2.1}$$

---

[1]For notational simplicity, we assume that one symbol is output per time unit. In this way, a time series can be labeled as a sequence.

J.M. Amigó, *Permutation Complexity in Dynamical Systems,*
Springer Series in Synergetics, DOI 10.1007/978-3-642-04084-9_2,
© Springer-Verlag Berlin Heidelberg 2010

This estimator of the probability of π converges with probability 1 to the true value in the limit of infinitely long time series, under the proviso that the underlying stochastic process is stationary or, at least, that the probability for $x_n < x_{n+k}$, $1 \leq k \leq L - 1$, does not depend on n [28]. Let us mention in passing that the ordinal pattern probability distributions have been calculated for some random processes and pattern lengths, like Gaussian, fractional Brownian, and autoregressive moving-average (ARMA) processes for $L \leq 4$ [30, 213]; see also [190].

The permutation entropy per symbol of order L of x_0^{N-1} is then defined as

$$h_L^*(x_0^{N-1}) = -\frac{1}{L} \sum_{\pi \in \mathcal{S}_L} \hat{p}(\pi) \log \hat{p}(\pi). \tag{2.2}$$

In the case of infinitely long sequences, one defines the permutation entropy of a sequence x_0^∞ as

$$h^*(x_0^\infty) = \lim_{L \to \infty} h_L^*(x_0^\infty), \tag{2.3}$$

provided the limit exists.

The general procedure followed so far is well known to the practitioners of non-linear time analysis: L is the *embedding dimension* and the *delay time T* is here 1 (since we take consecutive entries). As the window of size L slides along the time series x_0^∞, the vectors $\mathbf{x}_n = x_n^{n+L-1} \in \mathbb{R}^L$ describe the so-called *reconstructed trajectory* in the L-dimensional embedding space [1, 112, 166, 197]. The changes to be done when the sequences

$$x_n, x_{n+T}, \ldots, x_{n+(L-2)T}, x_{n+(L-1)T}, \tag{2.4}$$

have a delay time $T > 1$, are merely a matter of form but not of concept. Note that for deterministic sequences $x_n = f^n(x_0)$, $n \geq 0$, subsequence (2.4) is an orbit segment of f^T.

In general, h_L^* and h^* are defined for arbitrary-alphabet sequences whose symbols can be linearly ordered, while Shannon entropy applies to finite-alphabet sequences.[2] In practice all alphabets are finite because of the finite precision of the observation device and/or the finite real number representation of the computers. Such being the case, let $\mathbf{X} = \{X_n\}_{n \in \mathbb{N}_0}$ be the actual data source of the sequences x_0^∞, where now x_i are "discretized" values drawn from a finite alphabet S, and

$$h(\mathbf{X}) = -\lim_{n \to \infty} \frac{1}{L} \sum_{x_0, \ldots, x_{L-1} \in S} p(x_0, \ldots, x_{L-1}) \log p(x_0, \ldots, x_{L-1}),$$

---

[2]Real-valued data sources call for the concept of differential entropy [59].

its Shannon entropy. Usually, $h(\mathbf{X})$ is estimated by means of the so-called *plug-in*, *maximum likelihood*, or *naive estimator*

$$\hat{h}_L(x_0^{N-1}) = -\frac{1}{L}\sum \hat{p}(a_0 \ldots a_{L-1}) \log \hat{p}(a_0 \ldots a_{L-1}), \quad (2.5)$$

where the summation is over all blocks $a_0^{L-1} = a_0 \ldots a_{L-1} \in S^L$, and

$$\hat{p}(a_0 \ldots a_{L-1}) = \frac{\left|\{n : 0 \le n \le N - L, x_n^{n+L-1} = a_0^{L-1}\}\right|}{N - L + 1} \quad (2.6)$$

is the relative frequency of a_0^{L-1} in x_0^{N-1}.

Important for us is that if the process $\mathbf{X}$ is stationary and ergodic, then $h^*(x_0^\infty) = h(\mathbf{X})$ for a "typical" sequence (Chap. 6, Theorem 8). Therefore, in such cases $h_L^*(x_0^{N-1})$, with $L \ll N$, can be used as an estimator of $h(\mathbf{X})$ instead of (2.5). The numerical estimation of entropy via ordinal patterns will be discussed with more detail in Sect. 6.4, once the theoretical underpinnings of metrical permutation entropy of maps have been elucidated. At this point it suffices to advance that the computation is fast but the convergence is in general slow.

The slow convergence of h_L^* to the Shannon entropy seems to require great values of L for an accurate estimation. On the other hand, the superexponential growth of $|\mathcal{S}_L| = L!$ makes exhaustive sampling computationally unfeasible for, say, $L \gtrsim 12$, even if there would be enough data at our disposal. In Chap. 7 we shall learn sampling techniques that work pretty well in these cases. In practice, the estimation of both Shannon entropy and permutation entropy (or, for that matter, of any quantity involving the limit $L \to \infty$) suffers from *undersampling* when L becomes sufficiently large as compared to the length N of the sequence. Undersampling means that the observed relative frequencies (of blocks or ordinal patterns) are no longer good estimators of the corresponding probabilities, simply because the samples are too small to be statistically significant. The following first-order correction due to finite sample effects was proposed by Herzel [93]:

$$\hat{h}_L(x_0^{N-1}) \longleftarrow \hat{h}_L(x_0^{N-1}) - \frac{M_1}{2M_2}, \quad (2.7)$$

where M_1 is the number of words a_0^{L-1} with positive probabilities and M_2 is the number of samples ($M_2 = N - L + 1$ when the sequence is sampled by means of overlapping sliding windows, see (2.6)). In principle, the samples should be independent, but as stated in [94], the results are also satisfactory when the words overlap. Other corrections have been discussed by Grassberger [88] (who generalizes (2.7)) and Schmitt et al. [181] (who exploit Shannon–McMillan–Breiman's theorem of asymptotic equidistribution). Sometimes extrapolation techniques perform fine when undersampling occurs. One of them [195, 6] calls for plotting the partial entropies h_L^* against $1/L$; if the graph exhibits a distinctive linear part (showing

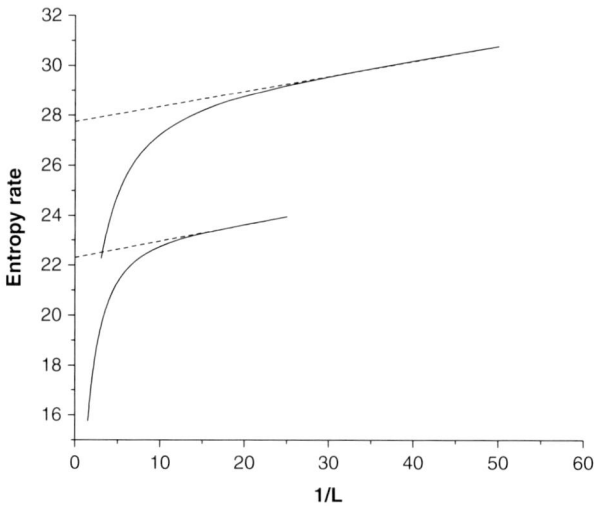

Fig. 2.1 Extrapolating the linear part (if any) of h_L^* vs $1/L$, over the undersampled values. The *continuous lines* correspond to entropy rates of finite order obtained from neurological time series

that h_L^*/L has already converged), then one extrapolates with a straight line this linear part till it intercepts the vertical axis ($1/L \to 0$), Fig. 2.1. See [127] for other methods to estimate the Shannon entropy and [167] for a review on entropy estimation.

Summing up, permutation entropy ("counting ordinal patterns") provides a conceptually simple and computationally fast method to estimate Shannon entropy. When compared to the usual block-based estimators ("counting blocks"), there is a difference that can be important in applications: the number of ordinal L-patterns does not depend on the alphabet. Specifically, the maximal number of length-L blocks (Shannon entropy) and length-L ordinal patterns (permutation entropy) grows with L as

$$|S|^L = e^{L \ln |S|} \quad \text{and} \quad L! \sim e^{L \ln L},$$

respectively, where S is the alphabet. It follows that if $|S|$ is very large, undersampling might set in earlier for block-based estimation than for ordinal pattern-based estimation. This occurs precisely with real-world or computer-generated data. Such an advantage has been reported in the literature, also in the computation of the *Rényi entropy*

$$h_{R_\alpha}(\mathbf{X}) = \lim_{L \to \infty} \frac{1}{L} \frac{1}{1 - \alpha} \log \left(\sum_{x_0, \ldots, x_{L-1} \in S} p(x_0, \ldots, x_{L-1})^\alpha \right), \qquad (2.8)$$

where $\alpha \geq 0$, $\alpha \neq 1$ ($\lim_{\alpha \to 1} h_{R_\alpha}(\mathbf{X}) = h(\mathbf{X})$), and the *Tsallis entropy*

$$h_{T_q}(\mathbf{X}) = \lim_{L \to \infty} \frac{1}{L} \frac{1}{q-1} \sum_{x_0, \dots, x_{L-1} \in S} \left(p(x_0, \dots, x_{L-1}) - p(x_0, \dots, x_{L-1})^q \right), \quad (2.9)$$

where $q \in \mathbb{R}$, $q \neq 1$ [213]. When $p(x_0, \dots, x_{L-1})$ is replaced in (2.8) and (2.9) by $p(\pi)$, $\pi \in S_L$ (or estimated by the relative frequency $\hat{p}(\pi)$), one speaks of the *Rényi permutation entropy* and the *Tsallis permutation entropy*, respectively. To complete the picture, let us add that the situation reverses when the alphabet comprises few symbols. But in this case, Lempel–Ziv complexity (specifically, LZ-76) can be a better choice than block counting [6]; see [82] for the entropy estimation in binary sequences.

Although less used than the "Shannon permutation entropy" h^*, one can also define the *topological permutation entropy* or *permutation capacity*,

$$h_0^*(x_0^\infty) = -\lim_{L \to \infty} h_{0,L}^*(x_0^\infty), \quad (2.10)$$

provided the limit exists, where the rate of finite order is given as

$$h_{0,L}^*(x_0^\infty) = -\frac{1}{L} \log N(L), \quad (2.11)$$

$N(L)$ being the number of distinct ordinal patterns defined by sliding windows x_n^{n+L-1} of size L. That is, we just count now how many different L-patterns are realized, instead of computing the relative frequency of those L-patterns. It follows that h_0^* is an upper bound of h^*. When the sequences x_0^∞ are seen as outputs of an information source $\mathbf{X}$, then $N(L)$ stands for the number of admissible L-patterns in the messages that $\mathbf{X}$ can emit, and one speaks of the permutation capacity or topological permutation entropy of $\mathbf{X}$ (Chap. 7).

The ordinal pattern-based approach to Shannon entropy can also be extended to the metric and topological entropy of maps; see Chaps. 7 and 8. The situation is specially simple for one-dimensional, piecewise monotone interval maps $f:I \to I$. In this case, we only need to numerically estimate the probabilities $\mu(P_\pi)$ of the admissible L-patterns ($P_\pi \neq \emptyset$), or just the number of distinct admissible patterns, to get an estimate of the metric or topological entropy of f, respectively (see (1.30), (1.31), and (1.32)). Thus, the estimation of $h_\mu^*(f)$ and $h_{\mathrm{top}}^*(f)$ boils down again to counting ordinal L-patterns. The computation of $h_{\mathrm{top}}^*(f)$ is also simpler than for its standard counterpart. The higher dimensional case will also be considered in Chaps. 6 and 7.

2.2 Permutation Complexity

Although complexity, (pseudo-)randomness, disorder, irregularity, typicality, etc., are terms that have been introduced eventually in different settings to mean more or less the same dynamical behavior, complexity is the preferred one when there is no measure (or probability) involved. In fact, Bandt and Pompe introduced permutation entropy in [28] via (2.1), (2.2), and (2.3) as a "natural complexity measure for time series." The time series can be the output of a random process or an orbit of a dynamical system. By analyzing the complexity of a signal (if no other information available), we are inquiring into the complexity of the source. An axiomatic characterization of complexity was proposed in [163].

The measurement of complexity and its eventual time variation is an issue of utmost important in the analysis of biomedical, economic, physical, and technical time series. Think of the forecasting of transitions to abnormal health conditions, financial crashes, severe weather, earthquakes, etc. Over the years, a battery of methods has been proposed and developed with this purpose or adapted from other fields like information theory and networks. Let us mention some of these methods (see also the references therein):

- Cross-correlation sum analysis [111]
- Lempel–Ziv complexity [208, 196, 90, 6, 78]
- Mutual information [90]
- Nonlinear cross-prediction analysis [183]
- Recurrence plots [73, 144, 200] and recurrence quantification analysis [81]
- Relative entropy [180]
- Statistical complexity [56, 143] (statistical complexity was introduced by Crutchfield and collaborators within a theory called computational mechanics [60, 185, 24])
- Statistical tests in the reconstructed phase space [120]
- Topological methods [209]

Permutation entropy and other related quantities are specially well suited to measure the complexity of random and deterministic dynamical systems for several reasons.

First of all, permutation entropy in its different variants involves counting ordinal patterns. With the exception of a few cases, the number of ordinal L-patterns realized by a map f increases with L. Therefore, the (logarithm of the) rate of this increasing is a natural measure (as stated by Bandt and Pompe) for quantifying the complexity of a deterministic time series or, more generally, of a dynamical system. In the metric variant, each admissible L-pattern contributes to the entropy a term containing its relative frequency or probability, respectively. In the topological variant, all such patterns make the same contribution to the entropy; formally, they are assigned the same probability. Since random, unconstrained processes have no forbidden patterns with probability 1 (hence, they have a superexponential growth of admissible ordinal patterns with length), their complexity, as measured by the permutation entropy, is infinite. At the other end, a periodic or quasiperiodic dynamic has vanishing or negligible permutation entropy. Complex systems lie between order and randomness. From a practical point of view, we can characterize them as having a positive, finite permutation entropy. Both metric and topological permutation entropies increase as the sequence "looks" more random.

Second, unlike other proposals for complexity measures, permutation entropy applies in principle both to finite-alphabet and arbitrary-alphabet sequences, albeit it is more interesting in the second case.

Technically we are assuming that the limits involved in the corresponding definitions (like (2.3) and (2.10)) converge. In practice, limits have to be estimated using a finite number of terms—real sequences are finite anyway. What we mean is that the actual tools of permutation complexity are going to be the permutation entropy rates of finite order, like $h_L^*(x_0^{N-1})$ and $h_{0,L}^*(x_0^{N-1})$, and other related quantities based on finite-length ordinal patterns, like probability distributions, information-theoretical tools (relative entropy, mutual information, etc.), complexity functionals. Moreover, since the maximal value of $h_L^*(x_0^{N-1})$ and $h_{0,L}^*(x_0^{N-1})$ is $\log L!$, we can eventually divide both entropy rates by $\log L!$ to obtain dimensionless quantities ranging between the two non-complex extremes: 0 (order) and 1 (randomness).

Finally, permutation entropy rates of finite order are computationally fast for the pattern lengths used in practice ($3 \leq L \leq 7$)—also for the Rényi (2.8) and Tsallis (2.9) permutation entropies. This allows calculation in real time, which is a significant advantage in applications. We come back to this point in the next chapters.

Application of ordinal patterns and permutation entropy to complexity analysis of data has been reported in different fields. For instance

- biomedical series [116, 45, 118]
- financial series [146, 147]
- physical series [28]
- statistical series [30, 146, 147, 212]

Let us underline at this point that the application by Keller [116] of ordinal patterns to electroencephalogram (EEG) data from children with epileptic disorders dates from about the same time as permutation entropy was formulated [28].

Similarly, one of the first applications of permutation entropy was the detection of dynamical changes in time series and, in particular, epileptic seizure detection from EEGs by Cao et al. [45]. Regarding the second application, the authors analyzed continuous EEG measurements recorded intracranially (also called depth EEG) with typically 28 electrodes. Figure 2.2 shows the normalized permutation entropy rate of order $L = 5$ for three different patients. Each signal is more than 5 h long, with a sample frequency of 200 Hz and time delay 3 (i.e., only every third entry in the EEG signal is taken into account, what amounts to sampling the signal with frequency 200/3 Hz). According to [45], the change of permutation complexity in all these cases indicates that the dynamics of the brain first becomes more regular right after the seizure, then its irregularity increases as it approaches the normal state.

Since these and other pioneering works, ordinal analysis of time series has remain a popular technique. In some cases, ordinal analysis has been incorporated into more general schemes, such as the *method of recurrence plots*, introduced by Eckmann et al. [73] to visualize the recurrences of dynamical systems. This method, which is being used to analyze virtually any natural data [144], is based on the *recurrence matrix* of a scalar or vectorial trajectory $(x_i)_{i=0}^{N-1}$ of a system in its state space S, defined as

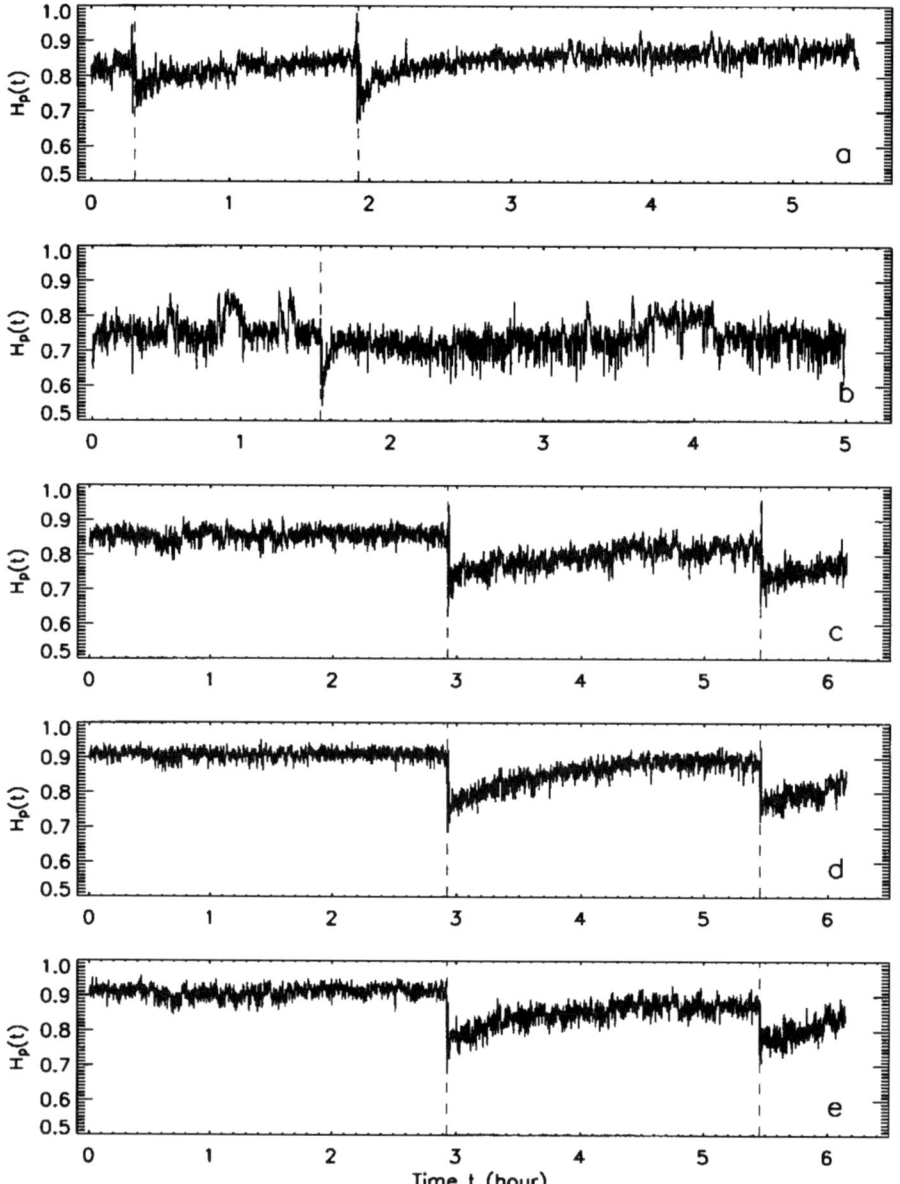

Fig. 2.2 [Reproduced with permission from [45].] Variation of the normalized h_5^* with time for EEG signals of (**a**) patient 1, channel 1, (**b**) patient 2, channel 1, and (**c**)–(**e**) patient 3, channels 1–3

$$\mathbf{R}_{i,j}(\varepsilon) = H(\varepsilon - \left\| x_i - x_j \right\|), \quad i,j = 0,\ldots,N-1, \qquad (2.12)$$

where ε is a threshold distance, $H(\,\cdot\,)$ is the *Heaviside function* ($H(x) = 0$ if $x < 0$ and $H(x) = 1$ otherwise), and $\|\cdot\|$ is a norm in S. Instead of using spatial closeness as in (2.12), *ordinal patterns recurrence plots* are based on the ordinal patterns $\pi(i)$ realized by the sequences x_i^{i+L-1}, $0 \le i \le N - L$. If $\delta(\pi, \pi') = 1$ for $\pi = \pi' \in \mathcal{S}_L$, and $\delta(\pi, \pi') = 0$ otherwise, set

$$\mathbf{R}_{i,j}(L) = \delta(\pi(i), \pi(j)), \tag{2.13}$$

$\pi(i), \pi(j) \in \mathcal{S}_L$, $0 \le i, j \le N - L$. According to [144], the main advantage of (2.13) is its robustness against non-stationary data.

To distinguish the kind of complexity captured by the tools of ordinal analysis— ordinal patterns, permutation entropy, permutation entropy rates of finite order, and other quantities based on order relations—we propose to call it *permutation complexity*. Therefore, permutation complexity has to do with the ordinal structure of data obtained from deterministic or random dynamical systems. These also include spatially extended systems, like the ones we shall consider in Chap. 10.

2.3 Estimation of Control Parameters from Symbolic Sequences

The basis of permutation complexity is the relation between order and dynamics. This relation is specially strong on one-dimensional intervals, where order and metric are intertwined, leading to such interesting results as Sarkovskii's theorem [179, 150]. It is therefore not surprising that the study of the ordinal structure of time series provides valuable information on the underlying dynamical system. In this section we learn how to recover the "control" parameter of a unimodal map from itineraries. The relationship between the itineraries of parametric unimodal maps and the value of the parameter that controls a particular dynamics was shown in [153, 203, 5].

Let $\mathcal{U}$ be the class of unimodal maps on an interval $I = [a, b] \subset \mathbb{R}$. A map $f : I \to I$ is *unimodal* if it is continuous, has a single turning point (called hereafter the *critical point*) x_c in I, and is monotone increasing on the left of x_c and decreasing on the right. The class $\mathcal{U}$ includes maps defined in a parametric way, say, $f_\nu(x) = \varphi(\nu, x)$, where $x \in I$, $\nu \in J \subset \mathbb{R}$ will be called the *control parameter*, and φ is a map on $I \times J$.

The class $\mathcal{U}$ includes the *logistic family* $g_\nu : [0, 1] \to [0, 1]$,

$$g_\nu(x) = \nu x(1 - x), \tag{2.14}$$

where $0 \le \nu \le 4$, and the *tent family* $\Lambda_\nu : [0, 1] \to [0, 1]$,

$$\Lambda_\nu(x) = \begin{cases} x/\nu & \text{if } 0 \le x \le \nu, \\ (1 - x)/(1 - \nu) & \text{if } \nu \le x \le 1, \end{cases} \tag{2.15}$$

where $0 < \nu < 1$; see Fig. 2.3. In particular, g_4 is the logistic map (1.19) and $\Lambda_{1/2}$ the symmetric tent map (1.17). The critical point of g_ν does not depend on ν: $x_c = \frac{1}{2}$

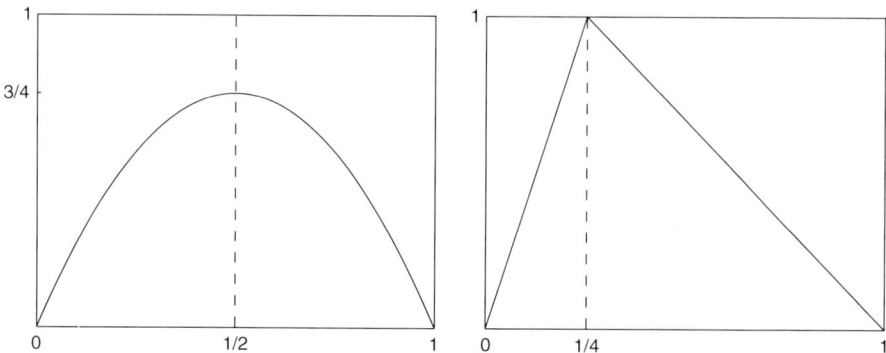

Fig. 2.3 Graphs of the logistic map g_v with $v = 3$ (*left*) and the tent map Λ_v with $v = 0.25$ (*right*)

for all v. On the opposite side, the critical point of Λ_v coincides with the parameter value: $x_c = v$. As usual in the literature, we will also refer to g_v and Λ_v just as the logistic and tent maps, respectively, when the parameter v is thought to be fixed. Note that Λ_v preserves the Lebesgue measure for all $v \in (0, 1)$.

For $f \in \mathcal{U}$, let $\Phi(x)$ be the itinerary of $x \in [a, b]$ with respect to the partition $\{A_0, A_1\}$, with $A_0 = [a, x_c)$ and $A_1 = [x_c, b]$. Specifically,

$$\Phi(x) = \Phi_0(x), \Phi_1(x), \ldots, \Phi_n(x), \ldots = (\Phi_i(x))_{i=0}^{\infty}, \qquad (2.16)$$

where

$$\Phi_n(x) = \begin{cases} 0 & \text{if } f^n(x) < x_c, \\ 1 & \text{if } f^n(x) \geq x_c. \end{cases}$$

As a result, any orbit $\mathcal{O}_f(x)$ can be encoded into a binary sequence. Whenever convenient, we will write $\Phi(f, x)$ instead of $\Phi(x)$ to make clear which unimodal map is generating the itinerary of x.

An interesting aspect of the binary sequences $\Phi(x)$ is that they can be endowed with a *signed lexicographical order* (sometimes called *Gray ordering*) $\leq$ that is equivalent to the order in $[a, b]$ in the following weakened sense:

(E1) If $x < y$, then $\Phi(x) \leq \Phi(y)$.
(E2) If $\Phi(x) < \Phi(y)$, then $x < y$.

A sufficient condition for $x < y$ if and only if $\Phi(x) \leq \Phi(y)$ is given in [57, Theorem II.5.4]. The order between binary sequences is defined as follows. Given $\Phi(x) \neq \Phi(y)$, let $i_{\min}$ be the first index such that $\Phi_i(x) \neq \Phi_i(y)$, $i \geq 0$. Depending on $i_{\min}$, three cases can occur:

(O1) If $i_{\min} = 0$, then $\Phi(x) < \Phi(y)$ iff $\Phi_0(x) < \Phi_0(y)$.

(O2) If $i_{\min} > 0$ and $\{\Phi_i(x){:}0 \leq i < i_{\min}\}$ contains an *even* number of 1's, then
$\Phi(x) < \Phi(y)$ iff $\Phi_{i_{\min}}(x) < \Phi_{i_{\min}}(y)$.

(O3) If $i_{\min} > 0$ and $\{\Phi_i(x){:}0 \leq i < i_{\min}\}$ contains an *odd* number of 1's, then
$\Phi(x) < \Phi(y)$ iff $\Phi_{i_{\min}}(x) > \Phi_{i_{\min}}(y)$.

Given $x, f_v(x), \ldots, f_v^{L-1}(x)$, suppose that their corresponding itineraries, namely,

$$(\Phi_i(x))_{i=0}^{\infty}, (\Phi_i(x))_{i=1}^{\infty}, \ldots, (\Phi_i(x))_{i=L-1}^{\infty},$$

are all different. Then, according to (E1)–(E2),

$$f^{\pi_0}(x) < \cdots < f^{\pi_{L-1}}(x) \Leftrightarrow (\Phi_i(x))_{i=\pi_0}^{\infty} < \cdots < (\Phi_i(x))_{i=\pi_{L-1}}^{\infty}. \qquad (2.17)$$

Before proceeding further, let us point out that this setting can be extended to l-*modal maps*, i.e., continuous and piecewise strictly monotone self-maps of compact intervals with l local maxima, which map endpoints to endpoints. For the applications we will discuss, it is sufficient to consider only unimodal maps ($l = 1$).

In some applications, one is confronted with the following task: given the "sharp" orbit $\mathcal{O}_{f_v}(x_0)$ of $x_0 \in [a, b]$ under $f_v \in \mathcal{U}$, find the value of the parameter v. In practice, the exact values of $\mathcal{O}_{f_v}(x_0)$ are seldom known because of the finite precision of real number computation, so one has only access to a (finite segment of a) "coarse-grained" orbit $(\hat{x}_i)_{i=0}^{\infty}$, where $\hat{x}_i$ is an approximation to $x_i = f_v^i(x_0)$. In some chaos-based cryptosystems, the situation is even worse: the plaintext (i.e., the message to be encrypted prior to its transmission or storage) is encoded via the symbolic sequences (2.16) of a chaotic map $f_v \in \mathcal{U}$, the value v being part of the secret key of the cipher (see, e.g., [131]). Therefore, the cryptanalyst has eventually only access to the binary code $\Phi(f_v, x)$ (via a so-called chosen-text attack) to recover the control parameter v. D. Arroyo has shown how to recover v with the aid of the ordinal patterns of f_v and their itineraries $\Phi(f_v, x)$, if f_v is ergodic with respect to its natural measure μ_v for all values of v [21].

For simplicity, the estimation of v from the symbolic sequences $\Phi(f_v, x)$ will be illustrated using the tent map Λ_v, which is chaotic for all $v \in (0, 1)$. Since the natural invariant measure of Λ_v is the Lebesgue measure, the probability that x is of type $\pi \in \mathcal{S}_L$ when drawn uniformly from $[0, 1]$ equals the length of $P_{\pi} = \{x \in [0, 1]{:}x$ defines $\pi\}$ (as in (1.27)). By ergodicity, the relative frequency of π in an orbit of Λ_v coincides with the length of P_{π}, except possibly for a set of initial conditions with length zero. For the tent map, the length of the sets P_{π} can be determined analytically. The simplest case corresponds to the L-pattern $\langle 0, 1, \ldots, L - 1\rangle$:

$$P_{\langle 0,1,\ldots,L-1\rangle} = (0, \phi_L(v)),$$

where $\phi_L(v)$ is the leftmost intersection of Λ_v^{L-1} and Λ_v^{L-2}. Therefore, the length of $P_{\langle 0,1,\ldots,L-1\rangle}$, hence the probability that x is of type $\langle 0, 1, \ldots, L - 1\rangle$ when drawn uniformly from the interval $[0, 1]$ is $\phi_L(v)$.

In order to calculate $\phi_L(v)$, use

$$\Lambda_v^n(x) = \begin{cases} x/v^n & \text{if } 0 \le x \le v^n, \\ (v^{n-1} - x)/v^{n-1}(1-v) & \text{if } v^n \le x \le v^{n-1}. \end{cases}$$

Equating Λ_v^{L-1} and Λ_v^{L-2}, it follows

$$\phi_L(v) = \frac{v^{L-2}}{2-v}. \tag{2.18}$$

Note that this function is 1-to-1 in the interval $0 \le v \le 1$ for $L \ge 2$, with $\phi_2(0) = \frac{1}{2}$, $\phi_L(0) = 0$ for $L \ge 3$, and $\phi_L(1) = 1$ for $L \ge 2$. This fact allows to determine v from $\phi_L(v)$. Furthermore, from the equation

$$\frac{d}{dv}\phi_L(v) = \frac{v^{L-3}}{(2-v)^2}[2(L-2) - (L-3)v] = \begin{cases} 0 & \text{if } v = 0, \\ L-1 & \text{if } v = 1, \end{cases} \tag{2.19}$$

it follows that $\phi_L(v)$ is a $\cup$-convex map on $0 \le v \le 1$ for $L \ge 2$ that converges to 0 on $0 \le v < 1$ (i.e., it "flattens") as $L \to \infty$. As a result, the higher the L the lower the precision with which v can be numerically read off from $\phi_L(v)$. Consequently, $L = 3, 4$ are the best choices for a quality estimation of v.

In more general terms, suppose that each $f_v \in \mathcal{U}$ is ergodic for $v \in J$ with the same invariant measure μ. Furthermore assume for the time being that $f_v(a) = a$ and $f_v(x) > x$ on a non-empty vicinity of a. Let (a, c) be the maximal interval in (a, x_c) such that $f_v(x) > x$. We claim that the interval

$$I_L^v = (a, c) \cap f_v^{-1}(a, c) \cap \cdots \cap f_v^{-(L-1)}(a, c)$$

coincides with $P_{\langle 0,1,\ldots,L-1 \rangle}$. Indeed, if $x \in I_L^v$, then $f_v^i(x) \in (a, c)$ for $0 \le i \le L-1$, and

$$x < f_v(x) \Rightarrow f_v(x) < f_v^2(x) \Rightarrow \cdots \Rightarrow f_v^{L-2}(x) < f_v^{L-1}(x).$$

Hence, $I_L^v \subset P_{\langle 0,1,\ldots,L-1 \rangle}$. Conversely, if $x \in P_{\langle 0,1,\ldots,L-1 \rangle}$, i.e.,

$$x < f_v(x) < f_v^2(x) < \cdots < f_v^{L-1}(x),$$

then $f_v^i(x) \in (a, c)$ for $0 \le i \le L-1$. Thus, $P_{\langle 0,1,\ldots,L-1 \rangle} \subset I_L^v$. This proves

$$I_L^v = P_{\langle 0,1,\ldots,L-1 \rangle}. \tag{2.20}$$

If otherwise $f_v(a) = a$ but $f_v(x) < x$ on a non-empty vicinity of a, then let (a, c) be the maximal interval in (a, x_c) such that $f_v(x) < x$. In this case, a similar reasoning (reversing the inequalities) shows that

$$I_L^v = P_{\langle L-1,L-2,\ldots,1,0 \rangle}. \tag{2.21}$$

Since the tent map, our workhorse in this section, complies with (2.20), we restrict attention to this case (similar arguments apply mutatis mutandis to case (2.21)). Because of ergodicity, the relative frequency at which a typical trajectory visits I_L^v is $\mu(I_L^v)$. If $\mu(I_L^v)$ happens to be different for each v, then $\mu(I_L^v)$ can be used to determine or estimate the control parameter v. In this case, the relative frequency of the ordinal pattern $\langle 0, 1, \ldots, L-1 \rangle$ in an orbit $\mathcal{O}_{f_v}(x)$ is just the number of times that $f_v^{i+j}(x)$ $\in (a, c)$ for $i \in \mathbb{N}_0$ and $j = 0, 1, \ldots, L-1$.

Figure 2.4 shows the relative frequencies of the ordinal patterns (a) $\langle 0, 1, 2, 3 \rangle$, (b) $\langle 0, 1, 3, 2 \rangle$, (c) $\langle 0, 3, 1, 2 \rangle$, and (d) $\langle 3, 0, 1, 2 \rangle$ found in a numerical simulation with the tent map. As expected, curve (a) approximates the function

$$\phi_4(v) = \frac{v^2}{2-v}$$

with great precision. Observe that a 1-to-2 functional relation between frequency and v, as it occurs in Fig. 2.4 (b)–(d), can also be acceptable, e.g., for cryptographic applications since it implies a reduction of the secret key space.

So far we have shown the possibility of recovering the control parameter v from the relative frequency of the pattern $\pi = \langle 0, 1, \ldots, L-1 \rangle$ (most conveniently for $L = 3, 4$), in a statistically significant sample of orbits of Λ_v. The ergodicity of Λ_v with respect to the Lebesgue measure on $[0, 1]$ and the 1-to-1 relation between v and

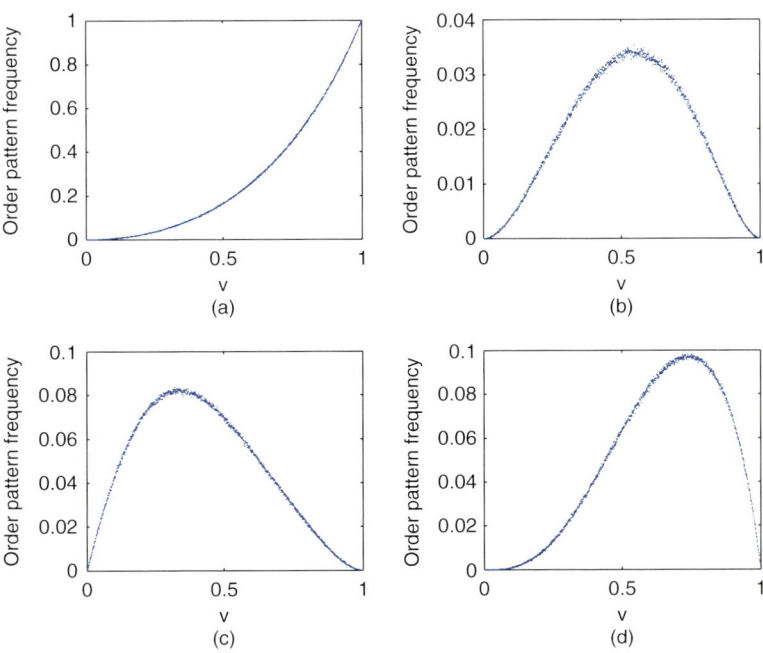

Fig. 2.4 Ordinal pattern frequencies for the tent map family. Here $L = 4$, and **(a)** $\pi = \langle 0, 1, 2, 3 \rangle$, **(b)** $\pi = \langle 0, 1, 3, 2 \rangle$, **(c)** $\pi = \langle 0, 3, 1, 2 \rangle$, and **(d)** $\pi = \langle 3, 0, 1, 2 \rangle$

the probability $\phi_L(v)$ of observing π were instrumental to achieve that goal. What about if we have access only to coarse-grained orbits $\Phi(\Lambda_v, x)$?

Let $b_0^{M-1} = b_0 b_1 \ldots b_{M-1}$, $b_i \in \{0, 1\}$, be the initial segment of length M of the symbolic sequence $\Phi(f_v, x)$. Take a sliding window of size $W < M$ along b_0^{M-1}. The result is $M - W + 1$ consecutive blocks of length W:

$$b_0^{W-1} = b_0 \ldots b_{W-1}, \ldots, b_i^{i+W-1} = b_i \ldots b_{i+W-1}, \ldots, b_{M-W}^{M-1} = b_{M-W} \ldots b_{M-1}.$$

The blocks b_i^{i+W-1}, $i = 0, 1, \ldots, M - W - L + 1$, define $M - W - L + 2$ ordinal patterns of length L. That is, if

$$b_{i+\pi_0}^{i+\pi_0+W-1} < b_{i+\pi_1}^{i+\pi_1+W-1} < \cdots < b_{i+\pi_{L-1}}^{i+\pi_{L-1}+W-1}, \tag{2.22}$$

then b_i^{i+W-1} is of type $\pi = \langle \pi_0, \pi_1, \ldots, \pi_{L-1} \rangle$. The order for finite sequences in (2.22) is defined the same way as for infinite sequences in (O1)–(O3).

Each block $b_i^{i+W-1} = b_i \ldots b_{i+W-1}$ locates $f_v^i(x)$ up to an uncertainty interval whose length goes to zero when $W, M \to \infty$:

$$f_v^i(x) \in A_{b_i} \cap f_v^{-1} A_{b_{i+1}} \cap \cdots \cap f_v^{-(W-1)} A_{b_{i+W-1}}.$$

This being the case, the ordinal patterns defined by, say, $x, f_v(x), f_v^2(x)$, and $b_0^{W-1}, b_1^W, b_2^{W+1}$ may be different as soon as two of the latter blocks overlap. Otherwise, both ordinal patterns will be the same because of (2.17). In sum, the relative frequencies of an ordinal L-pattern in the finite orbits $(f_v(x))_{i=0}^M$ and $(b_i^{i+W-1})_{i=0}^{M-W-L+1}$ will converge to each other in the limit $M \to \infty$, $W \to \infty$ ($W < M$). In practice, we expect the latter to be a good approximation of the former, at least for $L = 3, 4$, and W large enough, so that a good estimation of the control parameter is feasible.

Figure 2.5 shows the relative frequencies of the same 4-patterns as in Fig. 2.4 for the itineraries of the tent map family. Here $M = 10, 104$ and $W = 100$. Except for $v \simeq 0$ (an uninteresting region for cryptographic applications), the approximation is excellent. Some caveats related to the finite precision of the numerical simulations are discussed in [21]. In practical cases, the error in the estimation of the control parameter ranges between 10^{-3} and 10^{-4}. From the viewpoint of cryptographic applications, this amounts to a strong reduction of the key space, which compromises the security of the cipher.

The tent map family is a specimen of a more general family: unimodal, piecewise linear expanding Markov transformations (Annex A, Definition 9). Each topologically transitive transformation in this family (i.e., some power of its transition matrix is strictly positive) has a unique ergodic invariant measure, which furthermore is absolutely continuous with respect to the Lebesgue measure [134]. This measure can be calculated or numerically estimated by a variety of methods (Perron–Frobenius operator, Ulam's method, or just computation of long time averages) [105]. For the purpose envisaged in this chapter, an exact knowledge of the invariant measures is

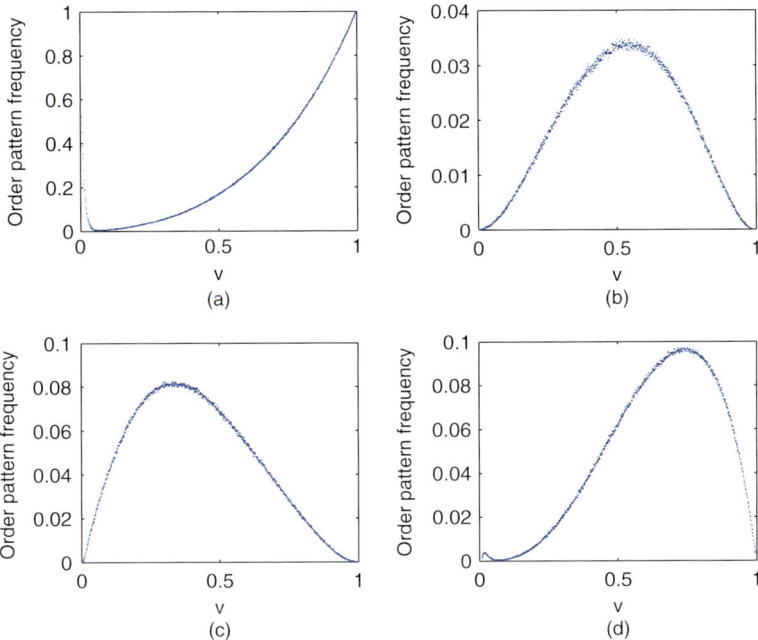

Fig. 2.5 Ordinal pattern frequencies of the tent map family using itineraries. Here $L = 4$, $W = 100$, $M = 10104$, and (**a**) $\pi = \langle 0, 1, 2, 3 \rangle$, (**b**) $\pi = \langle 0, 1, 3, 2 \rangle$, (**c**) $\pi = \langle 0, 3, 1, 2 \rangle$, and (**d**) $\pi = \langle 3, 0, 1, 2 \rangle$

not necessary, since the relative frequencies of the ordinal patterns can be calculated with numerical simulations. The important features are ergodicity, so that the statistical properties of the sharp and coarse-grained orbits do not depend on the initial conditions, and the absolute continuity of the (unique) invariant measure, which guarantees that it is accessible by numerical methods.

2.4 Characterizing Synchronization

As a last application, we are going to summarize the work of R. Monetti et al. [159] on characterizing synchronization in time series using ordinal patterns (therein called "symbols") and some related probability distributions.

Remember that $\mathcal{S}_L$ is a group with respect to the product of ordinal patterns (1.25). This being the case, given $\pi, \pi' \in \mathcal{S}_L$ there always exists a unique $\tau = \tau(\pi, \pi') \in \mathcal{S}_L$, called *transcription* from the *source pattern* π to the *target pattern* π', such that

$$\tau \circ \pi = \pi', \tag{2.23}$$

where (see (1.25))

$$\tau \circ \pi = \left\langle \pi_{\tau_0}, \pi_{\tau_1}, \ldots, \pi_{\tau_{L-1}} \right\rangle.$$

It follows that τ is a transcription from π to π' if and only if τ^{-1} is a transcription from π' to π.

As the source pattern π and the target pattern π' vary over $\mathcal{S}_L$, their transcription varies according to $\tau(\pi, \pi') = \pi' \circ \pi^{-1}$. Note that different pairs (π, π') can share the same transcription. As an example in $\mathcal{S}_3$, $\tau(\pi, \pi') = \langle 0, 2, 1 \rangle$ for

$$(\pi, \pi') = (012, 021), (021, 012), (120, 102), (102, 120), (201, 210), (210, 201)$$

(angular parentheses omitted for brevity). More generally, given $\tau \in \mathcal{S}_L$, there exist $L!$ pairs $(\pi, \pi') \in \mathcal{S}_L \times \mathcal{S}_L$ such that τ is the transcript from π to π'.

Another concept we need is that of order of an element. We say that the *order* of $\pi \in \mathcal{S}_L$ is $\mathrm{ord}(\pi) \in \mathbb{N}$ if $\mathrm{ord}(\pi)$ is the minimal number of times we have to multiply π by itself to obtain the identity permutation $\langle 0, 1, \ldots, L-1 \rangle$ (this is the only permutation whose order is 1).

The group $\mathcal{S}_L$ can be partitioned into non-overlapping sets of transcriptions according to their order. In mathematical notation, $\mathcal{S}_L = \cup_{1 \leq i \leq L!} \mathcal{C}_i$, where

$$\mathcal{C}_i = \mathcal{C}_i(L) = \{\tau \in \mathcal{S}_L : \mathrm{ord}(\tau) = i\}.$$

For obvious reasons, the sets $\mathcal{C}_i$ are called *order classes*. From $\mathrm{ord}(\tau^{-1}) = \mathrm{ord}(\tau)$, it follows that $\tau \in \mathcal{C}_i$ if and only if $\tau^{-1} \in \mathcal{C}_i$. Note that $\mathcal{C}_1(L) = \{\langle 0, 1, \ldots, L-1 \rangle\}$. The authors of [159] propose to measure the complexity of a transcription between a source and a target pattern by its order.

A permutation of the form $i_1 \mapsto i_2 \mapsto \cdots \mapsto i_n \mapsto i_1$ is called a *cycle* (or cyclic permutation) *of length* n and denoted by $(i_1, i_2, \ldots, i_n)$. The order of a cycle of length n is trivially n. It is also trivial that any permutation of $\{0, 1, \ldots, L-1\}$ can be written as the product of disjoint cyclic permutations. It follows that the order of any transcription (or any permutation for that matter) is the least common multiple (lcm) of the lengths of its decomposition into cycles. In particular, given L there are ordinal patterns $\tau \in \mathcal{C}_i(L)$ of orders $1 \leq i \leq L$ (just take $\tau = (0, \ldots, i-1)(i)(i+1)\cdots(L-1)$). For $L+1 \leq i \leq L!$, a hypothetical decomposition $\tau = (i_1, \ldots, i_{n_1})(j_1, \ldots, j_{n_2}) \cdots (k_1, \ldots, k_{n_p})$, $\tau \in \mathcal{C}_i(L)$, has to fulfill the constraints (i) $n_1 + n_2 + \cdots + n_p = L$ and (ii) $\mathrm{lcm}\{n_1, n_2, \ldots, n_p\} = i$, which will not be the case in general. For example, for $L = 7$ and $i = 10$ or 12, we can choose $n_1 = 2$ and $n_2 = 5$, or $n_1 = 3$ and $n_2 = 4$, respectively. But for $L = 7$ and $i = 8, 9$, or 11, conditions (i) and (ii) cannot be simultaneously satisfied.

Let us next turn attention to the probability density of transcriptions. Consider source and target ordinal patterns generated by the time series of a coupled dynamics. Due to the symmetry property between source and target patterns pointed out above, it is irrelevant which one refers to which subsystem, any of the two possible assignments being fine. Let $\mathcal{S}_L^s$ and $\mathcal{S}_L^t$ be the state spaces comprising the corresponding admissible source and target patterns of length L, respectively, and let $\Omega_L(\tau)$ be the set of all pairs $(\pi_s, \pi_t) \in \mathcal{S}_L^s \times \mathcal{S}_L^t$ such that $\tau \in \mathcal{S}_L$ is a transcription from π_s to π_t, i.e.,

$$\Omega_L(\tau) = \{(\pi_s, \pi_t) \in \mathcal{S}_L^s \times \mathcal{S}_L^t : \tau \circ \pi_s = \pi_t\}.$$

The probability density of transcriptions $P_L(\tau)$, $\tau \in \mathcal{S}_L$, can be written as

$$P_L(\tau) = \sum_{(\pi_s, \pi_t) \in \Omega_L(\tau)} P^J(\pi_s, \pi_t),$$

where $P^J(\pi_s, \pi_t)$ is the joint probability density. Furthermore, let $P^s(\pi_s)$, $P^t(\pi_t)$ be the marginal probability densities of the patterns $\pi_s \in S_s$ and $\pi_t \in S_t$, respectively. The matrix $M(\pi_s, \pi_t) = P^s(\pi_s)P^t(\pi_t)$ is the probability density matrix of transcriptions for two independent sequences of lengths L. In this case, the corresponding probability density of transcriptions $P_L^{ind}(\tau)$ can be evaluated as follows:

$$P_L^{ind}(\tau) = \sum_{(\pi_s, \pi_t) \in \Omega_L(\tau)} M(\pi_s, \pi_t).$$

A natural measure to assess how much $P_L(\tau)$ deviates from $P_L^{ind}(\tau)$ is provided by the *relative entropy* or *Kullback–Leibler distance* (see Annex B, (B.3))

$$D(P_L \| P_L^{ind}) = \sum_{\tau \in \mathcal{S}_L} P_L(\tau) \log \frac{P_L(\tau)}{P_L^{ind}(\tau)}.$$

To circumvent the asymmetry of the relative entropy with respect to its arguments, one can take the harmonic mean of $D(P_L \| P_L^{ind})$ and $D(P_L^{ind} \| P_L)$,

$$S_{KL}(L) = \frac{D(P_L \| P_L^{ind})\, D(P_L^{ind} \| P_L)}{D(P_L \| P_L^{ind}) + D(P_L^{ind} \| P_L)}.$$

In contrast to the symmetrization via the arithmetic mean, the bound

$$S_{KL}(L) \leq \min\{D(P_L \| P_L^{ind}), D(P_L^{ind} \| P_L)\}$$

furnishes more general conditions for the symmetrized Kullback–Leibler distance to be finite. Moreover we shall write $S_{KL}^{\mathcal{C}}(L)$ when the Kullback–Leibler distance is calculated using the probability densities $P_{\mathcal{C}}$ of the order classes (see Fig. 2.7). Finally, if $P_L(\tau)$ and $P_L^{ind}(\tau)$ are obtained using only transcriptions from an order class $\mathcal{C}_i(L)$, then the notation will be $S_{KL}^i(L)$. The point in doing so is that the dynamics of coupled systems may lead to the extinction of order classes, a feature referred to as *saturation* in [159].

Let us apply the method to a bidirectionally coupled Rössler–Rössler system [175] defined by the following set of equations:

$$\dot{x}_{1,2} = -w_{1,2}y_{1,2} - z_{1,2} + k(x_{2,1} - x_{1,2}),$$
$$\dot{y}_{1,2} = w_{1,2}x_{1,2} + 0.165y_{1,2},$$
$$\dot{z}_{1,2} = 0.2 + z_{1,2}(x_{1,2} - 10).$$

Here $w_1 = 0.99$ and $w_2 = 0.95$ are the mismatch parameters and k is the coupling constant. All the time series were generated using a fourth-order Runge–Kutta method with time step $\Delta t = 10^{-3}$ and initial conditions: $x_1(0) = -0.4$, $y_1(0) = 0.6$, $z_1(0) = 5.8$, $x_2(0) = 0.8$, $y_2(0) = -2$, and $z_2(0) = -4$. This chaotic system exhibits a rich synchronization behavior that ranges from phase ($k \approx 0.036$) to lag ($k \approx 0.14$) and finally to complete synchronization as k increases [175]. In [159] the authors only study the x-components of the Rössler subsystems. Specifically, time series of length 2^{19} (about 775 orbits) were sampled with delay $T = 150$ and dimension L, to obtain delay vectors

$$(x(n\Delta t), x((n + T)\Delta t), \ldots, x((n + (L - 1)T)\Delta t))$$

from either subsystem. Following [80] the delay was chosen so as to minimize the mutual information (Annex B, (B.6)) of the coordinates $x_1(t)$ and $x_1(t + T\Delta t)$ for the uncoupled system ($k = 0$).

Figure 2.6(a) shows the symmetrized Kullback–Leibler distance $S_{KL}^{\mathcal{C}}(L)$ obtained using the probability density $P_{\mathcal{C}}$ of order classes for $L = 6$ and $L = 7$. Figure 2.6 (b)–(d) shows $S_{KL}(L)$ obtained with the probability density of transcriptions in all non-empty order classes for $L = 6$ ($\mathcal{C}_2$–$\mathcal{C}_6$ in subfigure (b)) and $L = 7$ ($\mathcal{C}_7$, $\mathcal{C}_{10}$, and $\mathcal{C}_{12}$ in subfigure (c) and $\mathcal{C}_2$–$\mathcal{C}_6$ in subfigure (d)). We comment first the salient features of $S_{KL}^{\mathcal{C}}(6)$ and $S_{KL}^{\mathcal{C}}(7)$.

The increase of $S_{KL}^{\mathcal{C}}$ at $k \approx 0.036$ is due to the transition from (almost) uncoupled dynamics to phase synchronization. For stronger coupling k, $S_{KL}^{\mathcal{C}}$ increases rather monotonically until $k \approx 0.11$. For $k \in [0.11, 0.145]$, $S_{KL}^{\mathcal{C}}$ displays strong fluctuations revealing the presence of "intermittent-lag synchronization." This particular synchronization regime is characterized by synchronization periods interrupted by bursts of non-synchronized activity [175, 34]. The strong fluctuations sharply vanish at the onset of lag synchronization ($k \approx 0.145$). Lag synchronization is defined by the condition $x_1(t + \delta t) = x_2(t)$, i.e., the coincidence of the time series when shifted in time by a constant time lag δt. Both curves, $S_{KL}^{\mathcal{C}}(6)$ and $S_{KL}^{\mathcal{C}}(7)$, increase monotonically in the interval $k \in [0.145, 0.30]$ reflecting stronger synchronization. This trend is only interrupted within the range $k \in [0.232, 0.256]$, where a period-5 window occurs.

The periodic windows are better observed in Fig. 2.6(b)–(d). In fact, all curves exhibit a peak at $k \approx 0.061$ that corresponds to a period-3 window [175]. $S_{KL}^{6}(6)$ and $S_{KL}^{12}(7)$ indicate a period-6 window at $k \approx 0.11$. It seems that this window was not reported before [159], probably due to its extremely small size ($k \in [0.1094, 0.1096]$). All curves show clear signatures of periodic behavior in the range $k \in [0.232, 0.256]$. Intermittent-lag synchronization is particularly reflected by the strong fluctuations observed in Fig. 2.6(b) and (c) for $S_{KL}^{6}(6)$ and $S_{KL}^{10}(7)$, which

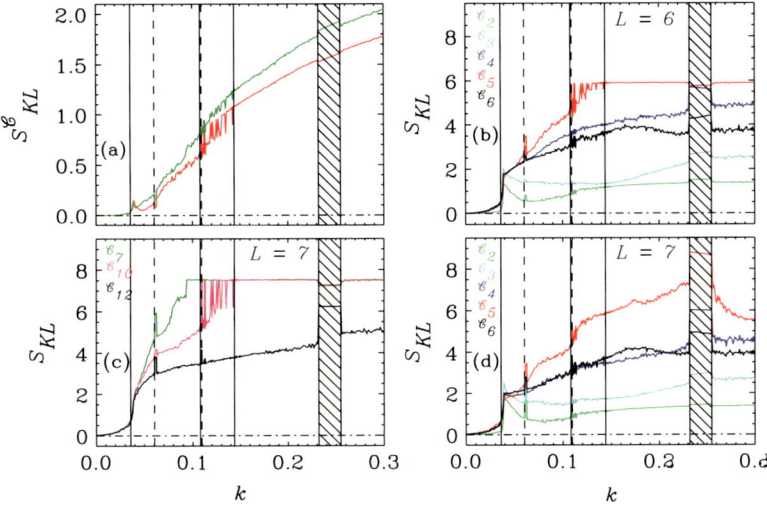

Fig. 2.6 [Reproduced with permission from [159].] (**a**) S_{KL}^{C} obtained using the probability density of the order classes for $L = 6$ (*lower curve*) and $L = 7$ (*upper curve*). (**b**)–(**d**) S_{KL} calculated with the probability density of transcriptions and sequence lengths shown in the plots. *Vertical full lines* from left to right locate the transitions to phase synchronization ($k \approx 0.036$), intermittent-lag synchronization ($k \approx 0.11$), and lag synchronization, respectively. The *vertical dashed lines* at $k \approx 0.061$ and $k \approx 0.11$ as well as the *hatched area* ($k \in [0.232, 0.256]$) indicate periodic windows

abruptly disappear at $k = 0.145$. Observe that different order classes provide complementary information of the coupled system. For instance, $S_{KL}^{5}(6)$ characterizes the intermittent-lag synchronization and the onset of lag synchronization better than $S_{KL}^{6}(6)$. In any case, these partial pieces of information add altogether to a global picture of the various synchronization stages.

Figure 2.6 also reveals that the Kullback–Leibler distance of some higher order classes saturates when the value of the coupling constant k increases. Indeed, Fig. 2.6(b) and (c) shows that the coupled dynamics lead to the extinction of order classes $\mathcal{C}_5(6)$, $\mathcal{C}_7(7)$, and $\mathcal{C}_{10}(7)$ at $k \approx 0.145$, $k \approx 0.09$, and $k \approx 0.145$, respectively.

Figure 2.7(a) and (b) shows the probability density $P_{\mathcal{C}}$ of the order classes for $L = 6$ and $L = 7$, respectively. Note that Fig. 2.6(a) displays the contrast between probability densities as in Fig. 2.7 and those of the independent processes. In particular, a vanishing contrast as for $k \approx 0.005$ indicates that the corresponding probability density $P_{\mathcal{C}}$ (which is clearly non-uniform) is similar to the probability density of transcriptions generated by two independent Rössler systems. In the vicinity of the transition to phase synchronization, $k \approx 0.039$, $P_{\mathcal{C}}$ deviates from the probability density of independent processes (Fig. 2.6(a)), and higher order classes dominate the coupled dynamics (Fig. 2.7). This trend is reversed when increasing k, and already at $k \approx 0.062$ (resp. $k \approx 0.074$), the class of order 2, $\mathcal{C}_2(6)$ (resp. $\mathcal{C}_2(7)$), is prevalent (except at $k = 0.299$ for $L = 6$, in which case $\mathcal{C}_1(6)$ prevails).

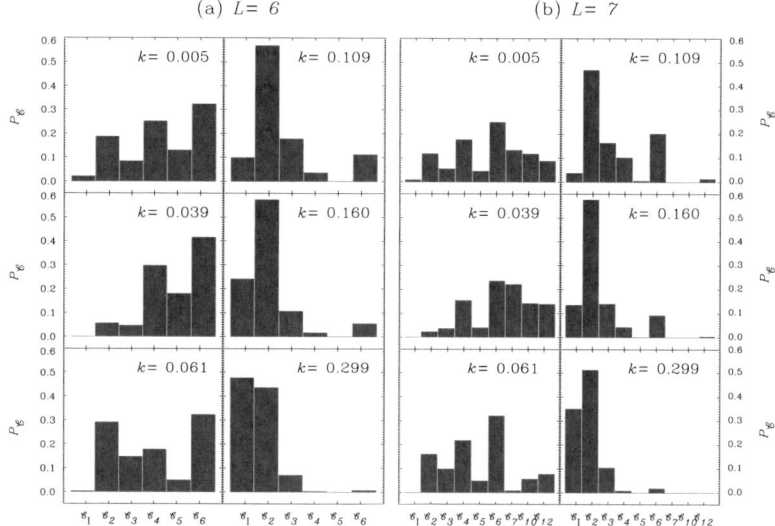

Fig. 2.7 [Reproduced with permission from [159].] (**a**) Probability density P_C of the feasible order classes for $L = 6$ and different values of the coupling constant k. (**b**) Idem for $L = 7$. Classes of orders 8, 9, and 11 are not allowed. Note that $C_1(L) = \{\langle 0, 1, \ldots, L - 1 \rangle\}$

If following [159] we agree to measure the complexity of a transcription by its order, then the probability density of order classes indicates how complex the relationship between the time series is. Figure 2.7 demonstrates that higher order transcriptions play an important role in the description of complex synchronization states such as phase synchronization ($k \geq 0.036$)—a regimen in which amplitudes remain chaotic and uncorrelated. As k increases, the probability densities of higher order classes decrease and some of them vanish, like $C_5(6)$, $C_7(7)$, and $C_{10}(7)$. In fact, simpler synchronization states such as intermittent-lag and lag synchronizations ($k > 0.11$) are predominantly described by lower order classes ($C_2(L)$ and $C_1(L)$). Clearly, the simplest synchronization state, namely complete synchronization, will only be described by the identity ($C_1(L)$).

Chapter 3
Ordinal Patterns

In this chapter we take a close look to the order relations and their consequences, mostly for dynamical systems defined by self-maps of one-dimensional intervals. More general situations will be considered and studied in detail in the following chapters.

Order has some interesting consequences in discrete-time dynamical systems. Just as one can derive sequences of symbol patterns from such a dynamic via coarse graining of the phase space, so it is also straightforward to obtain sequences of ordinal patterns if the phase space is linearly ordered. As we learnt in Sect. 1.2, not all ordinal patterns can be materialized by the orbits of a given dynamic under some mild mathematical assumptions. Furthermore, if an ordinal pattern of a given length is "forbidden," i.e., cannot occur, its absence pervades all longer patterns in form of more missing ordinal patterns. This cascade of outgrowth forbidden patterns grows super-exponentially (in fact, factorially) with the length, all its patterns sharing a common structure. Of course, forbidden and admissible ordinal patterns can be viewed as permutations; in combinatorial parlance, the admissible patterns are (the inverses of) those permutations avoiding the so-called forbidden root patterns in consecutive positions (see Sect. 3.4.2 for details). Let us mention that permutations avoiding general or consecutive patterns is a popular topic in combinatorics (see, e.g., [25, 74, 75]).

Forbidden *ordinal* patterns should not be mistaken for other sorts of forbidden patterns that may occur in dynamics with constraints. Forbidden patterns in symbol sequences occur, e.g., in Markov subshifts of finite type and, more generally, in random walks on oriented graphs. On the contrary, the existence of forbidden ordinal patterns does not entail necessarily any restriction on the patterns of the corresponding symbolic dynamics; the variability of *symbol* patterns is given by the statistical properties of the dynamics. As a matter of fact, the symbolic dynamics of one-dimensional chaotic maps are used to generate pseudo-random sequences, although all such maps have forbidden ordinal patterns.

J.M. Amigó, *Permutation Complexity in Dynamical Systems,*
Springer Series in Synergetics, DOI 10.1007/978-3-642-04084-9_3,
© Springer-Verlag Berlin Heidelberg 2010

3.1 Symbol Patterns

Before dealing with ordinal patterns, we are going to consider the symbol patterns defined by a symbolic dynamics with respect to a partition. The scope is to show that, under general conditions on the map, such symbol patterns have no restrictions, in contrast with the situation we will encounter when studying ordinal patterns.

Thus, let f be a measure-preserving map from a probability space $(\Omega, \mathcal{B}, \mu)$ to itself and $\alpha = \{A_0, \ldots, A_{|\alpha|-1}\}$ be a partition of $(\Omega, \mathcal{B}, \mu)$. Recall that the symbolic dynamics with respect to α, $\mathbf{X}^\alpha = \{X_n^\alpha\}_{n \in \mathbb{N}_0}$ with $X_n^\alpha : \Omega \to S = \{0, \ldots, |\alpha| - 1\}$, is defined as follows:

$$X_n^\alpha(x) = i_n \quad \text{if } f^n(x) \in A_{i_n}, \quad n \geq 0$$

(see (1.6)). The resulting sequence $(i_n)_{n \in \mathbb{N}_0}$ is called the *coded orbit* (or, sometimes, the (α, f)-name) of $x \in \Omega$.

In Sect. B.2.2 it is proven that if α is a generating partition for f, then $(\Omega, \mathcal{B}, \mu, f)$ is isomorphic [1] via the *coding map* $\Phi^\alpha : \Omega \to S^{\mathbb{N}_0}$,

$$(\Phi^\alpha(x))_n = X_n^\alpha(x),$$

to the one-sided shift $(S^{\mathbb{N}_0}, \mathcal{B}_\Pi(S), m, \Sigma)$, where $m = \mu \circ (\Phi^\alpha)^{-1}$ and $\Sigma \circ \Phi^\alpha = \Phi^\alpha \circ f$. Here $\mathcal{B}_\Pi(S)$ is the product sigma-algebra generated by the cylinder sets

$$C_{i_0,\ldots,i_n} = \{\mathbf{s} \in S^{\mathbb{N}_0} : s_0 = i_0, \ldots, s_n = i_n\},$$

$i_0, \ldots, i_n \in S$ (see Sect. A.2),

$$m(C_{i_0,i_1,\ldots,i_n}) = \mu(A_{i_0} \cap f^{-1}A_{i_1} \cap \cdots \cap f^{-n}A_{i_n},$$

and the partition

$$\{\Phi^\alpha(A_i) : i \in S\} = \{C_i : i \in S\}$$

is trivially generating for Σ. It follows that the coded orbits of f contain any arbitrary pattern. Indeed, given any *symbol pattern* of length $L \geq 1$, $i_0^{L-1} := i_0, i_1, \ldots, i_{L-1}$, where $i_n \in S$, choose

$$x \in \bigcap_{n=0}^{L-1} f^{-n}A_{i_n} = (\Phi^\alpha)^{-1}C_{i_0,\ldots,i_{L-1}}.$$

---

[1] The general definition of (metric) isomorphy or conjugacy between measure-preserving dynamical systems is given in Definition 12, Sect. A.1. The corresponding concept for continuous dynamical systems, that usually goes by the name of topological conjugacy, is given in Definition 25, Sect. B.3.

If the pattern has infinite length, $i_0^\infty = i_0, i_1, \ldots$, then there exits a unique point $x \in \Omega$ modulo 0 (i.e., possibly up to sets of measure 0), namely, $x = (\Phi^\alpha)^{-1}(\mathbf{s})$ with $\mathbf{s} = (i_0, i_1, \ldots) \in S^{\mathbb{N}_0}$, such that its coded orbit is precisely $\mathbf{s}$. Thus,

$$\bigcap_{n=0}^{\infty} f^{-n} A_{i_n} = \{x\}.$$

This means that Φ^α *separates points*: if $x_1 \neq x_2$ then $\Phi^\alpha(x_1) \neq \Phi^\alpha(x_2)$.

We conclude that if α is a generating partition for f, then the coded orbits $\Phi^\alpha(x)$, $x \in \Omega$, define any finite or infinite symbol pattern (in the second case, modulo 0).

If $f : \Omega \rightarrow \Omega$ is an automorphism, all the above generalizes to two-sided sequences. Sufficient conditions for f to have a generating partition in such a case are given by Krieger's theorem : ergodicity and a finite entropy.

Example 1 Take $g : [0, 1] \rightarrow [0, 1]$ to be the logistic map $g(x) = 4x(1 - x)$ and

$$\alpha = \{A_0 = [0, \tfrac{1}{2}), A_1 = [\tfrac{1}{2}, 1]\}. \tag{3.1}$$

(It is irrelevant whether the midpoint $\tfrac{1}{2}$ belongs to the left or to the right partition element.) Then α is a generating partition (use, for example, the conjugacy between the logistic map and the symmetric tent map, Example 24). In this case, the coding map $\Phi^\alpha : [0, 1] \rightarrow \{0, 1\}^{\mathbb{N}_0}$,

$$(\Phi^\alpha(x))_n = \begin{cases} 0 & \text{if } g^n(x) \in [0, \tfrac{1}{2}), \\ 1 & \text{if } g^n(x) \in [\tfrac{1}{2}, 1], \end{cases}$$

is an isomorphism between $([0, 1], \mathcal{B}, \mu, g)$ and the $(\tfrac{1}{2}, \tfrac{1}{2})$-Bernoulli shift, where $\mathcal{B}$ is the Borel sigma-algebra of $[0, 1]$ and

$$\mu([a, b]) = \int_a^b \frac{dx}{\pi \sqrt{x(1 - x)}},$$

$[a, b] \subset [0, 1]$. For example,

$$m\{C_{0,0}\} = \mu\{x \in [0, 1] : x \in A_0, g(x) \in A_0\} = \int_0^{1/2 - \sqrt{2}/4} \frac{dx}{\pi \sqrt{x(1 - x)}} = \frac{1}{4},$$

$$m\{C_{0,1}\} = \mu\{x \in [0, 1] : x \in A_0, g(x) \in A_1\} = \int_{1/2 - \sqrt{2}/4}^{1/2} \frac{dx}{\pi \sqrt{x(1 - x)}} = \frac{1}{4},$$

$$m\{C_{1,0}\} = \mu\{x \in [0, 1] : x \in A_1, g(x) \in A_0\} = \int_{1/2}^{1/2 + \sqrt{2}/4} \frac{dx}{\pi \sqrt{x(1 - x)}} = \frac{1}{4},$$

$$m\{C_{1,1}\} = \mu\{x \in [0, 1] : x \in A_1, g(x) \in A_1\} = \int_{1/2 + \sqrt{2}/4}^{1} \frac{dx}{\pi \sqrt{x(1 - x)}} = \frac{1}{4}.$$

Exercise 1 Let $E_2 : [0, 1] \to [0, 1]$ be the dyadic map $x \mapsto 2x \bmod 1$ and $\phi : \{0, 1\}^{\mathbb{N}_0} \to [0, 1]$ the map

$$(x_0, x_1, \ldots, x_k, \ldots) \mapsto \sum_{k=0}^{\infty} x_k 2^{-(k+1)}.$$

Check that ϕ is the inverse (modulo 0) of the coding map Φ^α of E_2 with respect to partition (3.1).

Shift transformations have generating partitions (namely, the cylinder sets $C_{i_0,\ldots,i_{n-1}}$ of any given length $n \geq 1$), hence their trajectories realize any possible symbol sequence.

3.2 Order Relations

A relation $\leq$ defined on every pair of elements of a set Ω is said to be a *total* or *linear order* if $\leq$ is reflexive, antisymmetric, and transitive. A set Ω endowed with a total order $\leq$ is called a *totally* or *linearly ordered set* and will be denoted by $(\Omega, \leq)$. As usual, $x < y$ means henceforth $x \leq y$ and $x \neq y$. The product of the totally ordered sets $(\Omega_1, \leq), (\Omega_2, \leq), \ldots, (\Omega_n, \leq)$ is also totally ordered via the *product order* (also called *lexicographical* or *dictionary order*): if $(x^{(1)}, x^{(2)}, \ldots, x^{(n)}) \neq (y^{(1)}, y^{(2)}, \ldots, y^{(n)})$, then $(x^{(1)}, x^{(2)}, \ldots, x^{(n)}) < (y^{(1)}, y^{(2)}, \ldots, y^{(n)})$ if

(i) $x^{(1)} < y^{(1)}$ or
(ii) $x^{(i)} = y^{(i)}$ for $i = 1, \ldots, k \leq n - 1$ and $x^{(k+1)} < y^{(k+1)}$.

Other conventions (e.g., the signed lexicographic order we considered in Sect. 2.3) are of course possible. The product order generalizes straightforwardly to "infinite products" (i.e., sequence spaces).

Suppose now that $(x_n)_{n \in \mathbb{N}_0}$ is a sequence whose elements (symbols, letters,...) x_n belong to a set (state space, alphabet, etc.) S endowed with a total ordering $\leq$. We say that a length-L block (segment, word, etc.) $x_n^{n+L-1} = x_n, x_{n+1}, \ldots, x_{n+L-1}$ defines the *ordinal (L-)pattern* $\pi = \langle \pi_0, \ldots, \pi_{L-1} \rangle$ if

$$x_{n+\pi_0} < x_{n+\pi_1} < \cdots < x_{n+\pi_{L-1}},$$

where in case $x_i = x_j$, we agree to set $x_i < x_j$ if, say, $i < j$.

Alternatively we also say that the block x_n^{n+L-1} is of type π, or that π is realized by x_n^{n+L-1}, and write $\pi = \pi(x_n^{n+L-1})$. As in Sect. 1.2, the set of ordinal L-patterns will be denoted by $\mathcal{S}_L$. Remember from Sect. 1.2 too that $\mathcal{S}_L$ can be promoted to a group of order $L!$ if equipped with the product (1.25). Unlike in Sect. 2.4, the algebraic structure of $\mathcal{S}_L$ will not be exploited in the sequel.

Example 2 Suppose that $S = \{a, b, c\}$ with $a < b < c$, and that we observe the block $x_0^2 = c, a, a$. Then x_0^2 defines the ordinal pattern $\langle 1, 2, 0 \rangle$ since $x_1 = x_2 = a < c = x_0$ and $1 < 2$. Observe that the following blocks of length 3 are also of type $\langle 1, 2, 0 \rangle$: (i) c, b, b, (ii) c, a, b, and (iii) b, a, a.

In other words, $\pi(x_n^{n+L-1})$ is a permutation of $\{0, 1, \ldots, L-1\}$ that encapsulates the ups and downs of the elements $x_n, x_{n+1}, \ldots, x_{n+L-1}$ in the set S; in case that two elements are equal, we take by convention the first one also as the smaller. This qualitative information is shown in Fig. 3.1 for patterns of length 3.

Given the sequence $(x_n)_{n \in \mathbb{N}_0}$, we say that $\pi \in \mathcal{S}_L$ is an *allowed* or *admissible* L-*pattern* if π is realized by some substring of length L of $(x_n)_{n \in \mathbb{N}_0}$; otherwise, π is called a *forbidden* L-*pattern*.

Proposition 1 *1. If $\pi = \langle \pi_0, \ldots, \pi_L \rangle$ is an allowed $(L+1)$-pattern of $(x_n)_{n \in \mathbb{N}_0}$, and $\check{\pi}$ is the L-pattern obtained from π by deleting the entry L, then $\check{\pi}$ is an allowed L-pattern of $(x_n)_{n \in \mathbb{N}_0}$.*
2. If $\pi = \langle \pi_0, \ldots, \pi_{L-1} \rangle$ is a forbidden L-pattern of $(x_n)_{n \in \mathbb{N}_0}$ and $\hat{\pi}$ is the $(L+1)$-pattern obtained from π by adding the entry L at any place, then $\hat{\pi}$ is a forbidden $(L+1)$-pattern of $(x_n)_{n \in \mathbb{N}_0}$.

Proof 1. If $\pi \in \mathcal{S}_{L+1}$ is allowed, this means that there exists a substring $x_n^{n+L} = x_n, x_{n+1}, \ldots, x_{n+L}$ of the sequence $(x_n)_{n \in \mathbb{N}_0}$ such that

$$x_{n+\pi_0} < x_{n+\pi_1} < \cdots < x_{n+\pi_L}. \tag{3.2}$$

Delete then x_{n+L} from (3.2) to show that the substring $x_n, x_{n+1}, \ldots, x_{n+L-1}$ is of type $\check{\pi} \in \mathcal{S}_L$.

2. Suppose by contradiction that the $(L+1)$-pattern $\hat{\pi} = \langle \hat{\pi}_0, \ldots, \hat{\pi}_L \rangle$ is allowed. Then, part 1 implies that the L-pattern obtained by removing from $\hat{\pi}$ the entry L, namely π, is an allowed pattern. $\square$

This general setting will crystallize in different ways as we move on. Let us advance three of them at this point.

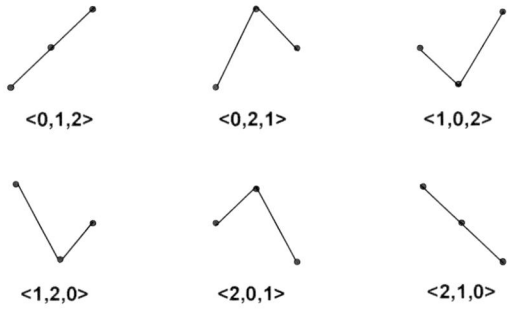

$\langle 0,1,2 \rangle$ $\langle 0,2,1 \rangle$ $\langle 1,0,2 \rangle$

$\langle 1,2,0 \rangle$ $\langle 2,0,1 \rangle$ $\langle 2,1,0 \rangle$

Fig. 3.1 Geometrical illustration of the six ordinal patterns of length 3

- The sequence $(x_n)_{n \in \mathbb{N}_0}$ can be the output of a finite-state stationary stochastic process. This corresponds to the usual information sources emitting a message composed by letters of a finite alphabet.
- The set S can be an interval $I \subseteq \mathbb{R}^q$, $q \geq 1$, and $(x_n)_{n \in \mathbb{N}_0}$ the output of a univariate ($q = 1$) or multivariate ($q > 1$) random process taking values in I.
- Still other possibility is that $(x_n)_{n \in \mathbb{N}_0}$ is the orbit of x_0 under a map $f{:}I \to I$, I being as before a q-dimensional interval or a homeomorphic copy thereof. In this case it is customary to neglect periodic points whose periods are shorter than the pattern length L considered, so as all points in the block x_n^{n+L-1} are different.

In the following sections and chapters we are going to dwell on all these settings.

3.3 Ordinal Patterns Defined by Maps

In Sect. 3.1 we saw that the symbolic dynamics of maps defines any symbol pattern of any length, under rather general assumptions. In this section we shall see that the situation is not quite the same when considering ordinal patterns.

Let $(\Omega, \leq)$ be a totally ordered set and $f : \Omega \to \Omega$ a map. Given $x \in \Omega$, set $x_n = f^n(x)$ for $n \geq 0$. If x is not a periodic point of period less than $L \geq 2$, we can then associate with x an ordinal pattern of length L, as follows. We say that x *defines the ordinal pattern* $\pi = \langle \pi_0, \ldots, \pi_{L-1} \rangle \in \mathcal{S}_L$, if $\pi = \pi(x_0^{L-1})$, i.e.,

$$x_{\pi_0} < x_{\pi_1} < \cdots < x_{\pi_{L-1}},$$

or, equivalently,

$$f^{\pi_0}(x) < f^{\pi_1}(x) < \cdots < f^{\pi_{L-1}}(x). \tag{3.3}$$

Set

$$P_\pi = \{x \in \Omega : x \text{ defines } \pi \in \mathcal{S}_L\} \tag{3.4}$$

as in Sect. 1.2 and

$$\mathcal{P}_L = \{P_\pi \neq \emptyset : \pi \in \mathcal{S}_L\}. \tag{3.5}$$

Therefore, $|\mathcal{P}_L|$ is the number of distinct ordinal L-patterns realized by the points of Ω.

Proposition 2 *Let* $(\Omega, \mathcal{B}, \mu, f)$ *be a measure-preserving dynamical system. Then* $\mathcal{P}_L$ *is a finite partition of* Ω *for all* $L \geq 2$ *if and only if* f *is aperiodic.*

We say that $f : \Omega \to \Omega$ is *aperiodic*, if

$$\mu \left(\bigcup_{n \geq 1} \{x \in \Omega : f^n(x) = x\} \right) = 0. \tag{3.6}$$

Proof In order that $\mathcal{P}_L$ fails to be a finite partition of Ω, it must happen that the complement of the disjoint union $\cup \{P_\pi \in \mathcal{P}_L\}$, which comprises all periodic points of f of period $p \leq L$, has a positive measure. But this possibility is excluded by (3.6). □

In particular, if f is ergodic with respect to μ, then f is aperiodic unless Ω is a finite set modulo 0 [202].

The family of sets $\mathcal{P}_L$ has some elementary properties. In Sect. 1.2 we saw that when going from $\mathcal{P}_L$ to $\mathcal{P}_{L+1}$, each "mother set" $P_{\pi_{\text{mother}}}$, $\pi_{\text{mother}} = \langle \pi_0, \ldots, \pi_{L-1} \rangle$, decomposes into several "daughter sets" $P_{\pi_{\text{daughter}}} \in \mathcal{P}_{L+1}$, where

$$\pi_{\text{daughter}} \in \{ \langle L, \pi_0, \ldots, \pi_{L-1} \rangle, \langle \pi_0, \ldots, \pi_k, L, \pi_{k+1}, \ldots, \pi_{L-1} \rangle, \langle \pi_0, \ldots, \pi_{L-1}, L \rangle \},$$

$0 \leq k \leq L - 2$. Therefore, each mother set is the (disjoint) union of her daughter sets. Correspondingly, we speak of mother and daughter patterns. To go back from π_{daughter} to π_{mother}, just delete the entry L. In particular, two different mother intervals cannot give birth to the same daughter interval.

Proposition 3 *(1) $\mathcal{P}_{L+1}$ is a refinement of $\mathcal{P}_L$, i.e., each $P_\pi \in \mathcal{P}_L$ is the union of elements of $\mathcal{P}_{L+1}$.*
(2) For every $P_{\pi'} \in \mathcal{P}_{L+1}$ there is a $P_\pi \in \mathcal{P}_L$ such that $f(P_{\pi'}) \subset P_\pi$.

Proof Statement (1) is trivial because

$$P_\pi = \cup \{P_{\pi'} \in \mathcal{P}_{L+1} : \pi' \text{ is a daughter pattern of } \pi \}.$$

To prove (2), let $x \in f(P_{\pi'})$, i.e., $x = f(y)$ where $y \in \Omega$ satisfies

$$f^{\pi'_0}(y) < f^{\pi'_1}(y) < \cdots < f^{\pi'_L}(y). \tag{3.7}$$

Let π'_{n_k}, $0 \leq k \leq L - 1$, be an order-isomorphic relabeling of those L entries of the ordinal pattern $\pi' \in \mathcal{S}_{L+1}$ which are positive. From (3.7) it follows that

$$f^{\pi'_{n_0} - 1}(x) < f^{\pi'_{n_1} - 1}(x) < \cdots < f^{\pi'_{n_{L-1}} - 1}(x),$$

hence $x \in P_\pi$ with $\pi = \langle \pi'_{n_0} - 1, \ldots, \pi'_{n_{L-1}} - 1 \rangle \in \mathcal{S}_L$. In words, π is obtained from π' after deleting the entry 0 and subtracting 1 from the remaining entries. Therefore, $f(P_{\pi'}) \subset P_\pi$. □

Example 3 To illustrate Proposition 3 (1)–(2), consider the logistic map g and the intervals $P_\pi \in \mathcal{P}_3$, (1.29). Then (see Figs. 1.5 and 1.6),

$$g(P_{\langle 0,1,2 \rangle}) = g((0, \tfrac{1}{4})) = (0, \tfrac{3}{4}) = P_{\langle 0,1,2 \rangle} \cup P_{\langle 0,2,1 \rangle} \cup P_{\langle 2,0,1 \rangle} = P_{\langle 0,1 \rangle},$$

$$g(P_{\langle 0,2,1 \rangle}) = g((\tfrac{1}{4}, \tfrac{5-\sqrt{5}}{8})) = (\tfrac{3}{4}, \tfrac{5+\sqrt{5}}{8}) = P_{\langle 1,0,2 \rangle} \subset P_{\langle 1,0 \rangle},$$

$$g(P_{\langle 2,0,1 \rangle}) = g((\tfrac{5-\sqrt{5}}{8}, \tfrac{3}{4})) = (\tfrac{3}{4}, 1) = P_{\langle 1,0,2 \rangle} \cup P_{\langle 1,2,0 \rangle} \subset P_{\langle 1,0 \rangle},$$

$$g(P_{\langle 1,0,2 \rangle}) = g((\tfrac{3}{4}, \tfrac{5+\sqrt{5}}{8})) = (\tfrac{5-\sqrt{5}}{8}, \tfrac{3}{4}) = P_{\langle 2,0,1 \rangle} \subset P_{\langle 0,1 \rangle},$$

$$g(P_{\langle 1,2,0 \rangle}) = g((\tfrac{5+\sqrt{5}}{8}, 1)) = (0, \tfrac{5-\sqrt{5}}{8}) = P_{\langle 0,1,2 \rangle} \cup P_{\langle 0,2,1 \rangle} \subset P_{\langle 0,1 \rangle}.$$

Observe that $\mathcal{P}_3$ is a Markov partition for g (i.e., $g(P_\pi) \supset P_\sigma$, whenever $g(P_\pi) \cap P_\sigma \neq \emptyset$, $P_\pi, P_\sigma \in \mathcal{P}_3$) with transition matrix

$$A = \begin{pmatrix} 1 & 1 & 1 & 0 & 0 \\ 0 & 0 & 0 & 1 & 0 \\ 0 & 0 & 0 & 1 & 1 \\ 0 & 0 & 1 & 0 & 0 \\ 1 & 1 & 0 & 0 & 0 \end{pmatrix}$$

(see Definition 9 and (A.2)). Needless to say, the partitions $\mathcal{P}_L$ are not in general Markovian.

Exercise 2 (1) Let $f : [a, b] \to [a, b]$ be a boundary-anchored unimodal map with full range (i.e., $f(a) = f(b) = a$ and $f([a, b]) = [a, b]$). Show that $\mathcal{P}_2$ is a Markov partition for f.

(2) Let Λ be the symmetric tent map. Using the information on $\mathcal{P}_4$ provided in Example 13, Sect. 6.3, shows that $\Lambda(P_{\langle 2,3,0,1 \rangle}) \cap P_{\langle 1,2,3,0 \rangle} \neq \emptyset$ but $P_{\langle 1,2,3,0 \rangle} \not\subseteq \Lambda(P_{\langle 2,3,0,1 \rangle})$.

A plain difference between symbol patterns and ordinal patterns of length L is their cardinality: the former grow exponentially with L (exactly as N^L, where N is the number of symbols) while the latter do superexponentially,

$$|\mathcal{S}_L| = L! \sim e^{L(\ln L - 1) + (1/2)\ln 2\pi L}, \tag{3.8}$$

see (1.34). Although one can construct maps whose orbits realize any possible ordinal pattern (more on this at the end of Sect. 4.2), numerical simulations support the conjecture that the number of ordinal L-patterns realized in the orbits of maps, like symbol patterns, grows only exponentially with L for "well-behaved" maps. In fact, we saw in Sect. 1.2 that if I is a closed interval of $\mathbb{R}$ and $f : I \to I$ is piecewise monotone, then (see (1.33))

$$|\mathcal{P}_L| \sim e^{L h_{\mathrm{top}}(f)}, \tag{3.9}$$

where $h_{\mathrm{top}}(f)$ is the topological entropy of f. From (3.8) and (3.9) we conclude the following result.

Proposition 4 *If f is a piecewise monotone self-map on a finite interval $I \subset \mathbb{R}$, then there exists $L \geq 2$ such that $P_\pi = \emptyset$ for some $\pi \in \mathcal{S}_L$.*

Ordinal patterns that do not appear in any orbit of f are called *forbidden (ordinal)*
patterns for f, at variance with the *admissible* or *allowed patterns*, for which there
are sets of points that realize them.

3.4 Properties of the Ordinal Patterns

We examine in this section three basic properties of ordinal patterns: invariance
under order isomorphism, superexponential growth of the forbidden patterns with
the length, and robustness against noise.

3.4.1 Invariance Under Order Isomorphism

Since ordinal patterns are not related to measure-theoretical or topological properties, metrically or topologically conjugate dynamical systems need not have the
same allowed (and hence forbidden) patterns, unless the conjugacy preserves linear
order—supposing that both state spaces are linearly ordered. In general, this will not
be the case.

For instance, we saw in Sect. 1.2 that the logistic map has the forbidden 3-pattern
$\langle 2, 1, 0 \rangle$, i.e., there are no three consecutive points in any orbit of the logistic map,
forming a strictly decreasing trio (see Fig. 1.6). However, Fig. 3.2 shows that the
dyadic map $E_2 : x \mapsto 2x \pmod 1$, $0 \leq x \leq 1$, has no forbidden patterns of length
3, despite being isomorphic to the logistic map. The reason is simple: the isomorphism between these two maps is proved via the semi-conjugacy[2] $\varphi : [0, 1] \to [0, 1]$,
$\varphi(x) = \sin^2 \pi x$, which does not preserve order on account of being increasing on
$(0, \frac{1}{2})$ and decreasing on $(\frac{1}{2}, 1)$.

Definition 1 Given two totally ordered sets $(\Omega_1, \leq_1)$ and $(\Omega_2, \leq_2)$, two maps
$f_1 : \Omega_1 \to \Omega_1$ and $f_2 : \Omega_2 \to \Omega_2$, and an invertible map $\phi : \Omega_1 \to \Omega_2$ such that
$\phi \circ f_1 = f_2 \circ \phi$, we say that f_1 and f_2 are *order isomorphic* if ϕ is order-preserving
(i.e., $x \leq_1 y$ implies $\phi(x) \leq_2 \phi(y)$). The map ϕ is called an order isomorphism.

It is trivial that if $\phi : \Omega_1 \to \Omega_2$ is an order isomorphism, then $x \in \Omega_1$ and
$\phi(x) \in \Omega_2$ define the same ordinal L-patterns, for all $L \geq 2$, under the f_1- and
f_2-dynamics, respectively. In other words, order-isomorphic maps have the same
allowed and forbidden patterns of any length. We conclude that ordinal patterns are
not invariants of metric nor topological conjugacy, but of order isomorphy.

Example 4 (1) The logistic map g (1.19) and the symmetric tent map Λ (1.17) are
not only isomorphic but also order isomorphic. Indeed, the isomorphism

$$\phi : x \mapsto \sin^2 \left(\tfrac{\pi}{2} x \right),$$

---

[2] Definition 25.

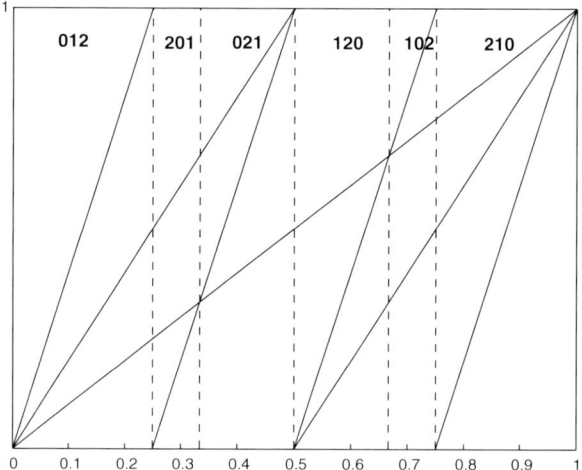

Fig. 3.2 All six 3-patterns are allowed for the shift map E_2:$x \mapsto 2x \pmod 1$: $P_{012} = \left(0, \frac{1}{4}\right)$, $P_{201} = \left(\frac{1}{4}, \frac{1}{3}\right)$, $P_{021} = \left(\frac{1}{3}, \frac{1}{2}\right)$, $P_{120} = \left(\frac{1}{2}, \frac{2}{3}\right)$, $P_{102} = \left(\frac{2}{3}, \frac{3}{4}\right)$, $P_{210} = \left(\frac{3}{4}, 1\right)$. A pattern $\langle \pi_0, \pi_1, \pi_2 \rangle$ has been shorthanded as $\pi_0 \pi_1 \pi_2$. Note that ordinal patterns are mirrored with respect to the central line $x = \frac{1}{2}$

(see Example 24) is strictly increasing and, hence, order preserving. This entails that allowed patterns for f correspond to allowed patterns for Λ in a one-to-one way.

(2) The same happens with the dyadic map $E_2 : x \mapsto 2x \pmod 1$, $0 \le x \le 1$, and the $\left(\frac{1}{2}, \frac{1}{2}\right)$-Bernoulli shift, since the isomorphism (modulo 0) $\phi_2 : \{0, 1\}^{\mathbb{N}_0} \to [0, 1]$,

$$\phi_2 : (x_1, x_2, \dots) \mapsto \sum_{k=1}^{\infty} x_k 2^{-k}$$

is order-preserving ($\{0, 1\}^{\mathbb{N}}$ endowed with the lexicographical order).

(3) The logistic map is isomorphic but not order isomorphic to the $\left(\frac{1}{2}, \frac{1}{2}\right)$-Bernoulli shift. Indeed, the corresponding isomorphy (actually, the coding map of Example 1) $\Phi^\alpha : [0, 1] \to \{0, 1\}^{\mathbb{N}_0}$ is not order preserving; e.g.,

$$\Phi^\alpha \left(\frac{1}{4}\right) = (0, 1^\infty) < \Phi^\alpha(\tfrac{3}{4}) = (1^\infty),$$

where binary strings are ordered lexicographically, while

$$\Phi^\alpha \left(\frac{1}{2}\right) = (1, 1, 0^\infty) > \Phi^\alpha(1) = (1, 0^\infty).$$

The forbidden ordinal patterns of the shift systems will be studied in Chap. 4.

On the other hand, if $\phi : \Omega_1 \to \Omega_2$ is order preserving but not one-to-one, then ordinal patterns are not necessarily invariant under ϕ. Let, for example, $\Omega_1 = \Omega_2 = $

$[0, 1] \times [0, 1] =: [0, 1]^2$ endowed with lexicographical order, $f : [0, 1]^2 \to [0, 1]^2$, $\phi : [0, 1]^2 \to [0, 1]$ the projection onto the first coordinate, $\mathbf{x}_0 = (x_0^{(1)}, x_0^{(2)}) \mapsto x_0^{(1)}$, and

$$(x_0^{(1)}, x_0^{(2)}) = \mathbf{x}_0 > f(\mathbf{x}_0) = (x_1^{(1)}, x_1^{(2)}),$$

so that $\mathbf{x}_0$ is of type $\langle 1, 0 \rangle$. If $x_0^{(1)} > x_1^{(1)}$, then $\phi(\mathbf{x}_0)$ is also of type $\langle 1, 0 \rangle$. But if $x_0^{(1)} = x_1^{(1)}$ (and $x_0^{(2)} > x_1^{(2)}$), then $\phi(\mathbf{x}_0)$ will be of type $\langle 0, 1 \rangle$ in virtue of the lexicographical convention (Sect. 3.2).

Proposition 5 *Given* $\Omega_1, \Omega_2 \subset \mathbb{R}$, *let* $f_i : \Omega_i \to \Omega_i$, $i = 1, 2$, *be topologically conjugate via a (continuous) map* $\phi : \Omega_1 \to \Omega_2$. *If* f_1 *is topologically transitive and, for all* $x \in \Omega_1$, *both* x *and* $\phi(x)$ *define the same ordinal pattern, then* ϕ *is order preserving.*

Proof Pick $x, x' \in \Omega_1$ such that $x < x'$. We must prove that $\phi(x) < \phi(x')$.

Because of continuity, for all $\varepsilon > 0$ there exists $0 < \delta < \frac{x'-x}{2}$ such that $|y - x| < \delta \Rightarrow |\phi(y) - \phi(x)| < \varepsilon$ and $|y' - x'| < \delta \Rightarrow |\phi(y') - \phi(x')| < \varepsilon$. Moreover, topological transitiveness implies that, given x, x' and δ as above, there exists $x_0 \in \Omega_1$, and positive integers $N = N(x, \delta), N' = N'(x', \delta)$ such that $|f_1^N(x_0) - x| < \delta$ and $|f_1^{N'}(x_0) - x'| < \delta$. Suppose without restriction $N < N' = N + k$, $k > 0$, and set $f_1^N(x_0) = y, f_1^{N'}(x_0) = y'$, hence $y' = f_1^k(y)$. By assumption, $y \in \Omega_1$ and $\phi(y) \in \Omega_2$ define the same ordinal $(k + 1)$-pattern, i.e.,

$$f_1^{\pi_0}(y) < \cdots < f_1^{\pi_k}(y) \Leftrightarrow f_2^{\pi_0}(\phi(y)) < \cdots < f_2^{\pi_k}(\phi(y)), \qquad (3.10)$$

where $0 \leq \pi_i \leq k$, and $\pi_i \neq \pi_j$ for $i \neq j$. Since $|y - x| < \delta$, $|f_1^k(y) - x'| < \delta$, and $\delta < \frac{x'-x}{2}$, we have $y < f_1^k(y) = y'$. From (3.10) it follows

$$\phi(y) < f_2^k(\phi(y)) = \phi(f_1^k(y)) = \phi(y').$$

By continuity, $\phi(y)$ and $\phi(y')$ can be made to lie arbitrarily close to $\phi(x)$ and $\phi(x')$. It follows $\phi(x) < \phi(x')$. $\square$

Finally, observe that the setting we are considering is more general than the setting of kneading theory [150] since our functions need not be continuous, but only piecewise continuous. Under some assumptions, the so-called kneading invariants completely characterize the order isomorphy of continuous, one-dimensional interval maps.

3.4.2 Growth of Forbidden Patterns with Length: Outgrowth Patterns

Forbidden ordinal patterns come in two flavors: *outgrowth* patterns and *root* patterns.

Outgrowth forbidden patterns appeared already in Sect. 1.2 when discussing the ordinal patterns of the logistic map: they are the patterns on the "trail" of a given forbidden pattern (see (1.36)). Consider now a general map $f : \Omega \to \Omega$. That $\pi = \langle \pi_0, \dots, \pi_{L-1} \rangle$ is forbidden for f means that the order relations

$$f^{\pi_0}(x) < f^{\pi_1}(x) < \cdots < f^{\pi_{L-1}}(x) \qquad (3.11)$$

cannot occur. This implies that the following $2(L+1)$ patterns of length $L+1$ are also forbidden for f:

Group I: $\langle L, \pi_0, \dots, \pi_{L-1} \rangle, \langle \pi_0, L, \pi_1, \dots, \pi_{L-1} \rangle, \dots, \langle \pi_0, \dots, \pi_{L-1}, L \rangle,$
Group II: $\langle 0, \pi_0 + 1, \dots, \pi_{L-1} + 1 \rangle, \langle \pi_0 + 1, 0, \pi_1 + 1, \dots, \pi_{L-1} + 1 \rangle,$
            $\dots, \langle \pi_0 + 1, \dots, \pi_{L-1} + 1, 0 \rangle.$
$$\qquad (3.12)$$

For suppose by contradiction that the pattern $\langle \pi_0, \dots, \pi_i, L, \pi_{i+1}, \dots, \pi_{L-1} \rangle$ is allowed. Then the inequalities

$$f^{\pi_0}(x) < \cdots < f^{\pi_i}(x) < f^L(x) < f^{\pi_{i+1}}(x) < \cdots < f^{\pi_{L-1}}(x)$$

would hold for some $x \in I$, hence (3.11) would occur for the same $x \in I$, contradicting the assumption that π is forbidden. Analogously, if $x \in I$ would realize the pattern $\langle \pi_0 + 1, \dots, \pi_i + 1, 0, \pi_{i+1} + 1, \dots, \pi_{L-1} + 1 \rangle$, then $f(x)$ would realize the pattern π—again a contradiction.

A weak form of the converse holds also true: if $\langle L, \pi_0, \dots, \pi_{L-1} \rangle, \langle \pi_0, L, \dots, \pi_{L-1} \rangle, \dots, \langle \pi_0, \dots, \pi_{L_0-1}, L \rangle \in \mathcal{S}_{L+1}$ are forbidden, then $\langle \pi_0, \dots, \pi_{L-1} \rangle \in \mathcal{S}_L$ is also forbidden.

Assume for the time being that the forbidden patterns (3.12), belonging to the "first generation," are all different. Then, proceeding similarly as before, we would find

$$2(L+1) \times 2(L+2) = 2^2(L+1)(L+2)$$

forbidden patterns of length $L+2$ in the second generation and, in general,

$$2^m(L+1) \cdots (L+m) = 2^m \frac{(L+m)!}{L!}$$

forbidden patterns of length $L+m$ in the mth generation, provided that all forbidden patterns up to (and including) the mth generation are different. Observe that all these forbidden patterns generated by π have the form

$$\langle *, \pi_0 + n, *, \pi_1 + n, *, \ldots, *, \pi_{L-1} + n, * \rangle \in \mathcal{S}_M. \tag{3.13}$$

Here $n = 0, 1, \ldots, M - L$, where $M - L \geq 1$ is the number of wildcards $* \in \{0, 1, \ldots, n - 1, L + n, \ldots, M - 1\}$ (with $* \in \{L, \ldots, M - 1\}$ if $n = 0$ and $* \in \{0, \ldots, M - L - 1\}$ if $n = M - L$). Forbidden M-patterns of the form (3.13), where $\pi = \langle \pi_0, \ldots, \pi_{L-1} \rangle$ is a forbidden pattern for f and $M > L$, are called *outgrowth (forbidden) patterns* of π. It is straightforward that if π' is an outgrowth pattern of π and π'' is an outgrowth pattern of π', then π'' is an outgrowth pattern of π.

A better upper bound on the number of outgrowth forbidden patterns of length M of π is obtained using the following reasoning. For fixed n, the number of outgrowth patterns of π of form (3.13) is $M!/L!$. This is because out of all possible permutations of the numbers $\{0, 1, \ldots, M - 1\}$, we only count those that have the entries $\{\pi_0 + n, \pi_1 + n, \ldots, \pi_{L-1} + n\}$ in the required order. Next, note that we have $M - L + 1$ choices for the value of n. Each choice generates a set of $M!/L!$ outgrowth patterns. These sets are not necessarily disjoint, but an upper bound on the size of their union, i.e., the set of all outgrowth forbidden patterns of length M of π, is given by

$$(M - L + 1)\frac{M!}{L!}.$$

Forbidden patterns that are not outgrowth patterns of other forbidden patterns of shorter length are called *root forbidden patterns*. They can be viewed as the root of the tree of forbidden patterns spanned by the outgrowth patterns they generate, branching taking place when going from one length (or generation) to the next. Therefore, they are instrumental in the study of the ordinal structure defined by a transformation—the remaining patterns, whether forbidden or allowed, follow from them. In view of (3.12), for proving that a forbidden L-pattern is a root pattern it suffices to show that it does not belong to group I nor to group II of a forbidden $(L - 1)$-pattern.

Example 5 Figure 3.3 depicts the graphs of the identity (main diagonal), the map $E_2 : x \to 2x \bmod 1, 0 \leq x \leq 1$, and its second and third iterates. The vertical dashed lines rise at the endpoints of the intervals $P_\pi \neq \emptyset$ of points x defining the allowed patterns $\pi \in \mathcal{S}_4$. We conclude that E_2 has 18 allowed 4-patterns, all consisting of a single component, and hence 6 forbidden 4-patterns, namely

$$\langle 0, 2, 3, 1 \rangle, \langle 1, 0, 2, 3 \rangle, \langle 1, 3, 2, 0 \rangle, \langle 2, 0, 1, 3 \rangle, \langle 3, 1, 0, 2 \rangle, \langle 3, 2, 0, 1 \rangle. \tag{3.14}$$

Since E_2 has no forbidden 3-patterns (see Fig. 3.2), we deduce that all these six forbidden 4-patterns are root patterns.   □

Given a permutation

$$\sigma = \begin{pmatrix} 0 & 1 & \ldots & M - 1 \\ \sigma_0 & \sigma_1 & \ldots & \sigma_{M-1} \end{pmatrix} = [\sigma_0, \ldots, \sigma_{M-1}],$$

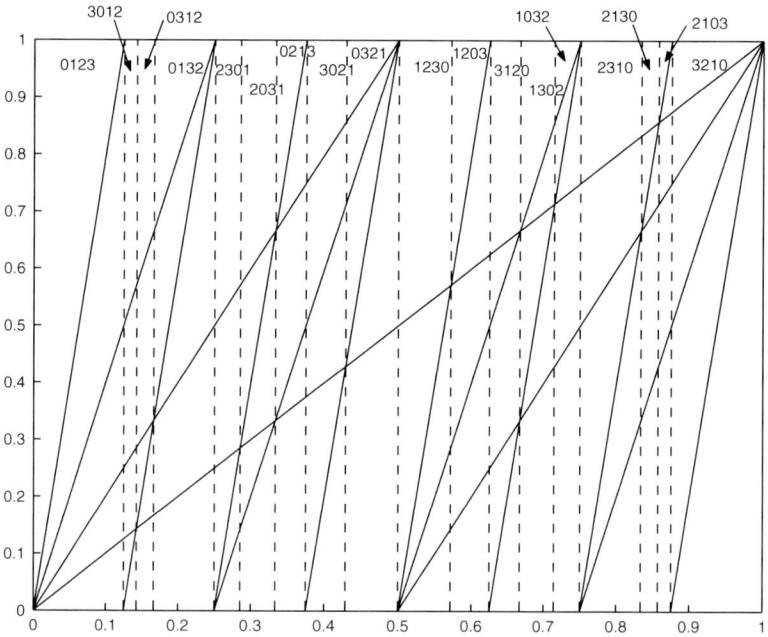

Fig. 3.3 The eighteen allowed 4-patterns of the map $E_2 : x \mapsto 2x$ (mod 1). For clarity, the allowed patterns have been written without angular parentheses nor separating commas. Note that the intervals P_π and the allowed patterns are mirrored with respect to the central line $x = 1/2$

we say that σ contains the *consecutive pattern* $\tau = [\tau_0, \ldots, \tau_{L-1}]$, $L < M$, if the sequence $\sigma_0, \ldots, \sigma_{M-1}$ contains a *consecutive* subsequence order isomorphic to the sequence $\tau_0, \ldots, \tau_{L-1}$. Alternatively, we say that σ avoids the *consecutive pattern* τ if it contains no consecutive subsequence order isomorphic to τ [74]. For instance, $\sigma = [5, 2, 0, 1, 4, 3]$ contains the consecutive pattern $\tau = [0, 2, 1]$ because σ contains the consecutive subsequence $1, 4, 3$ which is order isomorphic to $0, 2, 1$.

In order to apply results on pattern avoidance in combinatorics to forbidden ordinal patterns, recall from Sect. 1.2 that any ordinal pattern $\pi = \langle \pi_0, \ldots, \pi_{L-1} \rangle$ corresponds to the permutation $[\pi]^{-1} : \pi_i \mapsto i$, $0 \leq i \leq L - 1$, (1.23). Suppose furthermore that $\pi' = \langle \pi'_0, \ldots, \pi'_{M-1} \rangle$, $L < M$, is an outgrowth pattern of π, i.e., π' has the form (3.13). The permutation $[\pi'_0, \ldots, \pi'_{M-1}]^{-1} =: [\pi']^{-1}$ performs the substitutions

$$\ldots \quad \pi_0 + n \mapsto i_0, \ldots \quad \pi_1 + n \mapsto i_1, \ldots \quad \pi_{L-1} + n \mapsto i_{L-1}, \ldots,$$

where $n \in \{0, 1, \ldots, M - L\}$ and $0 \leq i_0 < i_1 < \cdots < i_{L-1} \leq M - 1$. Thus the sequence $i_0, i_1, \ldots, i_{L-1}$ is order isomorphic to $0, 1, \ldots, L - 1$. Note, furthermore, that $\pi_0, \ldots, \pi_{L-1}$ is a rearrangement of the consecutive sequence $0, \ldots, L - 1$, hence $\pi_0 + n, \ldots, \pi_{L-1} + n$ is a rearrangement of the consecutive sequence $n, \ldots, n + L - 1$. It follows that $\langle \pi'_0, \ldots, \pi'_{M-1} \rangle$ is an outgrowth pattern of $\langle \pi_0, \ldots, \pi_{L-1} \rangle$ if and only if the permutation $[\pi']^{-1}$ contains the permutation $[\pi]^{-1}$ as a consecutive pattern. Therefore the allowed patterns for f are the permutations that avoid all such consecutive subsequences for every forbidden root pattern of f.

Example 6 Take $\pi = \langle 2, 0, 1 \rangle$ to be a forbidden pattern for a certain function f. Then $\pi' = \langle 4, 2, 1, 5, 3, 0 \rangle$ is an outgrowth pattern of π because it contains the subsequence $4, 2, 3$ ($n = 2$). Equivalently, the permutation $[4, 2, 1, 5, 3, 0]^{-1} = [5, 2, 1, 4, 0, 3]$ contains the consecutive pattern $1, 4, 0$, which is order isomorphic to $[2, 0, 1]^{-1} = [1, 2, 0]$.

Let $\mathcal{S}^{\text{out}}(\pi)$ denote the family of outgrowth patterns of the forbidden pattern π,

$$\mathcal{S}_M^{\text{out}}(\pi) = \mathcal{S}^{\text{out}}(\pi) \cap \mathcal{S}_M$$
$$= \{\pi' \in \mathcal{S}_M : [\pi']^{-1} \text{ contains } [\pi]^{-1} \text{ as a consecutive pattern}\},$$

and

$$\mathcal{S}_M^{\text{avoid}}(\pi) = \mathcal{S}_M \backslash \mathcal{S}_M^{\text{out}}(\pi)$$
$$= \{\pi' \in \mathcal{S}_M : [\pi']^{-1} \text{ avoids } [\pi]^{-1} \text{ as a consecutive pattern}\}.$$

where $\backslash$ stands for set difference. The fact that some of the outgrowth patterns of a given length will be the same and that this depends on π makes the analytical calculation of $\left| \mathcal{S}_M^{\text{out}}(\pi) \right|$ extremely complicated. Yet, from [74] we know that there are constants $0 < c, d < 1$ such that

$$c^M M! < \left| \mathcal{S}_M^{\text{avoid}}(\pi) \right| < d^M M!$$

(for the first inequality, $L \geq 3$ is needed). This implies that

$$(1 - d^M) M! < \left| \mathcal{S}_M^{\text{out}}(\pi) \right| < (1 - c^M) M!. \tag{3.15}$$

This factorial growth with M is one of the mechanisms that make forbidden patterns a practical tool for detection of determinism in noisy time series. This topic will be addressed in detail in Chap. 9.

3.4.3 Robustness Against Noise in Deterministic Time Series

Determinism means functional dependence between the "current" value of a univariate or multivariate time series and some of its past values. In some theoretical models this dependence can involve infinitely many values, but we shall limit our attention to the more realistic processes with a finite number of dependent variables. Multivariate time series appear not only when the data source is vectorial but also when scalar deterministic processes are modeled as dynamical systems. Consider, for instance, a time process $y_{n+1} = g(y_n, y_{n-1}, \dots, y_{n-M+1})$, where g is a scalar self-map, the "memory" $M \geq 2$, and $(y_0, y_{-1}, \ldots, y_{-M+1}) \in \mathbb{R}^M$ is the initial condition. This process can be modeled as an M-dimensional dynamical system (or a multivariate process with memory one) via the change of variables

$$x_n^{(1)} = y_n, \, x_n^{(2)} = y_{n-1}, \, \ldots, \, x_n^{(M)} = y_{n-M+1},$$

so as

$$y_{n+1} = g(y_n, y_{n-1}, \ldots, y_{n-M+1}) \Leftrightarrow \mathbf{x}_{n+1} = \mathbf{f}(\mathbf{x}_n),$$

where

$$\mathbf{x}_n = (x_n^{(1)}, x_n^{(2)}, \ldots, x_n^{(M)}) \in \mathbb{R}^M,$$

and $\mathbf{f} : \mathbb{R}^M \to \mathbb{R}^M$ with

$$\mathbf{f}(\mathbf{x}_n) = (g(\mathbf{x}_n), x_n^{(1)}, x_n^{(2)}, \ldots, x_n^{(M-1)}) \in \mathbb{R}^M.$$

A similar strategy works out for vectorial maps. In sum, any deterministic time series can be considered as the orbit of a dynamical system of adequate dimensionality.

Exercise 3 Write the evolution process

$$x_{n+1} = f(x_n, x_{n-1}, y_{n-1}),$$
$$y_{n+1} = g(x_{n-1}, y_{n-2}),$$

as a five-dimensional dynamical system.

The perturbations that distort a deterministic time series during generation, transmission, observation, and/or measurement are generically referred to as *noise*. We elaborate next on the persistence of admissible and forbidden patterns when the observed data are "noisy," a property called robustness against noise. This property is essential for the applications of ordinal analysis since noise is ubiquitous in real data.

When modeling noise, there are two basic approaches:

- *Dynamical noise* is due to errors in the determination of the initial state and propagates with the dynamic. Thus, if we observe $y_0 = x_0 + \eta_0 \in \Omega \subset \mathbb{R}^q$ instead of the true initial state x_0, then the dynamical noise $(\eta_n)_{n \in \mathbb{N}_0}$ is defined as

$$y_n = f^n(y_0) = f^n(x_0 + \eta_0) = x_n + \eta_n,$$

where $\eta_n = f^n(x_0 + \eta_0) - f^n(x_0)$ depends on x_0 and η_0. Dynamical noise is detrimental to the predictability of the sequence $(f^n(x_0))_{n \in \mathbb{N}_0}$ when f exhibits sensitivity to initial conditions. This sensitivity is measured by its Lyapunov exponent(s) with respect to the natural invariant measure.

- *Observational* (or *additive*) *noise* adds a random fluctuation to the true value $x_n = f^n(x_0)$ in each iteration, that is, the observed value at "time" n is

$$z_n = x_n + \zeta_n,$$

where $\zeta = (\zeta_n)_{n \in \mathbb{N}_0}$ is an $\mathbb{R}^q$-valued random process that accounts for the different macroscopic and/or microscopic factors affecting the true value $f^n(x_0)$. If the random variables ζ_n are independent, then one says that ζ is *white noise*, otherwise the noise is *colored*. Since ordinal patterns depend only on arithmetical differences between observations close in time, the mean of the noise probability distribution is irrelevant. By the same token, we also expect that observational noises with similar variances and finite supports (or possibly thin-tailed distributions) will produce a similar structure of admissible and forbidden patterns. In numerical simulations, the support of the random variables ζ_n will be certainly bounded. White and colored noise are random time series, so random sequences can be viewed as consisting only of noise.

Dynamical noise belongs only to deterministic time series and is important in numerical simulations, whereas observational noise corrupts actual observations of experimental deterministic and random sequences.

Given a deterministic or random time series $\mathbf{x} = (x_n)_{n \in \mathbb{N}_0}$, we say that an ordinal pattern $\pi = \langle \pi_0, \pi_1, \ldots, \pi_{L-1} \rangle$ is *observable* or *visible* in $\mathbf{x}$ if $\mathbf{x}$ contains a length-L block $x_k^{k+L-1} = x_k, \ldots, x_{k+L-1}$ of type π, i.e., if $x_{k+\pi_0} < x_{k+\pi_1} < \cdots < x_{k+\pi_{L-1}}$. Otherwise, π is said to be *unobservable* or *missing* in $\mathbf{x}$. If $\mathbf{x}$ has been deterministically generated by f, then visible patterns are necessarily admissible for f, while forbidden patterns for f cannot be visible in $\mathbf{x}$ (nor in any other orbit of f for that matter). On the other hand, if π is missing in $\mathbf{x}$, this does not necessarily mean that π is forbidden for f—it might be visible in other orbit of f. Thus, forbidden patterns are a subset of the missing patterns. The same considerations apply to real, finite-length sequences.

We say that a visible (correspondingly, missing) ordinal L-pattern π in a deterministic time series $x_n = f^n(x_0)$ is *unconditionally robust* against dynamical or observational noise, if π is also visible (correspondingly, missing) in any perturbed time series $x_n + \xi_n$, $n \in \mathbb{N}_0$, where ξ_n is dynamical or observational noise, respectively. Likewise, we say that a visible (correspondingly, missing) ordinal L-pattern π in a deterministic time series $x_n = f^n(x_0)$ is *conditionally robust* against dynamical or observational noise, if π is also visible (correspondingly, missing) in any perturbed *finite* time series (or initial segment) $x_n + \xi_n$, $0 \leq n \leq N$, where ξ_n is, respectively, dynamical or observational noise with sufficiently small *amplitude* $A = A(x_0, N) = \max_{0 \leq n \leq N} \| \xi_n \|$.

Lemma 1 *Consider time series generated by a continuous self-map f of a closed interval $I \subset \mathbb{R}$.*

(1) Forbidden patterns are unconditionally robust against dynamical noise. (This is also true if f is not continuous.)

(2) Visible patterns are conditionally robust against dynamical noise.
(3) Visible and missing patterns are conditionally robust against observational noise.

Proof (1) If $\pi = \langle \pi_0, \ldots, \pi_{L-1} \rangle$ is a forbidden pattern for f, then π will be not visible in the sequence $(f^n(x_0))_{n \in \mathbb{N}_0}$ nor in the perturbed sequence $(f^n(x_0 + \eta_0))_{n \in \mathbb{N}_0}$ for any η_0 such that $x_0 + \eta_0 \in I$.

(2) An ordinal L-pattern π visible in $(f^n(x_0))_{n \in \mathbb{N}_0}$ will remain visible in a finite noisy sequence $y_n = f^n(x_0 + \eta_0) = x_n + \eta_n$, $0 \le n \le N$, only if $|\eta_0|$ is small enough. The size of $|\eta_0|$ will depend on the Lyapunov exponent (with respect to the natural invariant measure) of f.

(3) Consider the segment $x_n = f^n(x_0)$, $0 \le n \le N$, of the time series $(f(x_0))_{n \in \mathbb{N}_0}$, and suppose that

$$f^{\pi_0}(x_k) < f^{\pi_1}(x_k) < \cdots < f^{\pi_{L-1}}(x_k)$$

for some $k \in \{0, 1, , \ldots, N - L + 1\}$. Then

$$f^{\pi_0}(x_k) + \zeta_0 < f^{\pi_1}(x_k) + \zeta_1 < \cdots < f^{\pi_{L-1}}(x_k) + \zeta_{L-1}$$

holds also true as long as the perturbations ζ_i satisfy

$$\zeta_i < f^{\pi_{i+1}}(x_k) - f^{\pi_i}(x_k) + \zeta_{i+1}$$

for $i = 0, 1, \ldots, L - 2$.

From the result that visible patterns are robust against small observational noise, it follows that missing patterns (in particular, forbidden patterns) are likewise robust against small observational noise.   □

We conclude from Lemma 1 that visible patterns in univariate time series are conditionally robust however the kind of noise, whereas forbidden patterns are unconditionally robust against dynamical noise but conditionally robust against observational noise.

In case of multivariate time sequences ($I \subset \mathbb{R}^q$ with $q \ge 2$), property (1) of Lemma 1 remains the same, since the dimensionality of I does not enter in the proof. The situation is different with the conditional robustness. For example, suppose that $\mathbb{R}^q$ is lexicographically ordered, $x_k^{(1)} = x_{k+1}^{(1)}$ and $x_k^{(2)} < x_{k+1}^{(2)}$, hence $\mathbf{x}_k := (x_k^{(1)}, x_k^{(2)}) < (x_{k+1}^{(1)}, x_{k+1}^{(2)}) =: \mathbf{x}_{k+1}$. Then $\mathbf{x}_k + \zeta_k, \mathbf{x}_{k+1} + \zeta_{k+1}$ will not define the pattern $\pi = \langle 0, 1 \rangle$ if $\zeta_k^{(1)} > \zeta_{k+1}^{(1)}$, however, small their sizes are. A corresponding result holds for dynamical noise if, in the example above, the first component of $f^k(\mathbf{x}_0)$ can be made to increase or decrease by varying $\mathbf{x}_0$. In real cases though, in which time series are finite and maps have random-like properties, the coincidence of components is highly unlikely, at least if real numbers are represented with a high enough precision. We may infer that, although visible and missing patterns in multivariate sequences are not, in general, robust against observational nor dynamical noise, in practice they may be considered conditionally robust (as in the univariate case).

Conditional robustness has to do with the amplitude of the perturbation. What about the dependence of visible and missing patterns on the length N of the initial segment $(z_n)_{n=0}^{N-1}$ of a noisy time series $(z_n)_{n\in\mathbb{N}_0}$ of either type? Since an increase of N eventually transforms missing patterns of length $L < N$ into visible L-patterns, while visible patterns remain visible, it is clear that the number of missing L-patterns in time series contaminated by dynamical or observational noise will decrease with N. In other words, the longer the sequence, the higher the odds that some block z_n^{n+L-1} defines π. In the case of white noise only, $(z_n)_{n\in\mathbb{N}_0} = (\zeta_n)_{n\in\mathbb{N}_0}$, one can show that the decrease of missing ordinal L-patterns goes exponentially with N (see also Fig. 9.7).

If forbidden patterns were not robust against noise, they would be not useful in time-series analysis. The sort of applications we have in mind belong in the detection of determinism in univariate and multivariate time-series analysis, since (unconstrained) random real-valued time series have no forbidden patterns with probability 1. These and related issues will be discussed in Chap. 9.

Chapter 4
Ordinal Structure of the Shifts

Shift systems are dynamical systems which are used as universal models in information theory and stochastic processes. Besides they are interesting on its own because, in spite of their conceptual simplicity, they exhibit some of the intricacies of low-dimensional chaos, like sensitivity to initial conditions, strong mixing, and a dense set of periodic points.

In the last chapter we studied some general properties of the allowed and forbidden patterns associated with a dynamical system whose state space is linearly ordered. In this chapter we will be more specific and study the ordinal structure of the shift transformations. By ordinal structure we mean such properties as the length and number of the root forbidden patterns. Contrary to the generality of maps, we shall see that these issues can be ascertained with great detail for the shifts.

4.1 Ordinal Patterns and the Shift Maps

Let $E_N : [0, 1] \rightarrow [0, 1]$, $N \in \{2, 3, \dots\}$, be the shift or sawtooth map

$$E_N(x) = Nx \pmod 1 \tag{4.1}$$

(Fig. 4.1). Observe that if

$$x = \sum_{n=0}^{\infty} x_n \cdot N^{-(n+1)} =: 0. x_0 x_1 \dots x_n \dots,$$

$0 \leq x_n \leq N - 1$, is an N-ary expansion of $x \in [0, 1]$, then

$$Nx = \sum_{n=0}^{\infty} x_n \cdot N^{-n} = x_0 + \sum_{n=1}^{\infty} x_n \cdot N^{-n} = x_0 . x_1 x_2 \dots x_{n+1} \dots$$

and

J.M. Amigó, *Permutation Complexity in Dynamical Systems*,
Springer Series in Synergetics, DOI 10.1007/978-3-642-04084-9_4,
© Springer-Verlag Berlin Heidelberg 2010

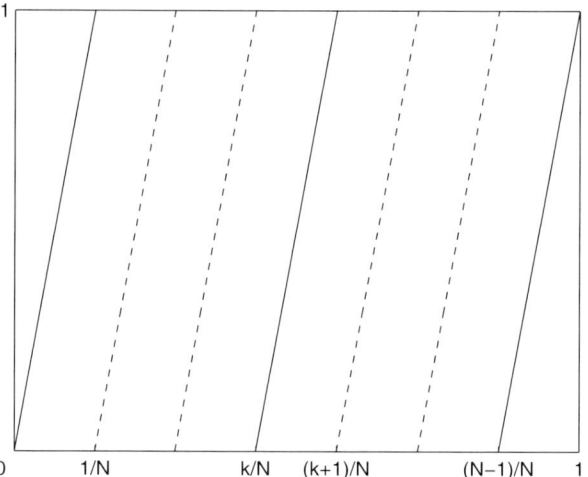

Fig. 4.1 The function $E_N(x) = Nx \bmod 1$. The figure shows only the first, the kth, and the last laps of the graph

$$E_N(0.x_0x_1\ldots x_n\ldots) = 0.x_1x_2\ldots x_{n+1}\ldots. \tag{4.2}$$

In other words, if we write $E_N(0.x_0x_1\ldots x_n\ldots) = 0.y_0y_1\ldots y_n\ldots$, then $y_n = x_{n+1}$ for $n \in \mathbb{N}_0$. This justifies the name "shift map" for E_N since it shifts the digits of the representation of x in base N, one position to the left (the first digit is deleted). Let us recall an N-ary expansion is not unique for some $x \in [0, 1]$ since

$$0.x_0\ldots x_{n-1}10^\infty = 0.x_0\ldots x_{n-1}0(N-1)^\infty,$$

where (as in Sect. 1.1.2) the upper symbol "∞" stands for indefinite repetition. But the set of points $x \in [0, 1]$ whose N-ary expansion ends with 10^∞ or $0(N-1)^\infty$ has zero Lebesgue measure, so such points can eventually be thought to have been removed from $[0, 1]$.

If we identify now an N-ary expansion $0.x_0x_1\ldots x_n\ldots$ of $x \in [0, 1]$ with the one-sided sequence $(x_0, x_1, \ldots, x_n, \ldots) \in S^{\mathbb{N}_0}$, $S = \{0, \ldots, N-1\}$, then action (4.2) translates into the action of the one-sided shift Σ on $S^{\mathbb{N}_0}$. Formally, if $\phi_N : S^{\mathbb{N}_0} \to [0, 1]$ is the map defined by

$$\phi_N : (x_n)_{n \in \mathbb{N}_0} \mapsto \sum_{n=0}^{\infty} x_n N^{-(n+1)}, \tag{4.3}$$

then ϕ_N is an order isomorphism modulo 0 between E_N and the one-sided shift Σ on $S^{\mathbb{N}_0}$, i.e.,

$$\phi_N \circ \Sigma = E_N \circ \phi_N, \tag{4.4}$$

the order of $S^{\mathbb{N}_0}$ being given by the lexicographical rule:

$$\mathbf{x} < \mathbf{x}' \Leftrightarrow \begin{cases} x_0 < x'_0, \\ \text{or} \\ x_0 = x'_0, \dots, x_{n-1} = x'_{n-1} \text{ and } x_n < x'_n \ (n \geq 1), \end{cases} \tag{4.5}$$

where $\mathbf{x} = (x_n)_{n \in \mathbb{N}_0}$ and $\mathbf{x}' = (x'_n)_{n \in \mathbb{N}_0}$. Observe that ϕ_N maps the cylinder set $C_{i_0 \dots i_n} = \{(x_n)_{n \in \mathbb{N}_0} : x_0 = i_0, \dots, x_n = i_n\}$ $(i_0, \dots, i_n \in S)$ to the interval

$$\left[\frac{i_0 N^n + \dots + i_n}{N^{n+1}}, \frac{i_0 N^n + \dots + i_n + 1}{N^{n+1}} \right].$$

Exercise 4 Let $\mathcal{B}$ be the Borel sigma-algebra on $[0, 1]$, λ the corresponding Lebesgue measure, and E_N the sawtooth map (4.1). Prove that the dynamical system $([0, 1], \mathcal{B}, \lambda, E_N)$ and the $(\frac{1}{N}, \dots, \frac{1}{N})$-Bernoulli one-sided shift are isomorphic (modulo 0).

Once we know that E_N and the one-sided shift Σ on N symbols are order-isomorphic (up to sets measure 0), it follows that they have the same forbidden patterns (see Sect. 3.4.1).

In general it is very difficult to work out the specifics of the forbidden patterns of a given map; the graphical methods can only help for small values of L. But we shall see next that the shifts and the signed shifts (to be defined in Chap. 5.) are an important exception. In particular, owing to the simple structure of one-sided and two-sided shifts, the structure of their admissible and forbidden patterns can be analyzed with great detail. By order isomorphy these conclusions will hold also for the sawtooth map family E_N (one-sided shifts), the baker map (two-sided shifts), and the logistic and symmetric tent maps (one-sided signed shifts), among others.

4.2 Forbidden Patterns for One-Sided Shifts

In Sect. 1.1.2 we saw that one-sided shifts Σ are continuous maps on the compact metric spaces $(\{0, 1, \dots, N - 1\}^{\mathbb{N}_0}, d)$, $N \geq 2$. Furthermore, if $\{0, 1, \dots, N-1\}^{\mathbb{N}_0}$ is lexicographically ordered (see (4.5)), then Σ is order-isomorphic (modulo 0) to E_N via map (4.3).

What is the structure of the allowed ordinal patterns for Σ? It is easy to convince oneself (see Example 7) that, given $\mathbf{x} = (x_0, \dots, x_{L-1}, \dots) \in \{0, 1, \dots, N - 1\}^{\mathbb{N}_0}$ of type $\pi \in S_L$, π can be decomposed into at most N blocks (separated by semicolons),

$$\langle \pi_0, \dots, \pi_{k_0-1}; \pi_{k_0}, \dots, \pi_{k_0+k_1-1}; \dots; \pi_{k_0+\dots+k_{N-2}}, \dots, \pi_{k_0+\dots+k_{N-2}+k_{N-1}-1} \rangle, \tag{4.6}$$

where $k_s \geq 0$ is the number of times the symbol $s \in \{0, 1, \ldots, N - 1\}$ appears in the segment $x_0^{L-1} = x_0, \ldots, x_{L-1}$ of $\mathbf{x}$ ($k_s = 0$ if none, with the corresponding block empty) and $k_0 + \cdots + k_{N-1} = L$. The entries $\pi_0, \ldots, \pi_{k_0-1}$ are the locations of the symbol 0 in x_0^{L-1}, the entries $\pi_{k_0}, \ldots, \pi_{k_0+k_1-1}$ are the locations of the symbol 1 in x_0^{L-1}, etc. For this reason, the first block will also be called the 0-block, and, in general, the $(s + 1)$th block,

$$\pi_{k_0+\cdots+k_{s-1}}, \ldots, \pi_{k_0+\cdots+k_{s-1}+k_s-1}, \tag{4.7}$$

$1 \leq s \leq N - 1$, will also be called the s-block. Decomposition (4.6) is sometimes called the *decomposition* of an allowed ordinal pattern $\pi \in S_L$ in s-blocks.

A (finite) subsequence of components of π of the form $\pi_i, \ldots, \pi_i + 1, \ldots, \pi_i + 2, \ldots$ (respectively, $\pi_i, \ldots, \pi_i - 1, \ldots, \pi_i - 2, \ldots$) will be called an *increasing* (respectively, *decreasing*) *subsequence*. Increasing or decreasing subsequences will be collectively called *monotone*. Observe that we use these concepts in a restrictive way.

We will see next that from the fact that allowed patterns for the one-sided shift must be decomposable as in (4.6), it is possible to deduce their structure.

Lemma 2 *The blocks in decomposition (4.6) obey the following basic restrictions.*

R1 *The first (leftmost) block, $\pi_0, \ldots, \pi_{k_0-1}$, contains the locations of the 0's in x_0^{L-1}. Each 0-run (i.e., a segment of two or more consecutive 0's contained in or intersected by x_0^{L-1}), if any, contributes an increasing subsequence of the same length as the 0-run. Solitary symbols 0's in x_0^{L-1}, if any, contribute components to the first block that do not form monotone subsequences.*

R2 *The last (rightmost) block, $\pi_{k_0+\cdots+k_{N-2}}, \ldots, \pi_{k_0+\cdots+k_{N-2}+k_{N-1}-1}$, contains the locations of the $(N-1)$'s in x_0^{L-1}. Each $(N-1)$-run contained in or intersected by x_0^{L-1}, if any, contributes a decreasing subsequence of the same length as the $(N - 1)$-run. Solitary symbols 1's in x_0^{L-1}, if any, contribute components to the last block that do not form monotone subsequences.*

R3 *Every intermediate block, $\pi_{k_0+\cdots+k_{j-1}}, \ldots, \pi_{k_0+\cdots+k_{j-1}+k_j-1}$, $1 \leq j \leq N - 2$, contains the locations of the j's in x_0^{L-1}. Each j-run contained in or intersected by x_0^{L-1}, if any, contributes a subsequence of the same length as the j-run that is increasing if the run is followed by a symbol $> j$, or decreasing if the run is followed by a symbol $< j$. Isolated symbols j's in x_0^{L-1}, if any, contribute components to the corresponding block that do not form monotone subsequences.*

R4 *If the entries $\pi_m \leq L - 2$ and $\pi_n \leq L - 2$ belong to the same block of $\pi \in S_L$ and π_m appears on the left of π_n (i.e., $0 \leq m < n \leq L - 1$), then $\pi_m + 1$ appears also on the left of $\pi_n + 1$ (i.e., $\pi_m + 1 = \pi_{m'}$, $\pi_n + 1 = \pi_{n'}$ and $0 \leq m' < n' \leq L - 1$), not necessarily in the same block.*

Proof **R1**) Consider a 0-run of length l in **x**:

$i =$	$n-1$	n	$n+1$	...	$n+l-1$	$n+l$
$\mathbf{x} =$	a	0	0	...	0	b

with $0 \le n$, $n + l \le L$, and $a, b > 0$. Hence the 0-block of $\pi(\mathbf{x})$ contains the increasing subsequence

$$\ldots, n, \ldots, n+1, \ldots, n+l-1, \ldots,$$

The "..." stands for entries proceeding from other 0-runs in **x**.

R2) Consider an $(N-1)$-run of length l in **x**:

$i =$	$n-1$	n	$n+1$	...	$n+l-1$	$n+l$
$\mathbf{x} =$	c	$N-1$	$N-1$	...	$N-1$	d

with $0 \le n$, $n + l \le L$, and $c, d < N - 1$. Hence the $(N-1)$-block of $\pi(\mathbf{x})$ contains the decreasing subsequence

$$\ldots, n+l-1, \ldots, n+1, \ldots, n, \ldots,$$

The "..." allows for entries proceeding from other $(N-1)$-runs in **x**.

R3) This restriction follows similarly to R1 for s-runs, $0 < s < N-1$, terminated with $b > s$ (increasing subsequences), and similarly to R2 for s-runs terminated with $d < s$ (decreasing subsequences).

R4) Since π_m and π_n belong to the same block and $\Sigma^{\pi_m}(\mathbf{x}) < \Sigma^{\pi_n}(\mathbf{x})$ for some $\mathbf{x} \in \{0, 1, \ldots, N-1\}^{\mathbb{N}_0}$, there exists $k \in \{0, 1, \ldots, N-1\}$ such that

$$\Sigma^{\pi_m}(\mathbf{x}) = (k, x_{\pi_m+1} \ldots) < (k, x_{\pi_n+1}, \ldots) = \Sigma^{\pi_n}(\mathbf{x}).$$

By the definition of lexicographical order, there are two possibilities: (i) $x_{\pi_m+1} < x_{\pi_n+1}$ and (ii) $x_{\pi_m+\kappa} = x_{\pi_n+\kappa}$ for $1 \le \kappa \le l-1$, $l \ge 2$, and $x_{\pi_m+l} < x_{\pi_n+l}$. In both cases,

$$\Sigma^{\pi_m+1}(\mathbf{x}) = (x_{\pi_m+1} \ldots) < (x_{\pi_n+1}, \ldots) = \Sigma^{\pi_n+1}(\mathbf{x})$$

and, hence, the entry $\pi_m + 1$ appears on the left of $\pi_n + 1$. $\square$

Example 7 Consider in $\{0, 1, 2\}^{\mathbb{N}_0}$ the sequence

$$\mathbf{x} = (2_0, 1_1, 1_2, 1_3, 2_4, 2_5, 0_6, 0_7, 1_8, 1_9, 0_{10}, 0_{11}, 2_{12}, 2_{13}, 2, 1, \ldots), \tag{4.8}$$

where a_k indicates that the entry $a \in \{0, 1, 2\}$ is at place k. Then **x** defines the ordinal pattern

$$\pi = \langle 6, 10, 7, 11; 9, 8, 1, 2, 3; 5, 0, 4, 13, 12 \rangle \in S_{14}.$$

The 0-block, $\pi_0^3 = 6, 10, 7, 11$, codifies the $k_0 = 4$ times the symbol 0 appears in x_0^{13}, grouped in two runs, x_6^7 and x_{10}^{11} (note the two increasing subsequences 6, 7 and 10, 11 in this block). The order results from

$$\Sigma_3^6(\mathbf{x}) = (0, 0, 1, \ldots) < \Sigma_3^{10}(\mathbf{x}) = (0, 0, 2, \ldots)$$
$$< \Sigma_3^7(\mathbf{x}) = (0, 1, 1, \ldots) < \Sigma_3^{11}(\mathbf{x}) = (0, 2, \ldots).$$

The 1-block, $\pi_4^8 = 9, 8, 1, 2, 3$, codifies the $k_1 = 5$ times the symbol 1 appears in x_0^{13}, grouped also in two runs: x_1^3, followed by the symbol $2 > 1$, and x_8^9, followed by the symbol $0 < 1$ (note the corresponding increasing subsequence 1, 2, 3, and decreasing subsequence 9, 8, in this block). The order results from

$$\Sigma_3^9(\mathbf{x}) = (1, 0, 0, \ldots) < \Sigma_3^8(\mathbf{x}) = (1, 1, 0, \ldots) < \Sigma_3^1(\mathbf{x}) = (1, 1, 1, \ldots) < \cdots$$

etc. Finally, the 2-block $\pi_9^{13} = 5, 0, 4, 13, 12$ codifies the $k_2 = 5$ appearances of the symbol 2 in x_0^{13}. The decreasing subsequences 5, 4 and 13, 12 come from the runs x_4^5 and x_{12}^{13}, respectively, where x_{12}^{13} is the intersection within x_0^{13} of a longer 2-run. The order results from

$$\Sigma_3^5(\mathbf{x}) = (2, 0, 0, \ldots) < \Sigma_3^0(\mathbf{x}) = (2, 1, 1, \ldots) < \Sigma_3^4(\mathbf{x}) = (2, 2, 0, \ldots) < \cdots.$$

The restriction R4 is easily checked to be fulfilled.

Observe that two sequences $\mathbf{x}, \mathbf{x}'$ with $x_0^{L-1} \neq x_0'^{L-1}$ may define the same ordinal L-pattern, while two sequences $\mathbf{y}, \mathbf{y}'$ with $y_0^{L-1} = y_0'^{L-1}$ may define different ordinal L-patterns (depending on $y_{L-2}, \ldots$ and $y'_{L-2}, \ldots$).

The restriction R4 implies some simple consequences for the relative locations of increasing and decreasing subsequences within the same block and their continuations (if any) outside the block.

Corollary 1 *In an allowed ordinal pattern $\pi \in S_L$, the following relations among its components hold.*

(A) *If $\pi_i, \pi_i + 1, \ldots, \pi_i + l - 1, 1 \leq l \leq L - 1$, is an increasing subsequence within the same block of $\pi \in S_L$ with $\pi_i + l < L$, then $\pi_i + l$ is on the right of $\pi_i + l - 1$ (i.e., $\pi_i + l - 1 = \pi_m$, $\pi_i + l = \pi_n$, and $m < n$).*
(B) *If $\pi_i, \pi_i - 1, \ldots, \pi_i - l + 1, 1 \leq l \leq L - 1$, is a decreasing subsequence within the same block of $\pi \in S_L$ with $\pi_i < L - 1$, then $\pi_i + 1$ is on the left of π_i (i.e., $\pi_i + 1 = \pi_j$ with $j < i$).*
(C) *If $\pi_i, \pi_i \pm 1, \ldots, \pi_i \pm l \mp 1$ and $\pi_j, \pi_j \pm 1, \ldots, \pi_j \pm h \mp 1, 1 \leq l, h \leq L - 1$, are two subsequences with the same monotonicity (upper signs for increasing, lower signs for decreasing subsequences) within the same block of $\pi \in S_L$, then they are fully separated or, if intertwined, then it may not happen that two or more entries of one of them are between two entries of the other.*

The proof is left as an easy exercise.

Theorem 1 *The one-sided shift on $N \geq 2$ symbols has no forbidden patterns of length $L \leq N + 1$.*

Proof If $L \leq N$ and $\pi = \langle \pi_0, \pi_1, \ldots, \pi_{L-1} \rangle$, then any "point" $\mathbf{x} \in \{0, 1, \ldots, N-1\}^{\mathbb{N}_0}$ with $x_{\pi_n} = n$, $0 \leq n \leq L - 1 \leq N - 1$, is trivially of type π:

$$\Sigma^{\pi_0}(\mathbf{x}) = (0, \ldots) < \Sigma^{\pi_1}(\mathbf{x}) = (1, \ldots) < \cdots < \Sigma^{\pi_{L-1}}(\mathbf{x}) = (L - 1, \ldots).$$

Thus, suppose $L = N + 1$ and note if $\mathbf{x} = (x_0, x_1, x_2, \ldots)$ is of type $\pi = \langle \pi_0, \pi_1, \ldots, \pi_N \rangle$, then the sequence $\bar{\mathbf{x}} = (N - 1 - x_0, N - 1 - x_1, N - 1 - x_2, \ldots)$ is of type $\pi_{\text{mirrored}} = \langle \pi_N, \pi_{N-1}, \ldots, \pi_1, \pi_0 \rangle$.

Given $\pi = \langle \pi_0, \pi_1, \ldots, \pi_N \rangle$, we can therefore assume, without loss of generality, that $\pi_0 < \pi_N$. Consider two cases.

- If $\pi_N \neq N$, then there is some $l \in \{1, 2, \ldots, N - 1\}$ such that $\pi_l = N$. In this case, the point $\mathbf{x} = (x_0, x_1, \ldots) \in \{0, 1, \ldots, N - 1\}^{\mathbb{N}_0}$, where

$$x_{\pi_0} = 0, \ x_{\pi_1} = 1, \ \ldots, \ x_{\pi_{l-1}} = l - 1, \ x_{\pi_l} = l - 1, \ x_{\pi_{l+1}} = l, \ \ldots,$$
$$x_{\pi_{N-1}} = N - 2, \ x_{\pi_N} = N - 1, \ x_{N+1} = x_{N+2} = N - 1$$

  is of type π. Indeed, it is enough to note that

$$\Sigma^{\pi_{l-1}}(\mathbf{x}) = (l - 1, x_{\pi_{l-1}+1}, \ldots) < (l - 1, N - 1, N - 1, \ldots)$$
$$= \Sigma^N(\mathbf{x}) = \Sigma^{\pi_l}(\mathbf{x}).$$

- If $\pi_N = N$, let us first assume that $\pi_0 \neq 0$. Then there is $k \in \{1, 2, \ldots, N-1\}$ such that $\pi_k + 1 = \pi_0$. In this case, the point $\mathbf{x} = (x_0, x_1, \ldots) \in \{0, 1, \ldots, N - 1\}^{\mathbb{N}_0}$, where

$$x_{\pi_0} = 0, \ \ldots, \ x_{\pi_k} = k, \ x_{\pi_{k+1}} = k, \ x_{\pi_{k+2}} = k + 1, \ \ldots,$$
$$x_{\pi_{N-1}} = N - 2, \ x_{\pi_N} = N - 1, \ x_{N+1} = N - 1$$

  is of type π. This is clear because

$$\Sigma^{\pi_k}(\mathbf{x}) = (k, 0, \ldots) < (k, x_{\pi_{k+1}+1}, \ldots) = \Sigma^{\pi_{k+1}}(\mathbf{x}).$$

  In the case that $\pi_0 = 0$, then there is $l \in \{1, 2, \ldots, N - 1\}$ such that $\pi_l = N - 1$. Now the sequence $\mathbf{x} = (x_0, x_1, \ldots) \in \{0, 1, \ldots, N - 1\}^{\mathbb{N}_0}$, where

$$x_{\pi_0} = 0, \ x_{\pi_1} = 1, \ \ldots, \ x_{\pi_{l-1}} = l - 1, \ x_{\pi_l} = l - 1, \ x_{\pi_{l+1}} = l, \ \ldots,$$
$$x_{\pi_{N-1}} = N - 2, \ x_{\pi_N} = N - 1$$

  is of type π, since

$$\Sigma^{\pi_{l-1}}(\mathbf{x}) = (l-1, x_{\pi_{l-1}+1}, \dots)$$
$$< (l-1, N-1, \dots) = \Sigma^{N-1}(\mathbf{x}) = \Sigma^{\pi_l}(\mathbf{x}). \ \square$$

Next we are going to show that the one-sided shift on N symbols has forbidden patterns (more specifically, forbidden *root* patterns) of any length $L \geq N+2$. In order to construct explicit instances, we need first to introduce some notation and definitions.

Consider a partition of the sequence $0, 1, \dots, L-1$ of the form

$$\overrightarrow{\mathbf{p}_1}, \overrightarrow{\mathbf{p}_2}, \dots, \overrightarrow{\mathbf{p}_d}, \dots, \overrightarrow{\mathbf{p}_D}, \tag{4.9}$$

where

$$\overrightarrow{\mathbf{p}_d} = e_d, e_d + 1, \dots, e_d + h_d - 1, \tag{4.10}$$

$1 \leq d \leq D, D \geq 2$, with (i) $h_d \geq 1$, $h_1 + \dots + h_D = L$, (ii) $e_1 = 0$, $e_D + h_D - 1 = L - 1$, and (iii) $e_d + h_d = e_{d+1}$ for $1 \leq d \leq D-1$, i.e., the *follower* of $\overrightarrow{\mathbf{p}_d}$, $e_d + h_d$, $d \leq D-1$, is the first element of p_{d+1}, namely, e_{d+1}. We call (4.9) a partition of $0, 1, \dots, L-1$ in D segments, (4.10) being an *increasing segment*, and denote by $\overleftarrow{\mathbf{p}_d}$ the *decreasing* or *reversed segment*

$$\overleftarrow{\mathbf{p}_d} = e_d + h_d - 1, \dots, e_d + 1, e_d.$$

We also call e_d the first element of $\overleftarrow{\mathbf{p}_d}$ and e_{d+1} the follower of $\overleftarrow{\mathbf{p}_d}$.

Since increasing and decreasing segments are nothing else but special cases of increasing and decreasing subsequences, respectively, the consequences (A)–(C) of restriction R4 apply as well. In the proof of the existence of forbidden root patterns below (Lemmas 3 and 4 and Theorem 2) we are going to use (A) and (B) in the following, particularized version (that will be also referred to as R4): *the follower (if any) of an increasing segment $\overrightarrow{\mathbf{p}_n}$ (correspondingly, decreasing segment $\overleftarrow{\mathbf{p}_n}$) in an allowed pattern π appears always to the right of $\overrightarrow{\mathbf{p}_n}$ (correspondingly, to the left of $\overleftarrow{\mathbf{p}_n}$).*

Definition 2 Consider partition (4.9) of $0, 1, \dots, L-1$ in segments.

1. We call

$$\pi = \langle \overrightarrow{\mathbf{p}_1}, \overrightarrow{\mathbf{p}_3}, \dots, \overleftarrow{\mathbf{p}_4}, \overleftarrow{\mathbf{p}_2} \rangle \quad \text{and} \quad \pi_{\text{mirrored}} = \langle \overrightarrow{\mathbf{p}_2}, \overrightarrow{\mathbf{p}_4}, \dots, \overleftarrow{\mathbf{p}_3}, \overleftarrow{\mathbf{p}_1} \rangle \tag{4.11}$$

   a tent *pattern of length L*.
2. We call

$$\pi = \langle \dots, \overleftarrow{\mathbf{p}_3}, \overleftarrow{\mathbf{p}_1}, \overrightarrow{\mathbf{p}_2}, \overrightarrow{\mathbf{p}_4}, \dots \rangle \quad \text{and} \quad \pi_{\text{mirrored}} = \langle \dots, \overleftarrow{\mathbf{p}_4}, \overleftarrow{\mathbf{p}_2}, \overrightarrow{\mathbf{p}_1}, \overrightarrow{\mathbf{p}_3}, \dots \rangle \tag{4.12}$$

   a spiraling *pattern of* length L.

Observe that the relation between partitions of $0, 1, \ldots, L - 1$ in segments and spiraling patterns of length L is one-to-one except when $\overrightarrow{\mathbf{p_1}} = 0$ ($h_1 = 1$). In this case, $\overleftarrow{\mathbf{p_1}}, \overrightarrow{\mathbf{p_2}} = 0, 1, \ldots, e_2 + h_2 - 1$ can be taken for $\overrightarrow{\mathbf{p_1'}} := 0, 1, \ldots, e_2 + h_2 - 1$ ($h_1' = h_2 + 1$).

Lemma 3 *If $N \geq 2$ is the number of symbols and π is a tent pattern with D segments, then π is forbidden if and only if $D \geq N + 2$.*

Proof Consider the tent pattern $\pi = \langle \overrightarrow{\mathbf{p_1}}, \overrightarrow{\mathbf{p_3}}, \ldots, \overleftarrow{\mathbf{p_4}}, \overleftarrow{\mathbf{p_2}} \rangle$. To begin with, the last entry $h_1 - 1$ of $\overrightarrow{\mathbf{p_1}}$ and the first entry e_3 of $\overrightarrow{\mathbf{p_3}}$ may not be in the same block, otherwise the R4 would be violated ($e_2 = h_1$ should be on the left of $e_3 + 1$ if $h_3 \geq 2$ or on the left of e_4 if $h_3 = 1$). Thus we separate them with a first semicolon:

$$\pi = \langle \overrightarrow{\mathbf{p_1}} ; \overrightarrow{\mathbf{p_3}}, \ldots, \overleftarrow{\mathbf{p_4}}, \overleftarrow{\mathbf{p_2}} \rangle.$$

Observe that the resulting leftmost block, $\overrightarrow{\mathbf{p_1}}$, complies with R1. Consider now the followers of $\overleftarrow{\mathbf{p_2}}$ and $\overleftarrow{\mathbf{p_4}}$ to conclude similarly that we need to separate these segments by a second semicolon:

$$\pi = \langle \overrightarrow{\mathbf{p_1}} ; \overrightarrow{\mathbf{p_3}}, \ldots, \overleftarrow{\mathbf{p_4}} ; \overleftarrow{\mathbf{p_2}} \rangle.$$

The resulting rightmost block satisfies R2.

The procedure continues along the same lines. In the kth step, R4 requires a kth semicolon between the segments $\overrightarrow{\mathbf{p_k}}$ and $\overrightarrow{\mathbf{p_{k+2}}}$, so that, if $D \geq N + 1$, the $(N - 1)$th semicolon will separate $\overrightarrow{\mathbf{p_{N-1}}}$ and $\overrightarrow{\mathbf{p_{N+1}}}$. All these intermediary blocks trivially fulfill R3.

In the particular case $D = N + 1$, the "central" block $\overrightarrow{\mathbf{p_N}\mathbf{p_{N+1}}}$ (N odd) or $\overrightarrow{\mathbf{p_{N+1}}\mathbf{p_N}}$ (N even) complies with R3 and R4, and hence π is allowed. A further segment $\overrightarrow{\mathbf{p_{N+2}}}$ would require an Nth semicolon to separate $\overrightarrow{\mathbf{p_N}}$ and $\overrightarrow{\mathbf{p_{N+1}}}$ in order not to violate R4.

The proof for π_{mirrored} is completely analogous.   □

Lemma 4 *If $N \geq 2$ is the number of symbols, π is a spiraling pattern with D segments, and $h_1 \geq 2$ (i.e., $\overrightarrow{\mathbf{p_1}} = 0, 1, \ldots$), then*

1. *π is forbidden if and only if (a) $D = N$ and $h_D \geq 2$ or (b) $D \geq N + 1$;*
2. *π is allowed if and only if (a') $D < N$ or (b') $D = N$ and $h_D = 1$.*

Part 2 of Lemma 4, which is the logical negation of part 1, has been explicitly formulated for further references.

Proof Consider the spiraling pattern (4.12). To begin with, the entries $h_1 - 1$ and $h_1 - 2$ of $\overleftarrow{\mathbf{p_1}} = h_1 - 1, \ldots, 1, 0$ may not be in the same block, otherwise R4 would be violated (e_2 should be on the left of $h_1 - 1$). Thus we separate them with a first semicolon:

$$\pi = \langle \ldots, \overleftarrow{\mathbf{p_3}}, h_1 - 1; h_1 - 2, \ldots, 1, 0, \overrightarrow{\mathbf{p_2}}, \overrightarrow{\mathbf{p_4}}, \ldots \rangle.$$

From here on, three possibilities can occur that we illustrate in a general step of even order. (i) If $\overrightarrow{\mathbf{p}_{2v}}$ consists of more than one element (i.e., $h_{2v} \geq 2$), then we apply R4 to $\overrightarrow{\mathbf{p}_{2v}}$ to conclude that we need a semicolon between $e_{2v} + h_{2v} - 2$ and $e_{2v} + h_2 - 1$ (since the follower of $\overrightarrow{\mathbf{p}_{2v}}$, i.e., the first entry of $\overleftarrow{\mathbf{p}_{2v+1}}$, is on the wrong side). (ii) If $\overrightarrow{\mathbf{p}_{2v}}$ consists of one element ($h_{2v} = 1$) and $\overrightarrow{\mathbf{p}_{2v-2}}$ consists of more than one element ($h_{2v-2} \geq 2$), then we apply R4 to the pair $\overrightarrow{\mathbf{p}_{2v}} = e_{2v}$ and $e_{2v-2} + h_{2v-2} - 1$, the last element of $\overrightarrow{\mathbf{p}_{2v-2}}$, which has been separated with a semicolon from the rest of elements in $\overrightarrow{\mathbf{p}_{2v-2}}$ two steps earlier. (iii) If both $\overrightarrow{\mathbf{p}_{2v}}$ and $\overrightarrow{\mathbf{p}_{2v-2}}$ consist of a single element ($h_{2v} = h_{2v-2} = 1$), apply R4 to the pair $\overrightarrow{\mathbf{p}_{2v-2}} = e_{2v-2} < \overrightarrow{\mathbf{p}_{2v}} = e_{2v}$ to infer the need for a semicolon separating them (since $e_{2v-2} + 1 = e_{2v-1}$, the first element of $\overleftarrow{\mathbf{p}_{2v-1}}$, is on the right of $e_{2v} + 1 = e_{2v+1}$, the first element of $\overleftarrow{\mathbf{p}_{2v+1}}$). As a general rule, we need one semicolon per segment $\overrightarrow{\mathbf{p}_{2v}}$ or $\overleftarrow{\mathbf{p}_{2v+1}}$ as long as there are still a posterior segment $\overleftarrow{\mathbf{p}_{2v+1}}$ or $\overrightarrow{\mathbf{p}_{2v+2}}$, respectively, on the "wrong" side. Note that all (intermediary) blocks ensued so far comply with R3.

Following this way, we run out of the $N - 1$ semicolons we may use (corresponding to the N symbols), after having considered the segment $\overrightarrow{\mathbf{p}_{N-1}}$. Yet if $D = N$ and $h_N \geq 2$, then $\overrightarrow{\mathbf{p}_N}$ will violate R1 if N is odd or R2 if N is even. If $D \geq N + 1$, then the segment $\overrightarrow{\mathbf{p}_{N+1}}$ will be on the wrong side of $\overrightarrow{\mathbf{p}_N}$ and the pattern will not comply with R4.

The proof for π_{mirrored} is completely analogous.   □

The constructive, stepwise procedure used in the proofs of Lemmas 3 and 4 can be used mutatis mutandis in general to decompose any ordinal pattern into well-formed (i.e., complying with R1–R4) blocks. For instance, one could start from the leftmost entry and move on rightward one entry at a time, inserting a semicolon between the current and the previous entry whenever necessary to enforce the restrictions R1–R4. Reciprocally, given a decomposition of an ordinal pattern π in s-blocks, one can easily construct a sequence $\mathbf{x} \in \{0, \ldots, N - 1\}^{\mathbb{N}_0}$ of type π.

Theorem 2 *The following patterns of length $L \geq N + 2$, together with their corresponding mirrored patterns, are forbidden root patterns.*

1. *The tent patterns with $N + 2$ segments*

$$\langle 0, \overrightarrow{\mathbf{p}_3}, \ldots, \overrightarrow{\mathbf{p}_N}, L - 1, \overleftarrow{\mathbf{p}_{N+1}}, \ldots, \overleftarrow{\mathbf{p}_2} \rangle \tag{4.13}$$

if N is odd or

$$\langle 0, \overrightarrow{\mathbf{p}_3}, \ldots, \overrightarrow{\mathbf{p}_{N+1}}, L - 1, \overleftarrow{\mathbf{p}_N}, \ldots, \overleftarrow{\mathbf{p}_2} \rangle \tag{4.14}$$

if N is even. Here $\overrightarrow{\mathbf{p}_1} = 0$ and $\overrightarrow{\mathbf{p}_{N+2}} = L - 1$.
2. *The spiraling pattern with $N + 1$ segments*

$$\langle L - 2, \overleftarrow{\mathbf{p}_{N-2}}, \ldots, \overleftarrow{\mathbf{p}_3}, 1, 0, \overrightarrow{\mathbf{p}_2}, \ldots, \overrightarrow{\mathbf{p}_{N-1}}, L - 1 \rangle \tag{4.15}$$

if N is odd or

$$\langle L-1, \overleftarrow{\mathbf{p}_{N-1}}, \ldots, \overleftarrow{\mathbf{p}_3}, 1, 0, \overrightarrow{\mathbf{p}_2}, \ldots, \overrightarrow{\mathbf{p}_{N-2}}, L-2\rangle, \tag{4.16}$$

if N is even. Here $\overrightarrow{\mathbf{p}_1} = 0, 1$, $\overrightarrow{\mathbf{p}_N} = L-2$, and $\overrightarrow{\mathbf{p}_{N+1}} = L-1$.

3. *The spiraling pattern with N segments*

$$\langle L-1, L-2, \overleftarrow{\mathbf{p}_{N-2}}, \ldots, \overleftarrow{\mathbf{p}_3}, 1, 0, \overrightarrow{\mathbf{p}_2}, \ldots, \overrightarrow{\mathbf{p}_{N-1}}\rangle \tag{4.17}$$

if N is odd or

$$\langle \overleftarrow{\mathbf{p}_{N-1}}, \ldots, \overleftarrow{\mathbf{p}_3}, 1, 0, \overrightarrow{\mathbf{p}_2}, \ldots, \overrightarrow{\mathbf{p}_{N-2}}, L-2, L-1\rangle, \tag{4.18}$$

if N is even. Here $\overrightarrow{\mathbf{p}_1} = 0, 1$, and $\overrightarrow{\mathbf{p}_N} = L-2, L-1$.

Of course, cases 2 and 3 are related to the two possibilities in Lemma 4.

Proof First of all, remember from Sect. 3.4.2, (3.12), that given a forbidden pattern

$$\langle \pi_0, \ldots, \pi_{L-2}\rangle \in \mathcal{S}_{L-1},$$

its outgrowth patterns of length L have the form (*group I*)

$$\langle L-1, \pi_0, \ldots, \pi_{L-2}\rangle, \langle \pi_0, L-1, \ldots, \pi_{L-2}\rangle, \ldots, \langle \pi_0, \ldots, \pi_{L-2}, L-1\rangle$$

or the form (*group II*)

$$\langle 0, \pi_0+1, \ldots, \pi_{L-2}+1\rangle, \langle \pi_0+1, 0, \ldots, \pi_{L-2}+1\rangle, \ldots, \langle \pi_0+1, \ldots, \pi_{L-2}+1, 0\rangle.$$

1. This case is trivial. Any tent pattern made out of $N+2$ segments is forbidden according to Lemma 3. Moreover, since the entries $L-1$ and 0 in patterns (4.13) and (4.14) are segments on their own, the number of segments D of these tent patterns will fall below the threshold value $D = N+2$ once $L-1$ (group I) or 0 (group II) are deleted.

2. Only (4.15) will be considered here, the proof for (4.16) and their mirrored patterns being completely analogous. That (4.15) is forbidden follows readily from Lemma 4 (b). To prove that π is also a root pattern, we need to show that it is not the outgrowth of any forbidden pattern of shorter length.

There are two possibilities. Suppose first that π is an outgrowth forbidden pattern of group I. Deletion of the entry $L-1$ yields then the spiraling pattern

$$\langle L-2, \overleftarrow{\mathbf{p}_{N-2}}, \ldots, \overleftarrow{\mathbf{p}_3}, 1, 0, \overrightarrow{\mathbf{p}_2}, \ldots, \overrightarrow{\mathbf{p}_{N-1}}\rangle,$$

which is allowed on account of having N segments, $h_1 = 2$, and a last segment $\overrightarrow{\mathbf{p}_N} = L-2$ of length 1 (Lemma 4 (b')).

Thus, suppose that π is an outgrowth forbidden pattern of group II. In this case, after removing the entry 0 and subtracting 1 from the remaining entries we are left with the pattern

$$\langle L-3, \overleftarrow{\mathbf{p}'_{N-2}}, \ldots, \overleftarrow{\mathbf{p}'_3}, 0, \overrightarrow{\mathbf{p}'_2}, \ldots, \overrightarrow{\mathbf{p}'_{N-1}}, L-2 \rangle, \qquad (4.19)$$

where $\overrightarrow{\mathbf{p}'_d} = e_d - 1, \ldots, e_d + h_d - 2, 2 \leq d \leq N+1$. Since $\overrightarrow{\mathbf{p}'_1} = 0 \, (h'_1 = h_1 - 1 = 1)$ and $\overrightarrow{\mathbf{p}'_2} = 1, \ldots \, (h'_2 = h_2 \geq 1)$, we can merge $\overrightarrow{\mathbf{p}'_1}$ and $\overrightarrow{\mathbf{p}'_2}$ into the new segment $\overrightarrow{\mathbf{p}''_1} := 0, 1, \ldots$, so that (4.19) is a spiraling pattern with $h''_1 \geq 2$ and the following N segments: $\overrightarrow{\mathbf{p}''_1}, \overrightarrow{\mathbf{p}'_3}, \ldots, \overrightarrow{\mathbf{p}'_{N-1}}, \overrightarrow{\mathbf{p}'_N} = L-3, \overrightarrow{\mathbf{p}'_{N+1}} = L-2$. According to Lemma 4 (b'), the ordinal pattern (4.19) is allowed.

3. This case uses Lemma 4 (a)–(a') instead. The proof proceeds similar to case 2.   □

Example 8 For $N = 2n+1$, Theorem 2 provides the following six forbidden patterns of minimal length $L = N + 2$:

$$\langle 0, 2, \ldots, 2n, 2n+2, 2n+1, \ldots, 3, 1 \rangle,$$
$$\langle 2n+1, 2n-1, \ldots, 1, 0, 2, \ldots, 2n, 2n+2 \rangle,$$
$$\langle 2n+2, 2n+1, \ldots, 1, 0, 2, \ldots, 2n-2, 2n \rangle,$$

and their mirrored patterns. For $N = 2n$, the six forbidden patterns of minimal length $L = N + 2$ provided by Theorem 2 are

$$\langle 0, 2, \ldots, 2n, 2n+1, \ldots, 3, 1 \rangle,$$
$$\langle 2n+1, 2n-1, \ldots, 1, 0, 2, \ldots, 2n-2, 2n \rangle,$$
$$\langle 2n-1, 2n-3 \ldots, 1, 0, 2, \ldots, 2n, 2n+1 \rangle,$$

and their mirrored patterns. In particular, for $N = 2$ we obtain the following minimal-length forbidden patterns:

$$\langle 0, 2, 3, 1 \rangle \qquad \langle 1, 3, 2, 0 \rangle,$$
$$\langle 3, 1, 0, 2 \rangle \qquad \langle 2, 0, 1, 3 \rangle,$$
$$\langle 1, 0, 2, 3 \rangle \qquad \langle 3, 2, 0, 1 \rangle.$$

Needless to say, these are the six 4-patterns we got in (3.14) by graphical means.

It was proven in [76] that the shift Σ_N has exactly six root forbidden L-patterns for each $L \geq N+2$, namely, those delivered by Theorem 2 after setting $\overrightarrow{\mathbf{p}_k} = k-1$ (respectively, $\overrightarrow{\mathbf{p}_k} = k$) in those segments not explicitly given in the tent patterns (4.13) and (4.14) (respectively, in the spiraling patterns (4.15), (4.16), (4.17), and (4.18)).

Corollary 2 *For every $K \geq 2$ there are self-maps on the interval $[0, 1]$ without forbidden patterns of length $L \leq K$.*

Proof Let $E_N : [0, 1] \to [0, 1]$ be the shift map $x \mapsto Nx \pmod 1$, $N = 2, 3, \ldots$. We know that E_N and Σ have the same allowed and forbidden patterns because they are order isomorphic (see (4.4)). Therefore if $N + 1 \leq K$, then E_N has no forbidden patterns of length $L \leq K$ because of Theorem 1. $\qquad \square$

It follows that *there exist n-dimensional interval maps without forbidden patterns*. For example, see Fig. 4.2, one can decompose $[0, 1]$ in infinite many half-open intervals (of vanishing length), $[0, 1] = \cup_{N=2}^{\infty} I_N$ and define on each I_N a properly scaled version of E_N, $\tilde{E}_N : I_N \rightarrow I_N$. In $\mathbb{R}^2$ one can repeat the said decomposition along the 1-axis and define on $I_N \times [0, 1]$ the function $(\tilde{E}_N, \mathrm{Id})$, where Id denotes the identity. Proposition 4 shows that adding some natural assumption, like piecewise monotonicity, can make all the difference.

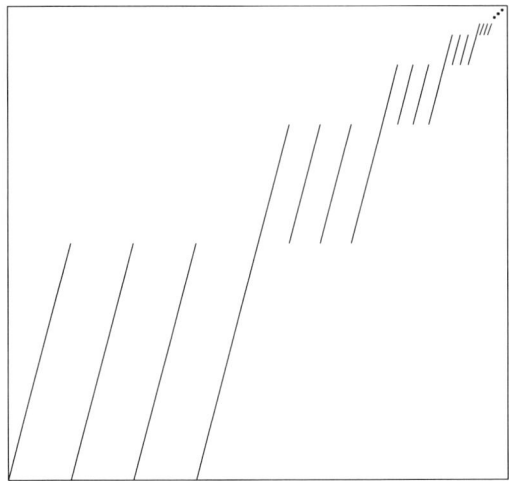

Fig. 4.2 A map with infinitely many monotonicity intervals and no forbidden patterns

4.3 Forbidden Patterns for Two-Sided Shifts

Consider now the bisequence space, $\{0, 1, \ldots, N-1\}^{\mathbb{Z}}$, equipped with the following lexicographical order. With the notation $\mathbf{x}^-$ for the *left sequence* $(x_{-n})_{n \in \mathbb{N}}$ of $\mathbf{x} \in \{0, 1, \ldots, N-1\}^{\mathbb{Z}}$ and $\mathbf{x}^+$ for the *right sequence* $(x_n)_{n \in \mathbb{N}_0}$, we set

$$\mathbf{x} < \mathbf{x}' \Leftrightarrow \begin{cases} \mathbf{x}^+ < \mathbf{x}'^+ \\ \text{or} \\ \mathbf{x}^- < \mathbf{x}'^- & \text{if } \mathbf{x}^+ = \mathbf{x}'^+ \end{cases}, \qquad (4.20)$$

where $\mathbf{x} = (\mathbf{x}^-, \mathbf{x}^+)$, $\mathbf{x}' = (\mathbf{x}'^-, \mathbf{x}'^+)$, and $<$ between right (respectively, left) sequences denote lexicographical order in $\{0, 1, \ldots, N-1\}^{\mathbb{N}_0}$ (respectively, $\{0, 1, \ldots, N-1\}^{\mathbb{N}}$). If we map $\{0, 1, \ldots, N-1\}^{\mathbb{Z}}$ onto $[0, 1] \times [0, 1] \equiv [0, 1]^2$ via

$$(\mathbf{x}^-, \mathbf{x}^+) \mapsto \left(\sum_{n=1}^{\infty} x_{-n} N^{-n}, \sum_{n=0}^{\infty} x_n N^{-(n+1)} \right), \qquad (4.21)$$

we find that the lexicographical order (4.20) in $\{0, 1, \ldots, N-1\}^{\mathbb{Z}}$ corresponds to the usual lexicographical order in $[0, 1]^2$. In order for this map to be one-to-one, we have to dispose of the usual ambiguities in either direction.

In relation with the ordinal patterns defined by the orbits of two-sided sequences,

$$\Sigma^i(\mathbf{x}) < \Sigma^j(\mathbf{x})$$
$$\Leftrightarrow \begin{cases} (x_i, x_{i+1}, \ldots) < (x_j, x_{j+1}, \ldots) \\ \text{or} \\ (x_{i-1}, x_{i-2}, \ldots) < (x_{j-1}, x_{j-2}, \ldots) \text{ if } (x_i, x_{i+1}, \ldots) = (x_j, x_{j+1}, \ldots), \end{cases}$$

where $i, j \geq 0$, $i \neq j$. It follows that the "exceptional" condition $(x_i, x_{i+1}, \ldots) = (x_j, x_{j+1}, \ldots)$ occurs if and only if $\Sigma^{|i-j|}(\mathbf{x}^+) = \mathbf{x}^+$, i.e., when the right sequence $\mathbf{x}^+$ of $\mathbf{x} \in \{0, 1, \ldots, N-1\}^{\mathbb{Z}}$ is periodic from the entry $\min\{i, j\}$ on with period $p = |i - j|$.

Lemma 5 *One-sided and two-sided shifts on N symbols have the same admissible and forbidden ordinal patterns.*

Proof (i) Suppose that the one-sided sequence $\mathbf{x}^+ \in \{0, 1, \ldots, N-1\}^{\mathbb{N}_0}$ defines an ordinal L-pattern π, i.e.,

$$\Sigma^{\pi_0}(\mathbf{x}^+) < \Sigma^{\pi_1}(\mathbf{x}^+) < \cdots < \Sigma^{\pi_{L-1}}(\mathbf{x}^+).$$

Then, the two-sided sequences $\mathbf{x} = (\mathbf{x}^-, \mathbf{x}^+)$, with $\mathbf{x}^- \in \{0, 1, \ldots, N-1\}^{\mathbb{N}}$ arbitrary, define the same ordinal pattern.

(ii) Suppose now that the two-sided sequence $\mathbf{x} = (\mathbf{x}^-, \mathbf{x}^+) \in \{0, 1, \ldots, N-1\}^{\mathbb{Z}}$ defines an ordinal L-pattern π,

$$\Sigma^{\pi_0}(\mathbf{x}) < \Sigma^{\pi_1}(\mathbf{x}) < \cdots < \Sigma^{\pi_{L-1}}(\mathbf{x}). \qquad (4.22)$$

If $\mathbf{x}^+$ is not eventually periodic, then (4.22) implies

$$\Sigma^{\pi_0}(\mathbf{x}^+) < \Sigma^{\pi_1}(\mathbf{x}^+) < \cdots < \Sigma^{\pi_{L-1}}(\mathbf{x}^+),$$

hence the pattern π is realized by the one-sided sequence $\mathbf{x}^+$. If $\mathbf{x}^+$ is eventually periodic, say

$$\mathbf{x}^+ = (x_0, \ldots, x_{k-1}, (x_k, \ldots, x_{k+p-1})^{\infty}),$$

i.e., $(\mathbf{x}^+)_{k+np} = (\mathbf{x}^+)_k$ for $k \geq 0$ and every $n \in \mathbb{N}$, then there are two subcases.

(ii-a) If $L \leq k + 2p$, then the periodicity of $\mathbf{x}^+$ is not visible in the segment x_0^{L-1}, so the pattern π is realized by the one-sided pattern $\mathbf{x}^+$.

(ii-b) If $L = k+np+v$ with $n \geq 2$ and $v \geq 1$, then $\Sigma^{k+p+i}(\mathbf{x}) = \cdots = \Sigma^{k+np+i}(\mathbf{x})$
for $i = 0, \ldots, v - 1$, so their negative sequences $(\Sigma^{k+p+i}(\mathbf{x}))^-, \ldots, (\Sigma^{k+np+i}(\mathbf{x}))^-$
have to be compared before ordering them. In this case, the pattern π is realized by
the one-sided sequence

$$\tilde{\mathbf{x}}^+ = (x_0, \ldots, x_{k+np+v-1}, (\Sigma^{k+np+v-1}(\mathbf{x}))^-)$$
$$= (x_0, \ldots, x_{k+np+v-1}, x_{k+np+v-2}, \ldots, x_0, x_{-1}, \ldots).$$

From (i) and (ii) we deduce that one-sided and two-sided shifts on $N \geq 2$ symbols
have the same admissible ordinal patterns, hence they have also the same forbidden
patterns.   □

As a corollary of Lemma 5, together with Theorems 1 and 2, we obtain the following result.

Theorem 3 *The two-sided shift on N symbols has no forbidden patterns of length
$L \leq N + 1$ and has forbidden root patterns for $L \geq N + 2$.*

Example 9 Let $I^2 = [0, 1] \times [0, 1]$ endowed with the Lebesgue measure, and let
$B : I^2 \to I^2$ be the *baker map*,

$$B(\xi, \eta) = \begin{cases} (2\xi, \frac{1}{2}\eta), & 0 \leq \xi < \frac{1}{2}, \\ (2\xi - 1, \frac{1}{2}\eta + \frac{1}{2}), & \frac{1}{2} \leq \eta \leq 1. \end{cases}$$

A generating partition of B is $A_0 = [0, \frac{1}{2}) \times [0, 1]$ and $A_1 = [\frac{1}{2}, 1] \times [0, 1]$. For
Σ take the two-sided $\left(\frac{1}{2}, \frac{1}{2}\right)$-Bernoulli shift. Then B and Σ are isomorphic (mod 0)
via the coding map $\Phi : I^2 \to \{0, 1\}^{\mathbb{Z}}$, given by

$$\Phi(\xi, \eta) = (\ldots, x_{-1}, x_0, x_1, \ldots),$$

where $x_n = i_n$ if $B^n(\xi, \eta) \in A_{i_n}$, $n \in \mathbb{Z}$. Since Φ preserves order (in fact, Φ is the
inverse of the order-preserving map $(\mathbf{x}^-, \mathbf{x}^+) \mapsto (\sum_{n=0}^{\infty} x_{-n}2^{-(n+1)}, \sum_{n=1}^{\infty} x_n2^{-n}))$,
we conclude that the baker transformation has no forbidden patterns of length ≤ 3.
The forbidden 4-patterns of the baker map are the same as those of the one-sided
shift, see (3.14).

Chapter 5
Ordinal Structure of the Signed Shifts

Shift transformations are a special case of a more general family: signed shift transformations—a sort of state-dependent shifts. The tent map is the simplest and perhaps most popular representative of the signed shifts. In this chapter we are going to show that most of the results on the ordinal structure of the shifts can be generalized to the signed shifts. By order isomorphy, these results apply also to more interesting cases, like the signed sawtooth maps.

5.1 Ordinal Patterns and the Tent Map

In this section we mimic the strategy used in the previous chapter, in order to get a handle on the ordinal patterns of the symmetric tent map. We will also address an issue pointed out in Fig. 1.7, namely, the interval structure of the sets P_π defining the allowed ordinal patterns of the logistic map.

5.1.1 A State-Dependent Shift Approach to the Tent Map

Just as some important dynamical properties of the sawtooth map E_N (like density of periodic points, sensitivity to initial conditions, topological transitivity, and the structure of its admissible and forbidden ordinal patterns) can be easily studied in the sequence space with the help of the relevant order isomorphisms, the same happens with the symmetric tent map. Remember from Sect. 1.1.3 that the symmetric tent map $\Lambda:[0,1] \to [0,1]$ is given by

$$\Lambda(x) = 1 - |1 - 2x| = \begin{cases} 2x & 0 \leq x \leq \frac{1}{2} \\ 2(1-x) & \frac{1}{2} \leq x \leq 1 \end{cases}. \tag{5.1}$$

For $x \in [0, 1]$, write

$$x = \sum_{n=0}^{\infty} x_n 2^{-(n+1)} = 0.x_0 x_1 \ldots x_n \ldots,$$

J.M. Amigó, *Permutation Complexity in Dynamical Systems,*
Springer Series in Synergetics, DOI 10.1007/978-3-642-04084-9_5,
© Springer-Verlag Berlin Heidelberg 2010

$x_n \in \{0, 1\}$. If $0 \le x < 1/2$, then

$$\Lambda(x) = 2x = 0.x_1x_2 \ldots x_{n+1} \ldots ,$$

hence the action of Λ coincides with the action of the sawtooth map E_2. Otherwise, if $1/2 \le x \le 1$, then

$$\Lambda(x) = 2 - 2x \equiv 1 - 2x \bmod 1$$
$$= 1 - 0.x_2x_3 \ldots x_{n+1} \ldots$$

Introducing the *dual bit*

$$x^* = 1 - x = \begin{cases} 1 & \text{if } x = 0 \\ 0 & \text{if } x = 1 \end{cases} \tag{5.2}$$

(thus, $(x^*)^* = x$), we have

$$\Lambda(x) = 0.x_1^* x_2^* \ldots x_{n+1}^* \ldots$$

because

$$0.x_1x_2 \ldots x_{n+1} + \cdots + 0.x_1^* x_2^* \ldots x_{n+1}^* \cdots = 0.11 \ldots 1 \ldots = 1.$$

All in all,

$$\Lambda(0.x_0x_1 \ldots x_n \ldots) = \begin{cases} 0.x_1x_2 \ldots x_{n+1} \ldots & \text{if } x_0 = 0, \\ 0.x_1^* x_2^* \ldots x_{n+1}^* \ldots & \text{if } x_0 = 1. \end{cases} \tag{5.3}$$

Identify now the binary representation $0. x_0 x_1 \ldots x_n \ldots$, $x_n \in \{0, 1\}$, of a number $x \in [0, 1]$, with the sequence

$$(x_0, x_1, \ldots, x_n, \ldots) \in \{0, 1\}^{\mathbb{N}_0},$$

via the map $\phi_2 : \{0, 1\}^{\mathbb{N}_0} \to [0, 1]$ defined as in (4.3) with $N = 2$. Then action (5.3) translates into the following zeroth-state-dependent shift on $\{0, 1\}^{\mathbb{N}_0}$:

$$\Sigma_{(+,-)}(x_0, x_1, \ldots, x_n, \ldots) = \begin{cases} (x_1, x_2, \ldots, x_{n+1}, \ldots) & \text{if } x_0 = 0 \\ (x_1^*, x_2^*, \ldots, x_{n+1}^*, \ldots) & \text{if } x_0 = 1 \end{cases} \tag{5.4}$$

(the subscripts $(+, -)$ will be explained later). Observe that if we write

$$\mathbf{x}^* = (x_0^*, x_1^*, \ldots, x_n^*, \ldots),$$

then

$$\Sigma_{(+,-)}(\mathbf{x}) = \begin{cases} \Sigma_2(\mathbf{x}) & \text{if } x_0 = 0, \\ \Sigma_2(\mathbf{x}^*) & \text{if } x_0 = 1, \end{cases}$$

where Σ_2 is the usual one-sided shift on sequences of two symbols.

A method of visualizing how the orbits of $\mathbf{x}$ are generated by $\Sigma_{(+,-)}$ is the following. Take as way of illustration

$$\mathbf{x} = (0, 1, 1, 0, 0, 0, 1, 0, 1, 1, 0, 0, 1, 1, 1, 0, 1, 0, 0, 1, 1, 0, 1, \ldots), \qquad (5.5)$$

so as

$$
\begin{aligned}
\Sigma^1_{(+,-)}(\mathbf{x}) &= (1 \quad 1 \quad 0 \quad 0 \quad 0 \quad 1 \quad 0 \quad 1 \quad 1 \quad 0 \quad 0 \quad 1 \ldots) = \Sigma^1_2(\mathbf{x}) \\
\Sigma^2_{(+,-)}(\mathbf{x}) &= (0 \quad 1 \quad 1 \quad 1 \quad 0 \quad 1 \quad 0 \quad 0 \quad 1 \quad 1 \quad 0 \quad 0 \ldots) = \Sigma^2_2(\mathbf{x}^*) \\
\Sigma^3_{(+,-)}(\mathbf{x}) &= (1 \quad 1 \quad 1 \quad 0 \quad 1 \quad 0 \quad 0 \quad 1 \quad 1 \quad 0 \quad 0 \quad 0 \ldots) = \Sigma^3_2(\mathbf{x}^*) \\
\Sigma^4_{(+,-)}(\mathbf{x}) &= (0 \quad 0 \quad 1 \quad 0 \quad 1 \quad 1 \quad 0 \quad 0 \quad 1 \quad 1 \quad 1 \quad 0 \ldots) = \Sigma^4_2(\mathbf{x}) \\
\Sigma^5_{(+,-)}(\mathbf{x}) &= (0 \quad 1 \quad 0 \quad 1 \quad 1 \quad 0 \quad 0 \quad 1 \quad 1 \quad 1 \quad 0 \quad 1 \ldots) = \Sigma^5_2(\mathbf{x}) \\
\Sigma^6_{(+,-)}(\mathbf{x}) &= (1 \quad 0 \quad 1 \quad 1 \quad 0 \quad 0 \quad 1 \quad 1 \quad 1 \quad 0 \quad 1 \quad 0 \ldots) = \Sigma^6_2(\mathbf{x}) \\
\Sigma^7_{(+,-)}(\mathbf{x}) &= (1 \quad 0 \quad 0 \quad 1 \quad 1 \quad 0 \quad 0 \quad 0 \quad 1 \quad 0 \quad 1 \quad 1 \ldots) = \Sigma^7_2(\mathbf{x}^*) \\
\Sigma^8_{(+,-)}(\mathbf{x}) &= (1 \quad 1 \quad 0 \quad 0 \quad 1 \quad 1 \quad 1 \quad 0 \quad 1 \quad 0 \quad 0 \quad 1 \ldots) = \Sigma^8_2(\mathbf{x}) \\
\Sigma^9_{(+,-)}(\mathbf{x}) &= (0 \quad 1 \quad 1 \quad 0 \quad 0 \quad 0 \quad 1 \quad 0 \quad 1 \quad 1 \quad 0 \quad 0 \ldots) = \Sigma^9_2(\mathbf{x}^*) \\
\Sigma^{10}_{(+,-)}(\mathbf{x}) &= (1 \quad 1 \quad 0 \quad 0 \quad 0 \quad 1 \quad 0 \quad 1 \quad 1 \quad 0 \quad 0 \quad 1 \ldots) = \Sigma^{10}_2(\mathbf{x}^*)
\end{aligned}
$$

etc., that is,

$$\Sigma^i_{(+,-)}(\mathbf{x}) = \begin{cases} \Sigma^i_2(\mathbf{x}) & \text{for } i = 0, 1, 4, 5, 6, 8, \ldots, \\ \Sigma^i_2(\mathbf{x}^*) & \text{for } i = 2, 3, 7, 9, 10, \ldots. \end{cases}$$

Write now $\mathbf{x}^*$ directly under $\mathbf{x}$, and mark (for example, with an underline) the initial digit of $\Sigma^i_{(+,-)}(\mathbf{x}), i \geq 0$:

$i =$	0	1	2	3	4	5	6	7	8	9	10	11	12
$\mathbf{x} =$	0	1	1	0	0	0	1	0	1	1	0	0	1
$\mathbf{x}^* =$	1	0	0	1	1	1	0	1	0	0	1	1	0

$$(5.6)$$

That is, we set out from x_0, which is always underlined. If $x_0 = 0$, then go over to x_1 and underline it. If $x_0 = 1$, then go down to x_1^* and underline it. In general, if $x_i = 0$ or $x_i^* = 0$, go one step rightward on the same row and underline x_{i+1} or x_{i+1}^*, respectively. On the other hand, if $x_i = 1$ or $x_i^* = 1$, we go one step rightward on the other row and underline x_{i+1}^* or x_{i+1}, respectively. The L-pattern π defined by $\mathbf{x}$ can be found now by ordering all the sequences on the $\mathbf{x}$-row and $\mathbf{x}^*$-row starting with an underlined bit, for $0 \leq i \leq L - 1$.

If $\mathbf{x}$ is sequence (5.5), then the ordinal L-patterns of $\mathbf{x}$ under $\Sigma_{(+,-)}$ are obtained by comparing the shifts $\Sigma^i(\mathbf{x})$ for $i = 0, 1, 4, 5, 6, 8, \ldots$ with the shifts $\Sigma^j(\mathbf{x}^*)$ for $j \neq i$. In particular, $\mathbf{x}$ is of type

$$\pi = \langle 4, 5, 9, 0, 2; 7, 6, 10, 1, 8, 3 \rangle \in \mathcal{S}_{11} \tag{5.7}$$

under the action of $\Sigma_{(+,-)}$.

Rather than deriving at this point the structure of the allowed ordinal patterns for $\Sigma_{(+,-)}$ (or the tent map Λ for this matter), which follows from the general results of the next section, let us prove here a particular property of the allowed patterns for $\Sigma_{(+,-)}$.

Lemma 6 *The subsequence $n+2, \ldots, n+1, \ldots, n$ ($0 \leq n \leq L-3$) cannot appear in the entries of an allowed L-pattern for $\Sigma_{(+,-)}$. Thus, the allowed ordinal patterns of $\Sigma_{(+,-)}$ cannot contain decreasing subsequences of length 3.*

Proof We prove by contradiction that the order relation

$$\Sigma^2_{(+,-)}(\mathbf{x}) < \Sigma_{(+,-)}(\mathbf{x}) < \mathbf{x} \tag{5.8}$$

cannot hold true. If $x_0 = 0$ there is no way that $\Sigma_{(+,-)}(\mathbf{x}) \equiv \Sigma_2(\mathbf{x}) < \mathbf{x}$. Hence $\mathbf{x} = (1, x_1, x_2, \ldots)$ and

$$\Sigma_{(+,-)}(\mathbf{x}) \equiv \Sigma_2(\mathbf{x}^*) = (x_1^*, x_2^*, \ldots).$$

By the same token, if $x_1^* = 0$ there is no way that $\Sigma_{(+,-)}(\Sigma_{(+,-)}(\mathbf{x})) \equiv \Sigma^2_{(+,-)}(\mathbf{x}) < \Sigma_{(+,-)}(\mathbf{x})$. Hence

$$\mathbf{x} = (1, 0, x_2, \ldots), \quad \Sigma_{(+,-)}(\mathbf{x}) = (1, x_2^*, x_3^*, \ldots), \quad \Sigma^2_{(+,-)}(\mathbf{x}) = (x_2, x_3, \ldots).$$

From $\Sigma_{(+,-)}(\mathbf{x}) < \mathbf{x}$ it follows $x_2^* = 0$. In turn, from $\Sigma^2_{(+,-)}(\mathbf{x}) = (1, x_3, \ldots) < \Sigma_{(+,-)}(\mathbf{x}) = (1, 0, x_3^*, \ldots)$ it follows $x_3 = 0$. So far, we found that $\mathbf{x} = (1, 0, 1, 0, x_4, \ldots)$ (thus $\Sigma_{(+,-)}(\mathbf{x}) = (1, 0, 1, x_4^*, \ldots)$ and $\Sigma^2_{(+,-)}(\mathbf{x}) = (1, 0, x_4, \ldots)$).
A straightforward induction along these lines yields

$$\mathbf{x} = (1, 0, 1, 0, \ldots, 1, 0, \ldots) = ((1,0)^\infty),$$

which is the binary expansion of the rational number 2/3. Since $\Sigma^2_{(+,-)}(\mathbf{x}) = \Sigma_{(+,-)}(\mathbf{x}) = \mathbf{x}$ for this particular sequence (in other words, 2/3 is a fixed point of $\Sigma_{(+,-)}$), the statement follows by contradiction.   □

Exercise 5 Prove, using representation (5.4) that the symmetric tent map has dense periodic points, sensitive dependence on initial conditions, and is topologically transitive.

5.1.2 The Interval Structure of the Sets P_π

The points in state space Ω defining an ordinal L-pattern π under the action of a map $f:\Omega \to \Omega$ build the set P_π, (3.4). The sets $P_\pi \neq \emptyset$, $\pi \in \mathcal{S}_L$, build in turn the set $\mathcal{P}_L$, which build a finite partition of Ω under the condition set by Proposition 2. In this section we examine the "topology" of $P_\pi \in \mathcal{P}_L$ for some one-dimensional interval maps. For continuous maps, those sets are clearly open sets (hence, an enumerable union of disjoint open intervals), but no further dissection can be made. For the sawtooth map family $x \mapsto Nx$ mod 1, $N \geq 2$, it is easy to convince oneself that P_π consists of a single open or half-open interval for all admissible patterns $\pi \in \mathcal{S}_L$, $L \geq 2$ (see Figs. 3.2 and 3.3). For the logistic map, Figs. 1.5 and 1.6 show that all $P_\pi \in \mathcal{P}_L$ with $L = 2, 3$ consist of a single open interval, but from Fig. 1.7 it can be read that

$$P_{\langle 0,3,1,2 \rangle} \approx (0.09549, 0.11698) \cup (0.18826, 0.25),$$

$$P_{\langle 2,0,3,1 \rangle} \approx (0.34549, 0.41318) \cup (0.61126, 0.65451),$$

$$P_{\langle 1,2,3,0 \rangle} \approx (0.93301, 0.95048) \cup (0.96985, 1).$$

We claim the following.

Proposition 6 *For the logistic map and the symmetric tent map, all $P_\pi \neq \emptyset$ consist of one or two components.*

As stated in Example 4 (1), the logistic map g and the symmetric tent map Λ are order isomorphic. Specifically, $g(\phi(x)) = \phi(\Lambda(x))$, where $\phi(x) = \sin^2\left(\frac{\pi}{2}x\right)$, $0 \leq x \leq 1$, so that

$$g^n(\phi(x)) = g^m(\phi(x)) \Leftrightarrow \phi(\Lambda^n(x)) = \phi(\Lambda^m(x)) \Leftrightarrow \Lambda^n(x) = \Lambda^m(x).$$

Thus, the curves $y = g^n(x)$ and $y = g^m(x)$ cross at x_0 if and only if the piecewise straight lines $y = \Lambda^n(x)$ and $y = \Lambda^m(x)$ cross at $\phi^{-1}(x_0)$. Moreover, the iterates of Λ have not only a simple graphical representation (triangular waves with frequencies increasing as powers of 2) but also a scaling property that makes Λ handier for the proof of Proposition 6:

$$\begin{array}{ll} \Lambda^n(x) = \Lambda^{n-1}(2x), & 0 \leq x \leq \frac{1}{2}, \\ \Lambda^n(x) = \Lambda^{n-1}(2(1-x)), & \frac{1}{2} \leq x \leq 1. \end{array} \tag{5.9}$$

Therefore, the left-half part of the graphs $(x, \Lambda^0(x)), (x, \Lambda^1(x)), \ldots, (x, \Lambda^L(x))$ is a "squeezed" copy of the graphs $(x, \frac{x}{2}), (x, \Lambda^0(x)), \ldots, (x, \Lambda^{L-1}(x))$ on the interval $0 \leq x \leq \frac{1}{2}$; indeed, upon rescaling the X-axis by a factor $\frac{1}{2}$, we have $(x, \frac{x}{2}) \mapsto (\frac{x}{2}, \frac{x}{2})$ and $(x, \Lambda^l(x)) \mapsto (\frac{x}{2}, \Lambda^l(x)) = (\frac{x}{2}, \Lambda^{l+1}(\frac{x}{2}))$. The corresponding right-half parts require the squeezed copy of the graphs $(x, 1 - \frac{x}{2}), (x, \Lambda^0(x)), \ldots, (x, \Lambda^{L-1}(x))$ on $0 \leq x \leq \frac{1}{2}$ to be further mirrored with respect to the line $x = \frac{1}{2}$ (this is the transformation $(x, y) \mapsto (x, 1 - x)$); see Fig. 5.1 for further insights.

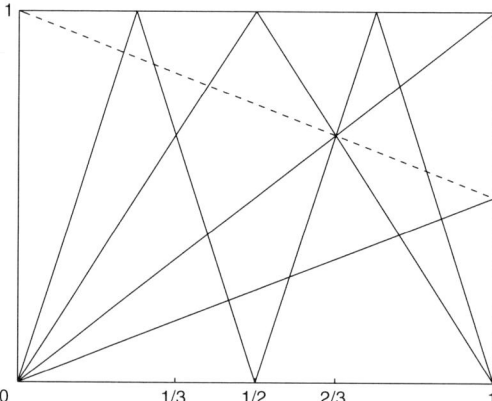

Fig. 5.1 If this figure is "opened" at the right side as a book put upside down, with the line $y = x/2$ only on the left page, the (*dashed*) line $y = 1 - x/2$ only on the right page, and the triangular waves $y = \Lambda(x)$, $y = \Lambda^2(x)$ on both, and the resulting graph is shrunk by a factor $1/2$ along the X-axis, then we get the graphs of $y = \Lambda^n(x)$, $0 \le n \le 3$. Alternatively, we can go from $\mathcal{P}_3^*$ to $\mathcal{P}_4^*$ just by going first rightward on the bottom page (containing $y = x/2$) of the closed book and then leftward on the top page (containing $y = 1 - x/2$)

Proof Proposition 6 follows from the considerations prior to Proposition 3 (remember the terminology mother and daughter intervals, here shortened to *mother* and *daughter*), together with the following facts.

The decomposition of a mother $P_{\pi_{\text{mother}}} \in \mathcal{P}_L$ into several daughters including two or more *twins* (disjoint subintervals with the same ordinal label) can only happen in intervals containing "vertex" or "bouncing-off" points x_v. As their name indicates, these points correspond to projections onto the X-axis of points at the bottom ($y = 0$) or at the ceiling ($y = 1$) of the unit square at which incoming (left) and outgoing (right) lines $y = \Lambda^l(x)$ meet, like $(\frac{1}{2}, 0)$ and $(\frac{1}{4}, 1)$ in Fig. 5.1. Possibly the most intuitive way to follow the growth of twins around vertex points uses the scaling property (5.9). If $0 < x_v < \frac{1}{2}$, consider the graphs of $y = \frac{x}{2}, y = \Lambda^0(x), \ldots, y = \Lambda^{L-1}(x)$ around $x = 2x_v$. If $2x_v \in P_{\langle \pi_0, \ldots, \pi_{L-1} \rangle}$, then the straight line $y = \frac{x}{2}$ generates (left to right) daughters of $P_{\pi_{\text{mother}}}$ (after squeezing) with labels $\pi_{\text{left}} = \langle \pi_0 + 1, \ldots, 0, \pi_k + 1, \ldots, \pi_{L-1} + 1 \rangle$, $\pi_{\text{central}} = \langle \pi_0 + 1, \ldots, \pi_k + 1, 0, \ldots, \pi_{L-1} + 1 \rangle$ and $\pi_{\text{right}} = \langle \pi_0 + 1, \ldots, 0, \pi_k + 1, \ldots, \pi_{L-1} + 1 \rangle = \pi_{\text{left}}$, with $x_v \in P_{\pi_{\text{central}}} \in \mathcal{P}_{L+1}$. Here k depends on the number of lines meeting at $(x_v, 0)$; if $k = 0$ or $L - 1$, then 0 is the first or last entry of the label, respectively. Hence, the set $P_{\pi_{\text{left}}} \cup P_{\pi_{\text{right}}} \in \mathcal{P}_{L+1}$ ($\pi_{\text{left}} = \pi_{\text{right}}$) consists of two disjoint interval components, one on each side of $P_{\pi_{\text{central}}}$. If, on the other hand, $\frac{1}{2} < x_v < 1$, consider the graphs of $y = 1 - \frac{x}{2}, y = \Lambda^0(x), \ldots, y = \Lambda^{L-1}(x)$ around $x = 2(1 - x_v)$. If $2(1 - x_v) \in P_{\langle \pi_0, \ldots, \pi_{L-1} \rangle}$, then the straight line $y = 1 - \frac{x}{2}$ generates daughters of $P_{\pi_{\text{mother}}}$ (after squeezing and mirroring) with labels $\pi_{\text{left}} = \langle \pi_0 + 1, \ldots, \pi_k + 1, 0, \ldots, \pi_{L-1} + 1 \rangle$, $\pi_{\text{central}} = \langle \pi_0 + 1, \ldots, 0, \pi_k + 1, \ldots, \pi_{L-1} + 1 \rangle$, and $\pi_{\text{right}} = \langle \pi_0 + 1, \ldots, \pi_k + 1, 0, \ldots, \pi_{L-1} + 1 \rangle = \pi_{\text{left}}$. As before, $P_{\pi_{\text{left}}} \cup P_{\pi_{\text{right}}}$ consists of two disjoint interval components, one on each side of $P_{\pi_{\text{central}}}$. Finally, for

$x_v = \frac{1}{2}$ the first set of graphs produces π_{left} and π_{central}, while the second produces π_{central} and $\pi_{\text{right}} = \pi_{\text{left}}$, with $x_v \in P_{\pi_{\text{central}}}$. This mechanism repeats again and again over all generations. After the step $L \to L + 1$, only the one-component daughters $P_{\pi_{\text{central}}}$, all of which contain some x_v, can in turn generate twins (two-component grand daughters); the corresponding two-component sisters $P_{\pi_{\text{left}}} \cup P_{\pi_{\text{right}}}$ cannot generate twins because they contain no vertex point. As a result, only one- or two-component intervals are possible, the latter forming a nested structure around some vertex points. From Fig. 5.1 it is clear that all such vertex points originate from $x = \frac{1}{2}, 1$ by squeezing and from $x = \frac{1}{4}$ by squeezing and mirroring.   □

Exercise 6 Discuss the interval structure of the sets P_π for the map $E_{-2}:x \mapsto -2x$ mod 1.

5.2 Ordinal Patterns and the Signed Shifts

The results of Sect. 5.1.1 can be generalized to a particular case of piecewise linear maps. Partition the unit interval $[0, 1]$ in $N \geq 2$ equal subintervals,

$$I_k = \left[\frac{k}{N}, \frac{k+1}{N}\right), \quad 0 \leq k \leq N - 2 \quad \text{and} \quad I_{N-1} = \left[\frac{N-1}{N}, 1\right]$$

(other choices regarding the endpoints are of course possible), and raise over I_k a "/-lap" of slope $+N$,

$$f(x) = Nx - k, \, x \in I_k,$$

or a "\-lap" of slope $-N$,

$$f(y) = k + 1 - Nx, \, x \in I_k.$$

A map of the unit interval whose graph consists of /-laps and \-laps of slopes $\pm N$, respectively, over the intervals I_k, $0 \leq k \leq N - 1$, will be called a *signed saw-tooth map*, the term "signed" referring to the fact that its laps can have positive or negative slope (see Fig. 5.2). We say that a signed sawtooth map f has *signature* $\sigma = (\sigma_0, \sigma_1, \ldots, \sigma_{N-1})$, where $\sigma_k \in \{+, -\}, 0 \leq k \leq N - 1$, to summarize that (the graph of) f has a /-lap over I_k whenever $\sigma_k = +$ and a \-lap whenever $\sigma_k = -$. In other words, the kth component of the signature gives the slope sign of the kth lap.

We have already met two important representatives of the signed sawtooth map family: the sawtooth map $E_N: x \mapsto Nx$ mod 1 ($\sigma = (+, \ldots, +)$) and the symmetric tent map Λ ($\sigma = (+, -)$).

Given a signature σ, define the *signed shift* $\Sigma_\sigma:\{0, \ldots, N-1\}^{\mathbb{N}_0} \to \{0, \ldots, N-1\}^{\mathbb{N}_0}$ as follows:

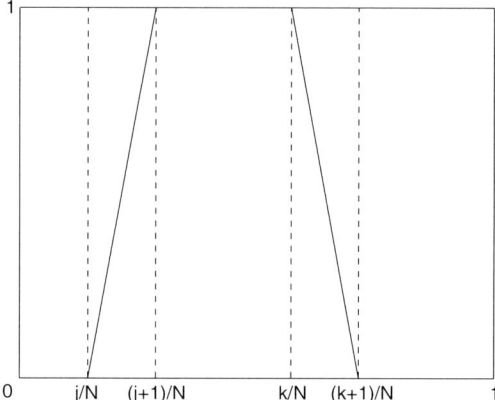

Fig. 5.2 The graph of a generic signed sawtooth map with slopes $\pm N$. The figure only depicts the jth lap, with positive slope, and the kth slope, with negative slope

$$\Sigma_\sigma(x_0, \ldots, x_n, \ldots) = \begin{cases} (x_1, \ldots, x_{n+1}, \ldots) & \text{if } x_0 = k, \sigma_k = +, \\ (N - 1 - x_1, \ldots, N - 1 - x_{n+1}, \ldots) & \text{if } x_0 = k, \sigma_k = -. \end{cases}$$

Therefore, if we define the *dual digit* of $k \in \{0, 1, \ldots, N - 1\}$ as

$$k^* = N - 1 - k, \tag{5.10}$$

(thus $(k^*)^* = k$), then

$$\Sigma_\sigma(\mathbf{x}) = \begin{cases} \Sigma_N(\mathbf{x}) & \text{if } x_0 = k \text{ and } \sigma_k = +, \\ \Sigma_N(\mathbf{x}^*) & \text{if } x_0 = k \text{ and } \sigma_k = -, \end{cases} \tag{5.11}$$

where

$$\mathbf{x}^* = (x_0^*, \ldots, x_n^*, \ldots) = (N - 1 - x_0, \ldots, N - 1 - x_n, \ldots)$$

is the *dual sequence* to $\mathbf{x} = (x_0, \ldots, x_n, \ldots) \in \{0, 1, \ldots, N - 1\}^{\mathbb{N}_0}$. In particular, if

$$N = 2\nu + 1,$$

then $\nu = (N-1)/2$ is "self-dual": $\nu^* = \nu$. Note that (5.10) generalizes the definition of dual bit, (5.2).

   Important for us is that if f is a signed sawtooth map with signature σ, then f and Σ_σ are order isomorphic via the map $\phi_N : \{0, 1, \ldots, N - 1\}^{\mathbb{N}_0} \to [0, 1]$ defined in (4.3). Observe that $\phi_N(0^\infty) = 0$, $\phi_N(1^\infty) = 1$, and

$$\frac{k}{N} \le \phi_N(\mathbf{x}) \le \frac{k + 1}{N} \quad \text{iff} \quad x_0 = k.$$

The technique described in Sect. 5.1 to keep track of the orbits of $\mathbf{x}$ under $\Sigma_{(+,-)}$ can be used for Σ_σ too. The number of symbols N goes in the definition of $\mathbf{x}^*$, while σ_k tells whether we have to jump from the current entry $x_i = k$ to x_{i+1}^* or from the current entry $x_i^* = k$ to x_{i+1} ($\sigma_k = -$), instead of remaining on the same line ($\sigma_k = +$), when underlining the entries of $\mathbf{x}$ in table (5.6).

Exercise 7 Check that

$$f(\phi_N(\mathbf{x})) = \phi_N(\Sigma_\sigma \mathbf{x}) = \begin{cases} \sum_{n=1}^{\infty} x_n N^{-n} & \text{if } x_0 = k \text{ and } \sigma_k = +, \\ 1 - \sum_{n=1}^{\infty} x_n N^{-n} & \text{if } x_0 = k \text{ and } \sigma_k = -. \end{cases}$$

We turn now to the ordinal patterns realized by a signed shift Σ_σ. Completely analogous to the case $\Sigma_{(+,\dots,+)} \equiv \Sigma_N$, Chap. 4, the allowed ordinal patterns for Σ_σ can also be decomposed into s-blocks, (4.6), where now the s-block (4.7) contains the locations of the symbol $s \in \{0,\dots,N-1\}$ in the segments $x_0^{L-1} := x_0,\dots,x_{L-1}$ of $\mathbf{x}$ and $(x^*)_0^{L-1} := x_0^*,\dots,x_{L-1}^*$ of $\mathbf{x}^*$, such that the zeroth component of $\Sigma_\sigma^i \mathbf{x}$, $0 \le i \le L-1$, is s (i.e., the locations of the symbol s which are underlined in the $\mathbf{x}$- or $\mathbf{x}^*$-row of table (5.6)). We shall presently see that each s-block consists basically of two kinds of subsequences: monotone ($\sigma_s = +$) or spiraling ($\sigma_s = -$), eventually intertwined by other subsequences of the same kind. Entries in an s-block not belonging to a subsequence will be referred to as solitary or single components or entries.

Theorem 4 *The non-empty blocks* $\pi_{k_0+\cdots+k_{s-1}},\dots,\pi_{k_0+\cdots+k_{s-1}+k_s-1}$, $0 \le s \le N-1$, *of* $\pi(\mathbf{x}) \in \mathcal{S}_L$ *fulfill the following basic restrictions:*

R*1   *If* $\sigma_s = +$, $0 < s < N-1$, *then the s-block is built by increasing subsequences,*

$$n,\dots,n+1,\dots,n+l-1 \tag{5.12}$$

($l \ge 2$) and/or decreasing subsequences,

$$n+l-1,\dots,n+1,\dots,n \tag{5.13}$$

($l \ge 2$) and/or solitary components ($l = 1$). If $\sigma_0 = +$, then the 0-block consists of increasing subsequences (5.12) and/or solitary components. If $\sigma_{N-1} = +$, then the $(N-1)$-block consists of decreasing subsequences (5.13) and/or solitary components.

R*2   *If* $\sigma_s = -$, $0 < s < N-1$, *then the s-block is built by even-length spiraling subsequences*

$$n+2l-2,\dots,n+2,\dots,n,\dots,n+1,\dots,n+3,\dots,n+2l-1 \tag{5.14}$$

with the entry $n+2l$ on an anterior block (if $n+2l \le L-1$) and/or the mirrored subsequences

$$n + 2l - 1, \ldots, n + 3, \ldots, n + 1, \ldots, n, \ldots, n + 2, \ldots, n + 2l - 2 \quad (5.15)$$

with the entry $n+2l$ on a posterior block (if $n+2l \leq L-1$) and/or odd-length spiraling subsequences

$$n + 2l, \ldots, n + 2l - 2, \ldots, n + 2, \ldots, n, \ldots, n + 1, \ldots, n + 3, \ldots, n + 2l - 1 \quad (5.16)$$

with the entry $n + 2l + 1$ on a posterior block (if $n + 2l + 1 \leq L - 1$) and/or the mirrored subsequences

$$n + 2l - 1, \ldots, n + 3, \ldots, n + 1, \ldots, n, \ldots, n + 2, \ldots, n + 2l - 2, \ldots, n + 2l \quad (5.17)$$

with the entry $n + 2l + 1$ on an anterior block (if $n + 2l + 1 \leq L - 1$) and/or solitary components. If $\sigma_0 = -$, then the first block consists of spiraling subsequences of the form (5.15) and/or (5.16) and/or solitary components. If $\sigma_{N-1} = -$, then the last block consists of spiraling subsequences of the form (5.14) and/or (5.17) and/or solitary components.

R*3   If (i) $\sigma_s = +$, (ii) the entries $m, n \leq L - 2$ belong to the s-block of $\pi(\mathbf{x})$, and (iii) m appears on the left of n, then $m + 1$ appears also on the left of $n + 1$ (not necessarily in the same block). If, on the other hand, (i) $\sigma_s = -$, (ii) the entries $m, n \leq L - 2$ belong to the s-block of $\pi(\mathbf{x})$, and (iii) m appears on the left of n, then $m + 1$ appears on the right of $n + 1$ (not necessarily in the same block).

Proof **R*1)** Let $s \in \{0, 1, \ldots, N - 1\}$ and consider an s-run of length $l \geq 2$ in the segment x_0^{L-1} of $\mathbf{x}$:

$i =$	$\ldots$	n	$n + 1$	$\ldots$	$n + l - 1$	$n + l$	$\ldots$
$\mathbf{x} =$	$\ldots$	s	s	$\ldots$	s	r	$\ldots$
$\mathbf{x}^* =$	$\ldots$	$N - 1 - s$	$N - 1 - s$	$\ldots$	$N - 1 - s$	$N - 1 - r$	$\ldots$

where $r \in \{0, 1, \ldots, N - 1\}$ and $r \neq s$. If (i) $s < N - 1$ and (ii) $x_{n+l} = r > s$, then this s-run contributes the increasing subsequence

$$n, \ldots, n + 1, \ldots, n + l - 1 \quad (5.18)$$

to the s-block of $\pi(\mathbf{x})$. If, on the other hand, (i) $s > 0$ and (ii) $x_{n+l} = r < s$, then the s-run contributes the decreasing subsequence

$$n + l - 1, \ldots, n + 1, \ldots, n. \quad (5.19)$$

The "$\ldots$" between the entries of these subsequences allow for entries eventually proceeding from other s-runs in $\mathbf{x}$ or $\mathbf{x}^*$ (see Example 7).

It follows that the 0-block can contain only increasing subsequences (and single entries not belonging to subsequences in the block), whereas the $(N-1)$-block can contain only decreasing subsequences (and single entries not belonging to subsequences in the block).

R*2) Consider an s-run of even length $2l$ in the segment x_0^{L-1} of $\mathbf{x}$. Thus,

$i =$	n	$n+1$	$\ldots$	$n+2l-2$	$n+2l-1$	$n+2l$
$\mathbf{x} =$	s	$N-1-1$	$\ldots$	s	$N-1-s$	r
$\mathbf{x}^* =$	$N-1-s$	s	$\ldots$	$N-1-s$	s	$N-1-r$

where $r \in \{0, 1, \ldots, N-1\}$ and $r \neq s$. Therefore, if (i) $s > 0$ and (ii) $x_{n+2l-1} = N-1-s < x_{n+2l}^* = N-1-r$, i.e., $r < s$, then the s-block of $\pi(\mathbf{x})$ will contain the spiraling subsequence

$$n+2l-2, \ldots, n+2, \ldots, n, \ldots, n+1, \ldots, n+3, \ldots, n+2l-1. \qquad (5.20)$$

Hence the entry $n+2l$ will appear in the r-block (provided $n+2l \leq L-1$), which precedes the s-block in $\pi(\mathbf{x})$ because $r < s$. If, on the other hand, (i) $s < L-1$ and (ii) $x_{n+2l-1} = N-1-s > x_{n+2l}^* = N-1-r$, i.e., $r > s$, then we obtain the mirrored, spiraling subsequence

$$n+2l-1, \ldots, n+3, \ldots, n+1, \ldots, n, \ldots, n+2, \ldots, n+2l-2, \qquad (5.21)$$

with the symbol $n+2l$ in a posterior block (provided $n+2l \leq L-1$), namely, on the r-block.

Consider now an s-run of odd length $2l+1$ in the segment x_0^{L-1} of $\mathbf{x}$. Thus,

$i =$	n	$n+1$	$\ldots$	$n+2l-1$	$n+2l$	$n+2l+1$
$\mathbf{x} =$	s	$N-1-s$	$\ldots$	$N-1-s$	s	$N-1-r$
$\mathbf{x}^* =$	$N-1-s$	s	$\ldots$	s	$N-1-s$	r

where $r \in \{0, 1, \ldots, N-1\}$ and $r \neq s$. Therefore, if (i) $s > 0$ and (ii) $x_{n+2l}^* = N-1-s < x_{n+2l+1} = N-1-r$, i.e., $r < s$, then the s-block of $\pi(\mathbf{x})$ will contain the spiraling subsequence

$$n+2l-1, \ldots, n+3, \ldots, n+1, \ldots, n, \ldots, n+2, \ldots, n+2l-2, \ldots, n+2l.$$

The entry $n+2l+1$ will appear on the r-block (provided $n+2l+1 \leq L-1$), which is on the left of the s-block because $r < s$. If, on the other hand, (i) $s < L-1$ and (ii) $x_{n+2l}^* = N-1-s > x_{n+2l+1} = N-1-r$, i.e., $r > s$, then we obtain the mirrored, spiraling subsequence

$$n+2l, \ldots, n+2l-2, \ldots, n+2, \ldots, n, \ldots, n+1, \ldots, n+3, \ldots, n+2l-1,$$

with the entry $n+2l+1$ in a block on the right of the s-block (provided $n+2l+1 \leq L-1$).

The corresponding results for the first ($s = 0$) and last ($s = N - 1$) blocks follow readily from these general results.

R*3) If m and n belong to the s-block, $\sigma_s = +$, and $\Sigma_\sigma^m(\mathbf{x}) < \Sigma_\sigma^n(\mathbf{x})$ for $\mathbf{x} \in \{0, 1, \ldots, N - 1\}^{\mathbb{N}_0}$, then

$$\Sigma_\sigma^m(\mathbf{x}) = (s, x_{m+1}, \ldots) < (s, x_{n+1}, \ldots) = \Sigma_\sigma^n(\mathbf{x}).$$

By the definition of lexicographical order, there are two possibilities: (i) $x_{m+1} < x_{n+1}$ or (ii) $x_{m+k} = x_{n+k}$ for $1 \leq k \leq l - 1$, $l \geq 2$, and $x_{m+l} < x_{n+l}$. In both cases,

$$\Sigma_\sigma^{m+1}(\mathbf{x}) = (x_{m+1}, \ldots) < (x_{n+1}, \ldots) = \Sigma_\sigma^{n+1}(\mathbf{x})$$

and, hence, the entry $m + 1$ appears on the left of $n + 1$ in $\pi(\mathbf{x})$.

If, on the other hand, m and n belong to the s-block, $\sigma_s = -$, and $\Sigma_\sigma^m(\mathbf{x}) < \Sigma_\sigma^n(\mathbf{x})$, then

$$\Sigma_\sigma^m(\mathbf{x}) = (s, x_{m+1}, \ldots) < (s, x_{n+1}, \ldots) = \Sigma_\sigma^n(\mathbf{x}).$$

As before, there are two possibilities: (i) $x_{m+1} < x_{n+1}$ and (ii) $x_{m+k} = x_{n+k}$ for $1 \leq k \leq l - 1$, $l \geq 2$, and $x_{m+l} < x_{n+l}$. In both cases,

$$\Sigma_\sigma^{m+1}(\mathbf{x}) = (N - 1 - x_{m+1}, \ldots) > (N - 1 - x_{n+1}, \ldots) = \Sigma_\sigma^{n+1}(\mathbf{x})$$

and, hence, the entry $m + 1$ appears on the right of $n + 1$ in $\pi(\mathbf{x})$. □

Conditions R*1–R*3 are not only necessary for an ordinal pattern to be allowed for Σ_σ, $\sigma = (\sigma_0, \ldots, \sigma_{N-1})$, but also sufficient. Indeed, given the s-block decomposition of $\pi \in S_L$ with each block satisfying the pertinent restrictions, then it is a simple matter to construct sequences $\mathbf{x} \in \{0, \ldots, N - 1\}^{\mathbb{N}_0}$ of type π. Furthermore, it is obvious that all L-patterns with $L \leq N$ are allowed for Σ_σ.

Corollary 3 *If $\pi = \langle \pi_0, \pi_1, \ldots, \pi_{L-1} \rangle$ is allowed (correspondingly, forbidden) for Σ_σ, $\sigma = (\sigma_0, \sigma_1, \ldots, \sigma_{N-1})$, then $\pi_{mirrored} = \langle \pi_{L-1}, \pi_{L-2}, \ldots, \pi_0 \rangle$ is allowed (correspondingly, forbidden) for $\Sigma_{\sigma_{mirrored}}$, where*

$$\sigma_{mirrored} := (\sigma_{N-1}, \sigma_{N-2}, \ldots, \sigma_0).$$

In the particular case $\sigma = \sigma_{mirrored}$, it follows that π is allowed (correspondingly, forbidden) for Σ_σ, iff $\pi_{mirrored}$ is also allowed (correspondingly, forbidden) for Σ_σ. These statements hold also true if "forbidden pattern" is replaced by "root forbidden pattern."

Proof The s-block structure of an allowed ordinal pattern is preserved under the transformation $\pi \mapsto \pi_{mirrored}$. Indeed, monotone subsequences transform into monotone subsequences (in particular, increasing subsequences of the 0-block transform in decreasing subsequences of the $(N - 1)$-block and vice versa), and spiraling subsequences go over to spiraling subsequences.

By the same token, mirrored outgrowth forbidden patterns for Σ_σ will be outgrowth forbidden patterns for $\Sigma_{\sigma_{\text{mirrored}}}$. It follows that $\pi \in \mathcal{S}_L$ is a root forbidden pattern for Σ_σ in the case $\sigma = \sigma_{\text{mirrored}}$, iff π_{mirrored} is also a root forbidden pattern for Σ_σ. □

Remark 1 If the first or last element of a monotone subsequence appearing in an s-block is assigned to the anterior or posterior block, respectively (if any), then the remaining subsequence preserves its increasing or decreasing character—or it becomes a single entry. If the leftmost or the rightmost element of a spiraling subsequence is assigned to the anterior or posterior block (if any), then the remaining subsequence preserves its spiraling character, eventually appearing also a new single entry in the same block. This implies that, when carrying out a decomposition of an ordinal L-pattern into s-blocks, $L \geq N$, we may assume without loss of generality that all s-blocks are non-empty.

For $\sigma_k = +, 0 \leq k \leq N - 1$, we recover from Theorem 4 the restrictions fulfilled by the allowed patterns for Σ_N (Lemma 2). In the case $\sigma = (+, -)$, considered in Sect. 5.1.1, there are only two symbols and two blocks in the decomposition of the ordinal patterns. Restrictions R*1 and R*2 entail then that $\pi = \langle 2, 1, 0 \rangle$ is forbidden for $\Sigma_{(+,-)}$ (Lemma 6). Indeed, $\pi_0, \pi_1 = 2, 1$ cannot occur in the 0-block because it is a decreasing sequence (R*1), hence $\pi = \langle 2; 1, 0 \rangle$; but then the entry 2 should appear on the right of $\pi_1, \pi_2 = 1, 0$ in order to form a spiraling subsequence (R*2); the restriction R*3 is also violated.

The five root forbidden 4-patterns for the logistic map (hence, for Λ and $\Sigma_{(+,-)}$) were found graphically in Sect. 1.2, (1.38). We check here that they do fail to satisfy the restrictions R*1–R*3:

- $\langle 0; 2, 3, 1 \rangle$ violates R*2; $\langle 0, 2; 3, 1 \rangle$ and $\langle 0, 2, 3; 1 \rangle$ violate R*3.
- $\langle 1; 0, 2, 3 \rangle$ violates R*3; $\langle 1, 0; 2, 3 \rangle$ and $\langle 1, 0, 2; 3 \rangle$ violate R*1.
- $\langle 1; 0, 3, 2 \rangle$ violates R*3; $\langle 1, 0; 3, 2 \rangle$ and $\langle 1, 0, 3; 2 \rangle$ violate R*1.
- $\langle 1; 3, 0, 2 \rangle$ violates R*3; $\langle 1, 0; 3, 2 \rangle$ and $\langle 1, 0, 3; 2 \rangle$ violate R*1.
- $\langle 3; 1, 2, 0 \rangle$ violates R*2; $\langle 3, 1; 2, 0 \rangle$ violates R*3 and $\langle 3, 1, 2; 0 \rangle$ violates R*1.

Exercise 8 Check that the allowed patterns for the logistic map, Fig. 1.7, comply with the restrictions (R*1)–(R*4).

Finally, let us prove that $\Sigma_{(+,-)}$ has root forbidden L-patterns for $L \geq 5$.

Theorem 5 *The patterns*

$$\pi = \langle 3, \ldots, L - 2, 0, 1, 2, L - 1 \rangle \in \mathcal{S}_L, \tag{5.22}$$

$L \geq 5$, *are root forbidden patterns for* $\Sigma_{(+,-)}$.

Proof Let us check that (5.22) is a forbidden pattern. First of all, $\pi_{L-5}, \pi_{L-4} = L - 2, 0$ cannot belong to the 0-block because $\pi_{L-5} + 1 = L - 1$ is not on the left of $\pi_{L-4} + 1 = 1$ (R*3). Hence

$$\pi = \langle 3, \ldots, L - 2; 0, 1, 2, L - 1 \rangle.$$

But $\pi_{L-4}, \pi_{L-3}, \pi_{L-2} = 0, 1, 2$ is not a spiraling subsequence, hence it violates R*2.

Furthermore, we claim that (5.22) is a root forbidden pattern. Otherwise, see (3.12), (i) π would be an outgrowth pattern of group I, i.e., the $(L-1)$-pattern obtained from π after removing the entry $L-1$,

$$\langle 3, \ldots, L-2, 0, 1, 2 \rangle \in \mathcal{S}_{L-1}, \tag{5.23}$$

would be forbidden or (ii) π would be an outgrowth pattern of group II, i.e., the $(L-1)$-pattern obtained from π after removing the entry 0 and subtracting 1 from each remaining entry,

$$\langle 2, \ldots, L-3, 0, 1, L-2 \rangle \in \mathcal{S}_{L-1}, \tag{5.24}$$

would be forbidden. But (5.23) admits the s-block decompositions

$$\langle 3, \ldots, L-2, 0; 1, 2 \rangle \text{ and } \langle 3, \ldots, L-2, 0, 1; 2 \rangle,$$

while (5.24) admits the decomposition

$$\langle 2, \ldots, L-3; 0, 1, L-2 \rangle.$$

$\square$

Exercise 9 Consider the eight cylinder sets $C_{i_0 i_1 i_2}$ of $\{0, 1\}^{\mathbb{N}_0}$. Check that the sequences of these sets are of the following types under $\Sigma_{(+,-)}$:

  (i) The sequences of C_{000} are of type $\langle 0, 1, 2 \rangle$.
  (ii) The sequences of C_{001} are also of type $\langle 0, 1, 2 \rangle$.
  (iii) The sequences $(0, 1, 0, 0, \ldots) \in C_{010}$ are of type $\langle 0, 1, 2 \rangle$, while the sequences $(0, 1, 0, 1, \ldots) \in C_{010}$ are of type $\langle 0, 2, 1 \rangle$.
  (iv) The sequences of C_{011} are of type $\langle 0, 2, 1 \rangle$ or $\langle 2, 0, 1 \rangle$.
  (v) The sequences of C_{100} are of type $\langle 2, 0, 1 \rangle$.
  (vi) The sequences of C_{101} are of type $\langle 1, 0, 2 \rangle$ or $\langle 2, 0, 1 \rangle$.
  (vii) The sequences of C_{110} are of type $\langle 1, 0, 2 \rangle$ or $\langle 1, 2, 0 \rangle$.
  (viii) The sequences of C_{111} are of type $\langle 1, 2, 0 \rangle$.

Among the signed sawtooth maps, those with signatures of alternating signs (we call them *alternating signatures*) have the special property of being continuous. The tent map is one of the two possibilities for $N = 2$. The next theorem generalizes the result that the tent map has a forbidden pattern already for $L = 3$.

Theorem 6 *Let Σ_σ be a shift with alternating signature $\sigma = (\sigma_0, \ldots, \sigma_{N-1})$.*

1. If N is even, then Σ_σ has forbidden L-patterns for $L \geq N + 1$.
2. If N is odd and $\sigma = (+, -, \ldots, -, +)$, then Σ_σ has forbidden L-patterns for $L \geq N + 1$.

3. *If N is odd and $\sigma = (-, +, \ldots, +, -)$, then (i) all ordinal $(N + 1)$-patterns are allowed for Σ_σ and (ii) Σ_σ has forbidden L-patterns for $L \geq N + 2$.*

In cases 2 and 3, along with a forbidden pattern $\pi \in \mathcal{S}_L$, π_{mirrored} will also be a forbidden pattern (Corollary 3).

Proof Remember that if Σ_σ has a forbidden pattern of length L_0, then its outgrowth patterns provide forbidden L-patterns for every $L \geq L_0$. Hence, we need only to exhibit forbidden patterns of the minimal lengths claimed in each case of Theorem 6.

1. Let $N \geq 2$ be even. There are two possibilities: (a) $\sigma_0 = +$ and $\sigma_{N-1} = -$ and (b) $\sigma_0 = -$ and $\sigma_{N-1} = +$. Since the signatures of these cases are mirrored from each other, we need to consider only one of them (Corollary 3), say (b).

A forbidden pattern of length $L = N + 1$ can be constructed attending to the positive signs of σ, together with the first and last negative signs, as follows. Take the entry $\pi_0 = 0$ for $\sigma_0 = -$,

$$\pi = \langle 0, \ldots \rangle,$$

the decreasing subsequence $\pi_{2k-1}, \pi_{2k} = 2k, 2k - 1$ for $\sigma_{2k-1} = +$, $1 \leq k \leq N/2 - 1$,

$$\pi = \langle 0, 2, 1, \ldots, 2k, 2k - 1, \ldots, N - 2, N - 3, \ldots \rangle,$$

and the increasing subsequence $\pi_{N-1}, \pi_N = N - 1, N$ for $\sigma_{N-1} = -$,

$$\pi = \langle 0, 2, 1, \ldots, 2k, 2k - 1, \ldots, N - 2, N - 3, N - 1, N \rangle \in \mathcal{S}_{N+1}.$$

(For $N = 2$, $\pi = \langle 0, 1, 2 \rangle \in \mathcal{S}_3$.) Then R*3 requires a first semicolon between $\pi_0 = 0$ and $\pi_1 = 2$, a second semicolon between $\pi_1 = 2$ and $\pi_2 = 1, \ldots$, and an $(N - 1)$th semicolon (the maximal number allowed) between $\pi_{N-2} = N - 3$ and $\pi_{N-1} = N - 1$. Still the increasing subsequence $\pi_{N-1}, \pi_N = N - 1, N$ in the last block ($\sigma_{N-1} = +$) violates R*1.

2. Let $N \geq 3$ be odd and $\sigma_0 = \sigma_{N-1} = +$. A forbidden pattern of length $L = N + 1$ can then be constructed attending to positive signs of σ. Take the decreasing subsequence $\pi_0, \pi_1 = 1, 0$ for $\sigma_0 = +$,

$$\pi = \langle 1, 0, \ldots \rangle,$$

the decreasing subsequence $\pi_{2k}, \pi_{2k+1} = 2k+1, 2k$ for $\sigma_{2k} = +$, $1 \leq k \leq (N-1)/2$,

$$\pi = \langle 1, 0, 3, 2, \ldots, 2k + 1, 2k, \ldots, N - 2, N - 3, \ldots \rangle,$$

and the increasing subsequence $\pi_{N-1}, \pi_N = N - 1, N$ for $\sigma_{N-1} = +$,

$$\pi = \langle 1, 0, 3, 2, \ldots, 2k + 1, 2k, \ldots, N - 2, N - 3, N - 1, N \rangle \in \mathcal{S}_{N+1}.$$

(For $N = 3$, $\pi = \langle 1, 0, 2, 3 \rangle \in \mathcal{S}_4$.) Then, R*3 requires a first semicolon between $\pi_0 = 1$ and $\pi_1 = 0$, a second semicolon between $\pi_1 = 0$ and $\pi_2 = 3, \ldots$, and an $(N - 1)$th semicolon (the maximal number allowed) between $\pi_{N-2} = N - 3$ and $\pi_{N-1} = N - 1$. Hence we are left with the increasing subsequence $\pi_{N-1}, \pi_N = N - 1, N$ in the last block ($\sigma_{N-1} = +$), what violates R*1.

3. Finally, let $N \geq 3$ be odd and $\sigma_0 = \sigma_{N-1} = -$.

(i) Let us prove that all ordinal $(N + 1)$-patterns are allowed for $\Sigma_{(-,+,\ldots,+,-)}$. Given $\pi \in \mathcal{S}_{N+1}$, there are three possibilities: (a) $N = \pi_0$, (b) $N = \pi_n$ with $1 \leq n \leq N - 1$, or (c) $N = \pi_N$. In the first case, π admits the allowed decomposition

$$\pi = \langle N, \pi_1; \pi_2; \ldots; \pi_k; \ldots; \pi_N \rangle .$$

In the second case, π admits the decomposition

$$\pi = \langle \pi_0; \pi_1; \ldots; \pi_{n-1}; N, \pi_{n+1}; \ldots; \pi_N \rangle$$

both if $\sigma_n = +$ or $\sigma_n = -$. In the third case, π admits the decomposition

$$\pi = \langle \pi_0; \pi_1; \ldots; \pi_k; \ldots; \pi_{N-1}, N \rangle .$$

(ii) A forbidden pattern of length $L = N + 2$ can be constructed attending to the blocks with negative sign. Let first $N = 5$ mod 4, so that the central sign of σ is $\sigma_{(N-1)/2} = -$. Take the increasing subsequence $\pi_0, \pi_1 = 0, 1$ for $\sigma_0 = -$,

$$\pi = \langle 0, 1, \ldots \rangle ,$$

the decreasing subsequence $\pi_N, \pi_{N+1} = 3, 2$ for $\sigma_{N-1} = -$,

$$\pi = \langle 0, 1, \ldots, 3, 2 \rangle ,$$

the increasing subsequence $\pi_2, \pi_3 = 4, 5$ for $\sigma_2 = -$,

$$\pi = \langle 0, 1, 4, 5, \ldots, 3, 2 \rangle ,$$

the decreasing subsequence $\pi_{N-2}, \pi_{N-1} = 7, 6$ for $\sigma_{N-3} = -$,

$$\pi = \langle 0, 1, 4, 5, \ldots, 7, 6, 3, 2 \rangle ,$$

and so on until arriving at the central block, $\sigma_{(N-1)/2} = -$, for which we take $\pi_{(N-1)/2}, \pi_{(N+1)/2}, \pi_{(N+3)/2} = N - 1, N + 1, N$,

$$\pi = \langle 0, 1, 4, 5, \ldots, N-1, N+1, N, \ldots, 7, 6, 3, 2 \rangle \in \mathcal{S}_{N+2}.$$

(For $N = 5$, $\pi = \langle 0, 1, 4, 6, 5, 3, 2 \rangle \in \mathcal{S}_7$.) Then, R*3 requires a first semicolon between $\pi_0 = 0$ and $\pi_1 = 1$, a second semicolon between $\pi_1 = 1$ and $\pi_2 = 4, \ldots$, an $((N+1)/2)$th semicolon between $\pi_{(N-1)/2} = N-1$ and $\pi_{(N+1)/2} = N+1$ or between $\pi_{(N+1)/2} = N+1$ and $\pi_{(N+3)/2} = N$ (since the central subsequence $N-1, N+1, N$ is not spiraling), $\ldots$, and an $(N-1)$th semicolon (the maximal number allowed) between $\pi_{N-1} = 6$ and $\pi_N = 3$. But the sequence $\pi_N, \pi_{N+1} = 3, 2$ in the last block ($\sigma_{N-1} = -$) violates R*3 because $\pi_N + 1 = 4$ is not on the right of $\pi_{N+1} + 1 = 3$.

In the case $N = 3 \bmod 4$, the central sign of σ is $\sigma_{(N-1)/2} = +$. The construction of a forbidden pattern of length $L = N+2$ follows the same assignment of entry pairs as before for σ_0, σ_{N-1}, σ_2, $\ldots$, $\sigma_{(N-3)/2}$, but takes $\pi_{(N+1)/2}, \pi_{(N+3)/2}, \pi_{(N+5)/2} = N+1, N, N-1$ for $\sigma_{(N+1)/2} = -$:

$$\pi = \langle 0, 1, 4, 5, \ldots, N-2, N+1, N, N-1, \ldots, 7, 6, 3, 2 \rangle \in \mathcal{S}_{N+2}.$$

(For $N = 3$, $\pi = \langle 0, 1, 4, 3, 2 \rangle \in \mathcal{S}_5$.) Then, R*3 requires a first semicolon between $\pi_0 = 0$ and $\pi_1 = 1$, a second semicolon between $\pi_1 = 1$ and $\pi_2 = 4$, $\ldots$, an $((N+1)/2)$th semicolon between $\pi_{(N-1)/2} = N-2$ and $\pi_{(N+1)/2} = N+1$ or between $\pi_{(N+1)/2} = N+1$ and $\pi_{(N+3)/2} = N$ (since the subsequence $\pi_{(N-1)/2}, \pi_{(N+1)/2}, \pi_{(N+3)/2} = N-2, N+1, N$ cannot belong to an s-block with positive sign because $\pi_{(N-1)/2} + 1 = N-1$ is not on the left of $\pi_{(N+3)/2} + 1 = N+1$), $\ldots$, and an $(N-1)$th semicolon (the maximal number allowed) between $\pi_{N-1} = 6$ and $\pi_N = 3$. But the sequence $\pi_N, \pi_{N+1} = 3, 2$ in the last block ($\sigma_{N-1} = -$) violates R*3 because $\pi_N + 1 = 4$ is not on the right of $\pi_{N+1} + 1 = 3$. □

A further signature with general features is $\sigma = (-, -, \ldots, -)$.

Theorem 7 *The shift Σ_σ with $\sigma_0 = \cdots = \sigma_{N-1} = -$, $N \geq 2$, has*

1. *allowed L-patterns for $L \leq N+1$ and*
2. *root forbidden L-patterns for $L \geq N+2$.*

Since $\sigma = (-, \ldots, -) = \sigma_{\mathrm{mirrored}}$, the number of root forbidden patterns for Σ_σ will be even (Corollary 3).

Proof 1. We need to consider only the case $L = N+1$, since all L-patterns with $L \leq N$ are trivially allowed. Given $\pi \in \mathcal{S}_{N+1}$, there are three possibilities: (i) $N = \pi_0$, (ii) $N = \pi_n$ with $1 \leq n \leq N-1$, or (iii) $N = \pi_N$. The decompositions (i)

$$\pi = \langle N, \pi_1; \pi_2; \ldots; \pi_k; \ldots; \pi_N \rangle,$$

(ii)

$$\pi = \langle \pi_0; \pi_1; \ldots; \pi_{n-1}; N, \pi_{n+1}; \ldots; \pi_N \rangle \quad \text{or} \quad \langle \pi_0; \pi_1; \ldots; \pi_{n-1}, N; \pi_{n+1}; \ldots; \pi_N \rangle$$

(since $\pi_{n-1}, \pi_n, \pi_{n+1} = \pi_{n-1}, N, \pi_{n+1}$ does not form a spiraling subsequence), and (iii)

$$\pi = \langle \pi_0; \pi_1; \ldots; \pi_k; \ldots; \pi_{N-1}, N \rangle,$$

show that any $\pi \in \mathcal{S}_{N+1}$ is allowed for Σ_σ, $\sigma_0 = \cdots = \sigma_{N-1} = -$.

2. Consider

$$\pi = \langle 0, 1, 2, \ldots, N-1, N, N+1 \rangle \in \mathcal{S}_{N+2}.$$

Then R*2 requires a first semicolon between 0 and 1, a second semicolon between 1 and 2, and an $(N-1)$th semicolon (the maximal number allowed) between $N-2$ and $N-1$. This leads to a last block $\pi_{N-1}, \pi_N, \pi_{N+1} = N-1, N, N+1$, which is not a spiraling subsequence. Hence π is forbidden.

The assumption that π is not a root forbidden pattern leads to the fact that π is outgrowth of the forbidden pattern

$$\langle 0, 1, 2, \ldots, N-1, N \rangle \in \mathcal{S}_{N+1},$$

whether π belongs to group I or II (3.12). But clearly this pattern admits the decomposition

$$\langle 0; 1; 2; \ldots; N-1, N \rangle,$$

with $N-1$ semicolons (the maximal number allowed). This contradiction shows that π is not an outgrowth forbidden pattern. Needless to say (Corollary 3),

$$\pi_{\text{mirrored}} = \langle N+1, N, N-1, \ldots, 2, 1, 0 \rangle$$

is also a root forbidden pattern.   □

To conclude this chapter, we consider briefly the existence of *root* forbidden patterns for the signed shifts on $N \geq 3$ symbols. For $\sigma = (+, \ldots, +)$ and $\sigma = (-, \ldots, -)$ we know that there exist root forbidden patterns for every $L \geq N+2$ (Theorems 2 and 7, respectively). The structure of the forbidden ordinal patterns depends, of course, on the signature of the signed shift envisaged, thus the construction of root forbidden patterns can only be done, in general, on a case-by-case basis.

To illustrate this point, consider the signed shifts (with mixed signs) on three symbols. Because of the relation between the allowed/forbidden patterns for Σ_σ and $\Sigma_{\sigma_{\text{mirrored}}}$, only the following four cases are really distinct:

Case a: $\sigma = (+, +, -)$, Case b: $\sigma = (+, -, +)$,
Case c: $\sigma = (+, -, -)$, Case d: $\sigma = (-, +, -)$.

These four cases were studied in [17]. There it is proven that all the signed shifts (a)–(d) have root forbidden L-patterns for $L \geq 5$. Furthermore, $\Sigma_{(+,-,+)}$ has two (root) forbidden 4-patterns, $\Sigma_{(+,-,-)}$ has one (root) forbidden 4-pattern, while $\Sigma_{(+,+,-)}$, $\Sigma_{(-,+,-)}$ have no forbidden 4-patterns. Of course, the same holds for any map order isomorphic to those signed shifts, in particular for the corresponding signed sawtooth maps.

Exercise 10 Check the following statements on root forbidden patterns for Σ_σ in the four cases a–d.

(a) The patterns

$$\pi = \langle 0, L-1, 2, 3, \ldots, L-2, 1 \rangle \in \mathcal{S}_L,$$

  $L \geq 5$, are root forbidden patterns for $\Sigma_{(+,+,-)}$.
(b) The patterns

$$\pi = \langle L-2, 0, L-4, \ldots, 3, 1, 2, 4, \ldots, L-3, L-1 \rangle \in \mathcal{S}_L$$

  if $L \geq 5$ is odd and

$$\pi = \langle L-1, L-3, \ldots, 3, 1, 2, 4, \ldots, L-4, 0, L-2 \rangle \in \mathcal{S}_L$$

  if $L \geq 6$ is even, together with their corresponding mirrored patterns, are root forbidden patterns for $\Sigma_{(+,-,+)}$. (If $L = 5$, then $\pi = \langle 3, 0, 1, 2, 4 \rangle$; if $L = 6$, then $\pi = \langle 5, 3, 1, 2, 0, 4 \rangle$.)
(c) The patterns

$$\pi = \langle 2, 1, 0, 3, 4 \rangle \in \mathcal{S}_5,$$
$$\pi = \langle L-3, \ldots, 4, 2, 1, 0, 3, 5, \ldots, L-4, L-2, L-1 \rangle \in \mathcal{S}_L$$

  for $L \geq 7$ odd, and

$$\pi = \langle L-1, L-2, L-4, \ldots, 4, 2, 1, 0, 3, 5, \ldots, L-3 \rangle \in \mathcal{S}_L$$

  for $L \geq 6$ even, are root forbidden patterns for $\Sigma_{(+,-,-)}$. Although $\sigma = (+,-,-) \neq \sigma_{mirrored} = (-,-,+)$, the mirrored patterns of these patterns are also root forbidden patterns for $\Sigma_{(+,-,-)}$.
(d) The patterns

$$\pi = \langle 0, 1, 4, 3, 2 \rangle \in \mathcal{S}_5,$$
$$\pi = \langle 0, 1, L-1, L-2, \ldots, 3, 2, 4, \ldots, L-3 \rangle \in \mathcal{S}_L$$

  if $L \geq 7$ is odd and

$$\pi = \langle 0, 1, L-1, L-2, \ldots, 4, 2, 3, \ldots, L-3 \rangle \in \mathcal{S}_L$$

if $L \geq 6$ is even, together with the corresponding mirrored patterns, are root forbidden patterns for $\Sigma_{(-,+,-)}$.

Exercise 11 Using signed sawtooth maps with alternating signature, construct a *continuous* map whose orbits realize all possible ordinal patterns (*hint*: the construction is similar to Fig. 4.2).

Chapter 6
Metric Permutation Entropy

The word "entropy" was coined by the German physicist R. Clausius (1822–1888), who introduced it in thermodynamics in 1865 to measure the amount of energy in a system that cannot produce work. The fact that the entropy of an isolated system never decreases constitutes the second law of thermodynamics and clearly shows the central role of entropy in many-particle physics. The direction of time is then explained as a consequence of the increase of entropy in all irreversible processes. Later on the concept of entropy was given a microscopic interpretation in the foundational works of L. Boltzmann (1844–1906) on gas kinetics and statistical mechanics [184]. The celebrated Boltzmann's equation reads in the usual physical notation

$$S = k_B \ln \Omega, \tag{6.1}$$

where here S is the entropy of the thermodynamical system, k_B is a physical constant (called Boltzmann's constant, $k_B = 1.3806504(24) \times 10^{-23}$ J/K) and Ω is the number of microscopic states consistent with the macroscopic constraints. In this realm, the entropy is a measure of the microscopic *disorder* of the system, the entropy being higher the more disordered the system.

In 1948 the word entropy came to the fore in the new context of information theory, coding theory, and cryptography through the seminal papers of C.E. Shannon[1] (1916–2001) [186]. This time, entropy measures the average *uncertainty* about the outcome of a random variable. More generally, the *entropy rate* measures the uncertainty per symbol (time unit, channel use, etc.) of a stationary stochastic process, eventually modeling an information source. Instead of associating entropy with uncertainty, one can alternatively speak of the average information gained by performing a random experiment. Entropy plays a paramount role in all information-related fields, being at the heart of the fundamental results.

---

[1] According to [64] "When Shannon had invented his quantity and consulted von Neumann on what to call it, von Neumann replied: 'Call it entropy. It is already in use under that name and besides, it will give you a great edge in debates because nobody knows what entropy is anyway.'"

J.M. Amigó, *Permutation Complexity in Dynamical Systems*,
Springer Series in Synergetics, DOI 10.1007/978-3-642-04084-9_6,
© Springer-Verlag Berlin Heidelberg 2010

Shannon's ideas, properly transformed, were incorporated by A.N. Kolmogorov (1903–1987) into ergodic theory in 1958 [126] to measure the *randomness* of deterministic dynamical systems. Kolmogorov's proposal was improved a short time later by Sinai [189]. The result became the most important invariant in the theory of discrete and continuous dynamical systems.

Since then the concept of entropy has evolved along different ways: Rényi entropy, topological entropy, sequence entropy, Tsallis entropy, directional entropy, permutation entropy, epsilon–tau entropy, etc. The basics of Shannon entropy, metric (Kolmogorov–Sinai or measure-theoretical) entropy, and topological entropy are systematized in Annex B.

Permutation entropy, both in the metric version (this chapter) and in the topological version (next chapter), was introduced by Bandt, Keller, and Pompe in [29] (see [28] as well). The main ingredient of permutation entropy is the ordinal patterns we studied in Chap. 3. As we shall see below, the definition of the metric permutation entropy of an information source is formally the same as Shannon's entropy, except for the fact that now probabilities refer not to length-*L* blocks of symbols but to the length-*L* ordinal patterns realized by them (assuming, of course, that those symbols can be ordered).

On defining the metric permutation entropy of maps, we depart from [29] to follow basically Kolmogorov's strategy: coarse-grain the state space with a partition, apply the definition of (in our case, permutation) entropy to the resulting symbolic dynamics, and then refine successively the original partition into the partition into separate points. Moreover, the partitions used may be taken to be product, uniform partitions, making possible the numerical estimation of metric permutation entropy under rather general conditions. Most importantly, we shall show that metric permutation entropy converges to the conventional metric entropy for ergodic self-maps of *n*-dimensional intervals.

6.1 The Metric Permutation Entropy of a Finite-State Process

Let $\mathbf{X} = \{X_n\}_{n\in\mathbb{N}_0}$ be a random process with finite state space S (see Annex A.3). We take without restriction $S = \{1, 2, \ldots, |S|\}$. As noted in Example 2, the relation between length-*L* words and length-*L* ordinal patterns is in general many-to-one. This is due to the fact that ordinal patterns do not take into account the sizes of the elements being compared, but only their relative order. The same happens with the *ranks* or *rank variables*, which are the outputs of a random process $\mathbf{R} = \{R_n\}_{n\in\mathbb{N}_0}$ subsidiary of $\mathbf{X}$, defined as follows:

$$R_n = |\{X_i, 0 \leq i \leq n : X_i \leq X_n\}| = \sum_{i=0}^{n} \delta(X_i \leq X_n),$$

where as usual the δ-function of a proposition is 1 if it holds and 0 otherwise. By definition, R_n is a *discrete* random variable with range $\{1, \ldots, n+1\}$, and the sequence

$\mathbf{R} = \{R_n\}_{n \in \mathbb{N}_0}$ builds a discrete-time, non-stationary stochastic process. The point about introducing rank variables is that the relation between length-L ordinal patterns $\pi(x_n^{n+L-1})$ and length-L ranks $r_n^{n+L-1} = r_n, r_{n+1}, \ldots, r_{n+L-1}$ is one-to-one. The many-to-one relation between X_0^{L-1} and R_0^{L-1} will be written as

$$R_0^{L-1} = \operatorname{rank}(X_0^{L-1}). \tag{6.2}$$

Ranks are specially useful in proofs.

Example 10 If, as in Example 2, $S = \{a, b, c\}$ with $a < b < c$ and $x_0^2 = c, a, a$, then $r_0^2 = 1, 1, 2$. All other words defining the same ordinal pattern $\pi(x_0^2) = \langle 1, 2, 0 \rangle$ define also the same rank variables:

$$r_0^2 = 1, 1, 2 = \operatorname{rank}(c, b, b) = \operatorname{rank}(c, a, b) = \operatorname{rank}(b, a, a).$$

Having defined the sibling concepts of ordinal patterns and rank variables of finite-alphabet sequences, we can proceed now very much the same way as we did when defining Shannon's entropy (rate) of stochastic processes or information sources in Sect. 1.1.1 (see also Annex B.1), this time though bookkeeping ordinal patterns instead of symbol blocks.

In this spirit, the *metric permutation entropy* of a stochastic process $\mathbf{X} = \{X_n\}_{n \in \mathbb{N}_0}$ is defined as

$$h^*(\mathbf{X}) = \lim_{L \to \infty} h^*(X_0^{L-1}), \tag{6.3}$$

provided the limit exists, where

$$h^*(X_0^{L-1}) = -\frac{1}{L} \sum_{x_0, \ldots, x_{L-1}} p(\pi(x_0^{L-1})) \log p(\pi(x_0^{L-1}))$$

is the *metric permutation entropy of order* $L \geq 2$ of $\mathbf{X}$. Here $p(\pi(x_0^{L-1}))$ is the probability for the length-L block $x_0^{L-1} = x_0, \ldots, x_{L-1}$ to be of type $\pi(x_0^{L-1}) \in S_L$. Alternatively,

$$h^*(X_0^{L-1}) = -\frac{1}{L} \sum_{r_0, \ldots, r_{L-1}} p(r_0^{L-1}) \log p(r_0^{L-1}) = h(R_0^{L-1}), \tag{6.4}$$

where $p(r_0^{L-1})$ is the probability for the block x_0^{L-1} to define the rank vector $r_0^{L-1} = r_0, \ldots, r_{L-1}$ (remember that the relation between $\pi(X_0^{L-1})$ and $R_0^{L-1} = \operatorname{rank}(X_0^{L-1})$ is one-to-one). In both cases,

$$h^*(\mathbf{X}) = h(\pi(\mathbf{X})) = h(\mathbf{R}),$$

where $h(\cdot)$ denotes the Shannon entropy of the corresponding stochastic process.

In case that the random process $\mathbf{X}$ is stationary, there is still a third way to look at its metric entropy permutation. If $(S^{\mathbb{N}_0}, \mathcal{B}_\Pi(S), m, \Sigma)$ is the sequence space model of $\mathbf{X}$ (see Annex A.3), then the non-empty cylinder sets

$$C_\pi = \{(x_n) \in S^{\mathbb{N}_0} : x_0^{L-1} \text{ is of type } \pi \in \mathcal{S}_L\}$$

build a partition of $(S^{\mathbb{N}_0}, \mathcal{B}_\Pi(S), m)$ with $m(C_\pi) = \Pr\{\pi(X_0^{L-1}) = \pi\} = \Pr\{R_0^{L-1} = r_0^{L-1}\}$, where $R_0^{L-1} = \text{rank}(X_0^{L-1})$, and $1 \le r_k \le k+1$ for $k = 0, \ldots, L-1$. Therefore

$$h^*(X_0^{L-1}) = -\frac{1}{L} \sum_{\pi \in \mathcal{S}_L} m(C_\pi) \log m(C_\pi). \tag{6.5}$$

As a result, the permutation entropy is sensitive to the measures of non-trivial order relationships observed in a word, as the Shannon entropy is sensitive to the measures of the different word values themselves.

When stationarity is important, as in (6.5), we call $\mathbf{X}$ an information source or just a source.

In the next lemma we use the conditional entropy of a random variable Y given another random variable X, $H(Y|X)$, which is the expected value of the entropies of the conditional distributions averaged over the conditioning variable X (see Annex B, (B.5)).

Lemma 7 *Given an ergodic source* $\mathbf{X} = \{X_n\}_{n \in \mathbb{N}_0}$, *the equality*

$$\lim_{k \to \infty} H(R_k^{k+l} | X_0^{k-1}) = \lim_{k \to \infty} H(X_k^{k+l} | X_0^{k-1})$$

holds for all $l \ge 0$.

That is, given a sufficiently long tail of previously observed symbols, the later ranks can be predicted virtually as well as the symbols themselves. Heuristically, this is because the rank of a late variable is sensitive effectively to the cumulative distribution function of the source, approximated by the normalized sum of X_0^{k-1}. In turn, this means that the information contained in R_k is the same as the information in X_k.

Proof Consider $R_k = \sum_{i=0}^{k} \delta(X_i \le X_k)$. For $a \in S = \{1, \ldots, |S|\}$ define the *sample frequency* of the letter a in the word x_0^k, $k \ge 0$, to be

$$\vartheta_k(a) = \frac{1}{k+1} \sum_{i=0}^{k} \delta(X_i = a).$$

With the help of $\vartheta_k(a)$ we may express R_k in terms of X_i, $0 \le i \le k$, namely,

$$R_k(X_k) = (k+1) \sum_{a=1}^{X_k} \vartheta_k(a),$$

where we assume the outcomes $X_0, \ldots, X_k$ to be known. Then, the identity

$$\Pr\{R_k = y\} = \sum_{q=1}^{|S|} \Pr\{X_k = q\}\delta\left(R_k(q) = y\right) \tag{6.6}$$

gives us the probability for observing some R_k with value $y \in \{1, \ldots, k+1\}$ by means of $\Pr\{X_k = q\}$, $1 \le q \le |S|$. Since, given X_0^{k-1} ($k \ge 1$), R_k is a deterministic function of the random variable X_k, i.e., $\Pr\{R_k = y | X_k = q\} = \delta(R_k(q) = y)$, (6.6) can be seen as an application of the law of total probability.

Without loss of generality, we may first rearrange the sum in (6.6) to consider only those symbol values q with non-zero $\Pr\{X_k = q\}$, summing to $N \le |S|$. Expand the sum,

$$\begin{aligned}
\Pr\left\{R_k = y\right\} = \; & \Pr\{X_k = 1\}\delta\left[y = (k+1)\vartheta_k(1)\right] \\
& + \Pr\{X_k = 2\}\delta\left[y = (k+1)(\vartheta_k(1) + \vartheta_k(2))\right] \\
& + \cdots + \Pr\{X_k = N\}\delta\left[y = (k+1)(\vartheta_k(1) + \cdots + \vartheta_k(N))\right].
\end{aligned}$$

Suppose all the relevant sample frequencies $\vartheta_k(1), \ldots, \vartheta_k(N)$ are greater than zero. This means that for any y, only a single one of the δ-functions can be non-zero, and hence we have a one-to-one transformation taking non-zero elements from the distribution $\Pr\{X_k\}$ without change into some bin for $\Pr\{R_k\}$. Since entropy is invariant to a renaming of the bins, and the remaining zero probability bins add nothing to the entropy, we conclude that, if $\vartheta_k(a) > 0$ for all a where the true probability $\Pr\{X_k = a\} > 0$ (i.e., $a = 1, \ldots, N$ after a hypothetical rearrangement), then $H(R_k | X_0^{k-1}) = H(X_k | X_0^{k-1})$ for $k \ge 1$. Because of the assumed ergodicity, we can make the probability that $\vartheta_k(a) = 0$ when $\Pr\{X_k = a\} > 0$ to be arbitrarily small by taking k to be sufficiently large, and the claim follows for $l = 0$.

This construction can be extended without change to words X_k^{k+l} of arbitrary length $l + 1 \ge 1$ via

$$\begin{aligned}
& \Pr\{R_k^{k+l} = y_0 \ldots y_l\} \\
& = \sum_{q_0, \ldots, q_l=1}^{N} \Pr\{X_k^{k+l} = q_0 \ldots q_l\}\delta(R_k(q_0) = y_0) \ldots \delta(R_{k+l}(q_l) = y_l).
\end{aligned}$$

Observe that if $\vartheta_k(a) > 0$ for $1 \le a \le N$, then the same happens with $\vartheta_{k+1}(a), \ldots, \vartheta_{k+l}(a)$ and $H(R_k^{k+l} | X_0^{k-1}) = H(X_k^{k+l} | X_0^{k-1})$ follows. Again, ergodicity guarantees that there exist realizations of X_0^{k+l} with sufficiently large k, whose sample frequencies fulfill the said condition. $\square$

Example 11 As way of illustration, suppose that $X_n = 0, 1$ are independent random variables with probability $\Pr\{X_n = 0\} = \Pr\{X_n = 1\} = \frac{1}{2}$. Given $x_0^{k-1} =$

$x_0 \ldots x_{k-1} \in \{0, 1\}^k$, set $N_0 = \left| \{ i : x_i = 0 \text{ in } x_0^{k-1} \} \right|$, $0 \leq N_0 \leq k$. Consider the case $l = 1$ in Lemma 1. There are two possibilities:

(i) $0 \leq N_0 \leq k - 1$. Then

$$
\begin{aligned}
x_k^{k+1} &= 0, 0 &\Rightarrow\quad r_k^{k+1} &= N_0 + 1, N_0 + 2, \\
x_k^{k+1} &= 0, 1 &\Rightarrow\quad r_k^{k+1} &= N_0 + 1, k + 2, \\
x_k^{k+1} &= 1, 0 &\Rightarrow\quad r_k^{k+1} &= k + 1, N_0 + 1, \\
x_k^{k+1} &= 1, 1 &\Rightarrow\quad r_k^{k+1} &= k + 1, k + 2.
\end{aligned}
$$

Each of these events has the joint probability

$$
\Pr\{N_0 = v, R_k^{k+1} = r_k^{k+1}\} = \frac{\binom{k}{v}}{2^k} \cdot \frac{1}{4} = \frac{1}{2^{k+2}} \binom{k}{v}
$$

and conditional probability

$$
\Pr\{R_k^{k+1} = r_k^{k+1} | N_0 = v\} = \frac{1}{4},
$$

where $0 \leq v \leq k - 1$ and $r_k^{k+1} = (v + 1, v + 2), (v + 1, k + 2), (k + 1, v + 1),$ or $(k + 1, k + 2)$.

(ii) $N_0 = k$. Then

$$
\begin{aligned}
x_k^{k+1} &= 0, 0 \,\&\, x_k^{k+1} = 0, 1 \,\&\, x_k^{k+1} = 1, 1 &\Rightarrow\quad r_k^{k+1} &= k + 1, k + 2, \\
x_k^{k+1} &= 1, 0 & \Rightarrow\quad r_k^{k+1} &= k + 1, k + 1.
\end{aligned}
$$

These events have the joint probabilities

$$
\Pr\left\{ N_0 = k, R_k^{k+1} = (k + 1, k + 2) \right\} = \frac{1}{2^k} \cdot \frac{1}{4} \cdot 3 = \frac{3}{2^{k+2}},
$$

$$
\Pr\left\{ N_0 = k, R_k^{k+1} = (k + 1, k + 1) \right\} = \frac{1}{2^k} \cdot \frac{1}{4} = \frac{1}{2^{k+2}}
$$

and conditional probabilities

$$
\Pr\left\{ R_k^{k+1} = (k + 1, k + 2) | N_0 = k \right\} = \frac{3}{4},
$$

$$
\Pr\left\{ R_k^{k+1} = (k + 1, k + 1) | N_0 = k \right\} = \frac{1}{4}.
$$

From Annex (B.5) and (i)–(ii), we get

$$H(R_k^{k+1}|X_0^{k-1}) = -4 \times \sum_{v=0}^{k-1} \frac{1}{2^{k+2}} \binom{k}{v} \log \frac{1}{4} - \frac{3}{2^{k+2}} \log \frac{3}{4} - \frac{1}{2^{k+2}} \log \frac{1}{4}$$

$$= 4 \times \frac{2}{2^{k+2}}(2^k - 1) + \frac{8}{2^{k+2}} - \frac{3}{2^{k+2}} \log 3$$

$$= 2\left(1 - \frac{3}{2^{k+3}} \log 3\right).$$

On the other hand, since the random variables X_n are independent,

$$H(X_k^{k+1}|X_0^{k-1}) = H(X_k^{k+1}) = 2.$$

It follows that $H(R_k^{k+1}|X_0^{k-1})$ and $H(X_k^{k+1}|X_0^{k-1})$ coincide in the limit $k \to \infty$, as guaranteed by Lemma 7.

With Lemma 7 in hand, we turn to the main result.

Theorem 8 *For a finite-alphabet ergodic source* **X**, *the permutation entropy exists and equals the metric entropy:* $h^*(\mathbf{X}) = h(\mathbf{X})$.

Proof We prove inequalities in both directions.

(a) $\limsup_{L \to \infty} h^*(X_0^{L-1}) \leq h(\mathbf{X})$. Given X_0^{L-1}, the corresponding rank variables are uniquely determined via $R_0^{L-1} = \mathrm{rank}(X_0^{L-1})$. By [59, Chap. 2, Exercise 5], $H(\varphi(Z)) \leq H(Z)$ for any discrete random variable Z and function φ, so $H(R_0^{L-1}) \leq H(X_0^{L-1})$ and thus (see (6.4)),

$$\limsup_{L \to \infty} h^*(X_0^{L-1}) = \limsup_{L \to \infty} h(R_0^{L-1}) \leq \limsup_{L \to \infty} h(X_0^{L-1}) = h(\mathbf{X}).$$

(b) $\liminf_{L \to \infty} h^*(X_0^{L-1}) \geq h(\mathbf{X})$. There are several ways to prove this inequality. Consider, for instance,

$$\liminf_{L \to \infty} h^*(X_0^{L-1})$$

$$= \liminf_{L \to \infty} \frac{1}{L} H(R_0^{L-1})$$

$$= \liminf_{L \to \infty} \frac{1}{L} \left(\left[H(R_{L-1}|R_0^{L-2}) + \cdots + H(R_{L^*+1}|R_0^{L^*}) \right] + H(R_0^{L^*}) \right)$$

for any $L^* < L - 1$, where we have applied the chain rule for entropy (B.9). As $R_1^k = \mathrm{rank}(X_1^k)$ we apply the data processing inequality $H(Y|\varphi(Z)) \geq H(Y|Z)$ [59] to all elements of the first term on the right-hand side:

$$\liminf_{L \to \infty} h(X_0^{L-1})$$

$$\geq \liminf_{L \to \infty} \frac{1}{L} \left(\left[H(R_{L-1}|X_0^{L-2}) + \cdots + H(R_{L^*+1}|X_0^{L^*}) \right] + H(R_0^{L^*}) \right).$$

By Lemma 7 with $l = 0$, for any $\varepsilon > 0$ there exists some L^* such that

$$\left| H(X_L|X_0^{L-1}) - H(R_L|X_0^{L-1}) \right| < \varepsilon$$

for $L > L^*$, so

$$
\liminf_{L \to \infty} h(X_0^{L-1})
$$
$$
> \liminf_{L \to \infty} \left(\frac{1}{L} \left[H(X_{L-1}|X_0^{L-2}) + \cdots + H(X_1|X_0) + H(X_0) \right] \right.
$$
$$
\left. + \frac{1}{L} \left[H(R_0^{L^*}) - H(X_0^{L^*}) \right] - \left(\frac{L - L^* - 1}{L} \right) \varepsilon \right)
$$
$$
= h(\mathbf{X}) - \varepsilon,
$$

since $H(X_0^{L^*}) = H(X_0) + H(X_1|X_0) + \cdots + H(X_{L^*}|X_0^{L^*-1})$ (B.9).
The existence of the limit and equality follows from (a) and (b).   □

Observe in the proof of Theorem 8 that the ergodicity hypothesis was used only in part (b) via Lemma 7, while part (a) is completely general. We highlight this particular result in the following corollary for further reference.

Corollary 4 *For finite-alphabet sources* $\mathbf{X}$,

$$\limsup_{L \to \infty} h^*(X_0^{L-1}) \le h(\mathbf{X})$$

holds.

In order to deal further with the general, nonergodic case, we appeal to the theorem on ergodic decompositions [114]: if Ω is a compact metrizable space and $T:(\Omega, \mathcal{B}, \mu) \to (\Omega, \mathcal{B}, \mu)$ is a continuous transformation, then there is a partition of Ω into T-invariant subsets Ω_w, each equipped with a sigma-algebra $\mathcal{B}_w$ and a probability measure μ_w, such that T acts ergodically on each probability space $(\Omega_w, \mathcal{B}_w, \mu_w)$, the indexing set being another probability space $(W, \mathcal{F}, \nu)$. Furthermore,

$$\mu(E) = \int_W \int_E d\mu_w d\nu(w) = \int_W \mu_w(E) d\nu(w) \quad (E \in \mathcal{B}).$$

The family $\{\mu_w : w \in W\}$ is called the *ergodic decomposition* of μ.

If Σ is the shift on the (compact, metric) sequence space $(S^{\mathbb{N}_0}, \mathcal{B}_\Pi(S), m)$, the indexing set can be taken to be $S^{\mathbb{N}_0}$, i.e.,

$$m(C) = \int_{S^{\mathbb{N}_0}} \int_C dm_s dm(s) = \int_{S^{\mathbb{N}_0}} m_s(C) dm(s) \quad (C \in \mathcal{B}_\Pi(S)), \qquad (6.7)$$

where $m_{\Sigma(s)} = m_s$ [89]. This result shows that any source which is not ergodic can be represented as a mixture of ergodic subsources. The next lemma states that such a decomposition holds also for the entropy.

Lemma 8 (Ergodic Decomposition of the Entropy) [89] *Let* $(S^{\mathbb{N}_0}, \mathcal{B}_\Pi(S), m, \Sigma)$ *be the sequence space model of a stationary finite-alphabet random process* $\mathbf{X} = \{X_n\}_{n \in \mathbb{N}_0}$. *Let* $\{m_s : s \in S^{\mathbb{N}_0}\}$ *be the ergodic decomposition of m. If* $h_{m_s}(\mathbf{X})$ *is m-integrable, then*

$$h(\mathbf{X}) = \int_{S^{\mathbb{N}_0}} h_{m_s}(\mathbf{X}) dm(s). \tag{6.8}$$

Theorem 9 *Under the assumptions of Lemma 8,*

$$\liminf_{L \to \infty} h^*(X_0^{L-1}) \geq h(\mathbf{X}) \tag{6.9}$$

for any finite-alphabet source $\mathbf{X}$.

Proof Fix $L \geq 2$. From (6.5) and (6.7),

$$h^*(X_0^{L-1}) = -\frac{1}{L} \sum_{\pi \in \mathcal{S}_L} \left(\int_{S^{\mathbb{N}_0}} m_s(C_\pi) dm(s) \right) \log \left(\int_{S^{\mathbb{N}_0}} m_s(C_\pi) dm(s) \right)$$

$$\geq -\frac{1}{L} \sum_{\pi \in \mathcal{S}_L} \left(\int_{S^{\mathbb{N}_0}} m_s(C_\pi) \log m_s(C_\pi) dm(s) \right) \tag{6.10}$$

$$= \int_{S^{\mathbb{N}_0}} \left(-\frac{1}{L} \sum_{\pi \in \mathcal{S}_L} m_s(C_\pi) \log m_s(C_\pi) \right) dm(s)$$

$$= \int_{S^{\mathbb{N}_0}} h^*_{m_s}(X_0^{L-1}) dm(s),$$

where in (6.10) we have used Jensen's inequality,

$$\Phi \left(\int_{S^{\mathbb{N}}} f d\mu \right) \leq \int_{S^{\mathbb{N}}} \Phi \circ f d\mu,$$

with $\Phi(t) = t \log t$ convex in $[0, \infty)$ and $f(s) = m_s(C_\pi) \geq 0$.

Therefore,

$$\liminf_{L\to\infty} h^*(X_0^{L-1}) \geq \liminf_{L\to\infty} \int_{S^{\mathbb{N}_0}} h^*_{m_s}(X_0^{L-1}) dm(s)$$

$$\geq \int_{S^{\mathbb{N}_0}} \left(\liminf_{L\to\infty} h^*_{m_s}(X_0^{L-1}) \right) dm(s) \qquad (6.11)$$

$$= \int_{S^{\mathbb{N}_0}} h^*_{m_s}(\mathbf{X}) dm(s),$$

where we have applied Fatou's lemma in (6.11) to the sequence of positive and (by hypothesis) m-measurable functions $h^*_{m_s}(X_0^{L-1})$. Observe that $h^*_{m_s}(\mathbf{X})$ exists for all $s \in S^{\mathbb{N}_0}$ (and is m-integrable as a function of s) since $h^*_{m_s}(\mathbf{X}) = h_{m_s}(\mathbf{X})$ by Theorem 8 ($\mathbf{X}$ is ergodic with respect to m_s). Therefore,

$$\lim_{L\to\infty} \inf h^*(X_0^{L-1}) \geq \int_{S^{\mathbb{N}_0}} h_{m_s}(\mathbf{X}) dm(s) = h(\mathbf{X})$$

by (6.8).   □

Corollary 4 and Theorem 9 yield the following result.

Corollary 5 *Under the assumptions of Lemma 8, $h^*(\mathbf{X}) = h(\mathbf{X})$ holds for any finite-alphabet source $\mathbf{X}$.*

6.2 Permutation Metric Entropy of Maps

In this section we shall use the previous results on finite-alphabet stochastic processes to show that the equality between permutation and metric entropies holds also for ergodic self-maps on domains homeomorphic to q-dimensional compact intervals.

We say that a set $D \subset \mathbb{R}^q$ is a (q-dimensional) *simple domain* if it is homeomorphic to a q-dimensional compact interval (hence D is compact). In particular, one-dimensional simple domains are close intervals. As a subset of $\mathbb{R}^q$, D is also ordered. Let D be a q-dimensional simple domain and $f:D \to D$ a μ-preserving map, with μ being a probability measure on $(D, \mathcal{B} \cap D)$ and $\mathcal{B}$ being the Borel sigma-algebra of $\mathbb{R}^q$. In order to define the permutation entropy of f, consider a q-dimensional compact interval $I \supset D$ and product partitions

$$\iota = \prod_{k=1}^{q} \{I_{1,k}, \ldots, I_{N_k,k}\} \qquad (6.12)$$

of I into $|\iota| = N_1 \cdots N_q$ subintervals of lengths $\Delta_{j,k}$, $1 \leq j \leq N_k$, in each coordinate k. As for the norm of ι (see (1.13)), the perhaps most popular are the *Euclidean norm*,

$$\|\iota\| = \max_{j_1,\ldots,j_q} \left(\sum_{k=1}^{q} \Delta_{j_k,k}^2 \right)^{1/2} =: \|\iota\|_2 \tag{6.13}$$

(i.e., $\|\iota\|_2$ is the longest diagonal of the bins $I_{j_1,1} \times \cdots \times I_{j_q,q} \in \iota$) and the *supremum norm*,

$$\|\iota\| = \max_{j,k} \Delta_{j,k} =: \|\iota\|_\infty . \tag{6.14}$$

For definiteness, the intervals are lexicographically ordered in each dimension, that is, points in $I_{j,k}$ are smaller than points in $I_{j+1,k}$ and, for the multiple dimensions, $I_{j,k} < I_{j,k+1}$, so there is an order relation between all the N partition elements, and we can enumerate them with a single index $i \in \{1,\ldots,|\iota|\}$:

$$\iota = \{I_i : 1 \leq i \leq |\iota|\}, \quad I_i < I_{i+1}$$

(i.e., points in I_i are smaller than points in I_{i+1}).

Below we shall consider refinements of product and general partitions. As usual we write $\alpha \leq \beta$ to mean that the partition β is a *refinement* of the partition α (of $(D, \mathcal{B} \cap D)$ or of any other measurable space for that matter), meaning that the elements of α are unions of the elements of β. By an *increasing sequence of partitions* we mean therefore a sequence of partitions, $(\alpha_n)_{n \in \mathbb{N}}$, such that $\alpha_n \leq \alpha_{n+1}$ for all n. If, as in the present case, the state space is a product space, then by a *product refinement* of partition (6.12) we mean any product partition of I obtained by subdividing some or all of the intervals $\{I_{1,k},\ldots,I_{N_k,k}\}$, $1 \leq k \leq q$.

Furthermore, let κ be the partition of D defined as

$$\kappa = \iota \cap D = \{I_i \cap D \neq \emptyset : 1 \leq i \leq |\iota|\} = \{K_j : 1 \leq j \leq |\kappa|\}.$$

In words, κ consists of all subintervals $I_i \in \iota$ contained in the interior of D, together with the overlaps with D of those I_i that intersect the boundary of D. Partitions κ of the form $\kappa = \iota \cap D$, where ι is a product partition and D a simple domain, will be called *quasi-product partitions*; if, moreover, ι is a box (i.e., uniform) partition, κ will be called a *quasi-box partition*. For simplicity, we set $\|\kappa\| = \|\iota\|$.

Next let $\mathbf{X}^\kappa = \{X_n^\kappa\}_{n \in \mathbb{N}_0}$ be the symbolic dynamics associated with $f : D \to D$ with respect to the partition κ:

$$X_n^\kappa(x) = j \quad \text{if} \quad f^n(x) \in K_j, \ n = 0, 1, \ldots .$$

Hence $\mathbf{X}^\kappa$ is a stationary, $|\kappa|$-state random process on $(D, \mathcal{B} \cap D, \mu)$ with alphabet $S^\kappa = \{1,\ldots,|\kappa|\}$.

Example 12 If $I = [0, 1]$ and $\kappa = \{K_j : 1 \leq j \leq 10^k\}$, with $K_j = [(j-1)10^{-k}, j10^{-k})$ for $1 \leq j \leq 10^k - 1$ and $K_{10^k} = [1 - 10^{-k}, 1]$, then $\mathbf{X}^\kappa$ can be written as follows: $X_n^\kappa(x) = \lfloor f^n(x) \cdot 10^k \rfloor + 1$ for $0 \leq x < 1$ and $X_n^\kappa(1) = 10^k$.

According to (B.16) (with $\alpha = \kappa$), the entropy of the symbolic dynamics $\mathbf{X}^\kappa$ equals the metric entropy of f with respect to κ:

$$h_\mu(f,\kappa) = h_\mu(\mathbf{X}^\kappa). \tag{6.15}$$

If we take now an increasing sequence of product refinements $\kappa \equiv \kappa_0 \leq \kappa_1 \leq \cdots$ such that $\|\kappa_n\| \to 0$, then we deduce from Theorem 25 that $h_\mu(f) = \lim_{n\to\infty} h_\mu(\mathbf{X}^{\kappa_n})$. This suggests to define the metric permutation of f as $h_\mu^*(f) = \lim_{n\to\infty} h_\mu^*(\mathbf{X}^{\kappa_n})$. The fact that the limit $n \to \infty$ proceeds by successive refinements of κ_0 and the way product partitions are being numbered guarantees that the order relations are preserved. This means, in particular, that if $X_k^{\kappa_n}(x) = i < j = X_{k+1}^{\kappa_n}(x)$ $(1 \leq i,j \leq |\kappa_n|)$, then $X_k^{\kappa_{n+1}}(x) = i' < j' = X_{k+1}^{\kappa_{n+1}}(x)$ $(1 \leq i',j' \leq |\kappa_{n+1}|)$ for all $x \in D$ and $k \in \mathbb{N}_0$. Thus $h_\mu^*(f)$ has a good chance to exist.

Definition 3 Given a measure-preserving dynamical system $(D, \mathcal{B} \cap D, \mu, f)$, and a lexicographically ordered, quasi-product partition κ_0 of $(D, \mathcal{B} \cap D, \mu)$, the metric permutation entropy of f with respect to the measure μ is defined by

$$h_\mu^*(f) = \lim_{n\to\infty} h_\mu^*(\mathbf{X}^{\kappa_n}) \tag{6.16}$$

(provided the limit exists), where $(\kappa_n)_{n\in\mathbb{N}}$ is a sequence of successive product refinements of κ_0 such that $\|\kappa_n\| \to 0$ and $\mathbf{X}^{\kappa_n}$ is the symbolic dynamics of f with respect to κ_n.

It is plain that this definition is independent from the auxiliary interval $I \supset D$ used to construct κ_0 and also independent from the particular collection of product refinements κ_n used, as long as $\|\kappa_n\| \to 0$. This being the case, we may take quasi-box partitions in (6.16).

One practical reason for using product partitions is that they make numerical calculations much easier. But most importantly, we claim that $\lim_{\|\alpha_n\|\to 0} h_\mu^*(\mathbf{X}^{\alpha_n})$ does not depend on the particular increasing sequence $(\alpha_n)_{n\in\mathbb{N}_0}$ of successive refinements of a general finite partition α_0 of $(D, \mathcal{B} \cap D, \mu)$, as long as (i) they converge to the point partition of D, $\epsilon = \{\{x\}:x \in D\}$, and (ii) the numbering of the elements of $\alpha_1, \alpha_2, \ldots$ preserves the order relations through the process of refinement. Condition (i) requires that α_n consists of connected sets for all n and $\lim_{n\to\infty} \|A\| = 0$ for all $A \in \alpha_n$. Condition (ii) means that if $A_i, A_j \in \alpha_n$ and $i < j$, then $i' < j'$ whenever $A_i \supset A_{i'}' \in \alpha_{n+1}$ and $A_j \supset A_{j'}' \in \alpha_{n+1}$ (this is automatically satisfied by the lexicographically ordered, product refinements ι_n).

Lemma 9 *Let $(D, \mathcal{B} \cap D, \mu, f)$ be a measure-preserving dynamical system, α_0 a finite partition of $(D, \mathcal{B} \cap D, \mu)$, and $(\alpha_n)_{n\in\mathbb{N}}$ a sequence of successive refinements of α_0 preserving the order relations and converging to the point partition. Then*

$$h_\mu^*(f) = \lim_{n\to\infty} h_\mu^*(\mathbf{X}^{\alpha_n}),$$

where $\mathbf{X}^{\alpha_n}$ is the symbolic dynamics of f with respect to the partition α_n.

Proof Roughly speaking, the increasing sequences $\cdots \leq \kappa_n \leq \kappa_{n+1} \leq \cdots$ and $\cdots \leq \alpha_n \leq \alpha_{n+1} \leq \cdots$ are equivalent in the sense that, given κ_n there is a partition α_m with $\|\alpha_m\| \lesssim \|\kappa_n\|$ which can resolve the orbits of f with the same precision as κ_n does—and reciprocally. Of course, the ordinal patterns of length $L = 2, 3, \ldots$ of a given orbit will be, in general, different, depending on the partitions used. Nevertheless, there will be a one-to-one relation between the ordinal L-patterns realized by $\mathbf{X}^{\alpha_n}$ and $\mathbf{X}^{\kappa_n}$ in the limit $n \to \infty$, and the same holds for the corresponding probabilities. Therefore,

$$\lim_{n\to\infty} h_\mu^*(\mathbf{X}^{\alpha_n}) = \lim_{n\to\infty} h_\mu^*(\mathbf{X}^{\kappa_n}) = h_\mu^*(f).$$

$\square$

The partitions $\mathcal{P}_L$, Eq. (3.5) build a sequence of successive refinements, but they do not preserve in general the order relations because their elements eventually decompose into different components. For the same reason, they cannot converge in general to the partition of D into separate points, ϵ, nor are their norms otherwise expected to vanish as $L \to \infty$.

Having shown that the metric permutation entropy does not depend on the partitions used in its calculation (with the provisos stated in Lemma 9), we turn to the main result of this chapter.

Theorem 10 *Let $f: D \to D$ be ergodic with respect to the measure μ, and suppose that $h_\mu^*(f)$ exists. Then $h_\mu^*(f) = h_\mu(f)$.*

Proof Let κ_0 be a quasi-box partition of $(D, \mathcal{B} \cap D, \mu)$ and $(\kappa_n)_{n\in\mathbb{N}}$ a sequence of successive product refinements of κ_0. Then,

$$h_\mu(f, \kappa_n) = h_\mu(\mathbf{X}^{\kappa_n})$$

by (6.15), where $\mathbf{X}^{\kappa_n} = \{X_k^{\kappa_n}\}_{k\in\mathbb{N}_0}$ is the symbolic dynamics of f with respect to the partition κ_n. Furthermore, $h_\mu(\mathbf{X}^{\kappa_n}) = h_\mu^*(\mathbf{X}^{\kappa_n})$ by Theorem 8, since $\mathbf{X}^\kappa$ is ergodic with respect to the measure μ if f is ergodic with respect to μ. Putting together, we have so far

$$h_\mu^*(f) = \lim_{n\to\infty} h_\mu^*(\mathbf{X}^{\kappa_n}) = \lim_{n\to\infty} h_\mu(\mathbf{X}^{\kappa_n}) = \lim_{n\to\infty} h_\mu(f, \kappa_n).$$

From Theorem 25 (Annex B) it follows then

$$\lim_{n\to\infty} h_\mu(f, \kappa_n) = h_\mu(f)$$

and we are done.

$\square$

If instead of Theorem 8, we use Corollary 5 in the previous proof for every process $\mathbf{X}^\kappa$, we conclude also $h_\mu^*(f) = h_\mu(f)$ for μ-preserving maps. This requires the technical assumption that $h_{m_s}(\mathbf{X}^\kappa)$ is m-integrable, where $\{m_s : s \in S^{\mathbb{N}_0}\}$, $S =$

$\{1, \ldots, |\kappa|\}$, is the ergodic decomposition of m, and m the shift-invariant measure of the sequence space model $(S^{\mathbb{N}_0}, \mathcal{B}_\Pi(S), m, \Sigma)$ of $\mathbf{X}^\kappa$—and this for every partition κ.

Theorem 11 *Let* $f: D \to D$ *be* μ-*preserving, and suppose that* $h_\mu^*(f) = \lim_{n \to \infty} h_\mu^*$ *$(\mathbf{X}^{\kappa_n})$ exists. Under the assumptions of Lemma 8 for each* $\mathbf{X}^{\kappa_n}$, *the equality* $h_\mu^*(f) = h_\mu(f)$ *holds.*

6.3 On the Definition of Metric Permutation Entropy for Maps

The original definition of permutation entropy by Bandt, Keller, and Pompe [29] was presented in Sect. 1.2. Recall that it involves closed *one-dimensional* intervals I, maps $f: I \to I$, and sets of the form

$$P_\pi = \left\{ x \in I : f^{\pi_0}(x) < f^{\pi_1}(x) < \cdots < f^{\pi_{L-1}}(x) \right\},$$

where $\pi = \langle \pi_0, \ldots, \pi_{L-1} \rangle \in \mathcal{S}_L$, $L \geq 2$. Recall once again that

$$\mathcal{P}_L = \{ P_\pi \neq \emptyset : \pi \in \mathcal{S}_L \}.$$

In most situations of interest, $\mathcal{P}_L$ will be a partition of $(I, \mathcal{B} \cap I, \mu)$, where $\mathcal{B}$ is the Borel sigma-algebra of $\mathbb{R}$ and μ is an f-invariant measure. This is going to be our setting throughout this section.

Bandt, Keller, and Pompe define then the metric permutation entropy of order L as[2]

$$h_\mu^{*\mathrm{BKP}}(f, L) = -\frac{1}{L-1} \sum_{\pi \in \mathcal{S}_L} \mu(P_\pi) \log \mu(P_\pi) \tag{6.17}$$

and the permutation entropy of f to be

$$h_\mu^{*\mathrm{BKP}}(f) = \lim_{L \to \infty} h_\mu^{*\mathrm{BKP}}(f, L), \tag{6.18}$$

provided the limit exists.

As compared to conventional entropy, $h_\mu^{*\mathrm{BKP}}(f)$ has at least one remarkable feature: it involves only one infinite limit over the length of the word, while $h_\mu(f)$ involves additionally a second infinite process, namely, a supremum over partitions—unless a generating partition is known. This fact can be rephrased by saying that the sequence $\mathcal{P}_L$ builds a "generator" for $h_\mu^{*\mathrm{BKP}}$.

Let us highlight at this point the main result concerning $h_\mu^{*\mathrm{BKP}}(f)$:

Theorem 12 [29] *If* $f: I \to I$ *is piecewise monotone, then* $h_\mu^{*\mathrm{BKP}}(f) = h_\mu(f)$.

---

[2] Bandt, Keller, and Pompe chose the factor $1/(L-1)$ instead of $1/L$ (see (1.30)) because $\pi(x_0^0)$ contributes nothing to the entropy. Of course, either choice yields the same limit when $L \to \infty$.

Example 13 For the symmetric tent map (1.17), the elements of $\mathcal{P}_2$ are

$$P_{\langle 0,1 \rangle} = (0, \tfrac{2}{3}), \qquad P_{\langle 1,0 \rangle} = (\tfrac{2}{3}, 1);$$

the elements of $\mathcal{P}_3$ are

$$P_{\langle 0,1,2 \rangle} = (0, \tfrac{1}{3}), \qquad P_{\langle 0,2,1 \rangle} = (\tfrac{1}{3}, \tfrac{2}{5}), \qquad P_{\langle 2,0,1 \rangle} = (\tfrac{2}{5}, \tfrac{2}{3}),$$

$$P_{\langle 1,0,2 \rangle} = (\tfrac{2}{3}, \tfrac{4}{5}), \qquad P_{\langle 1,2,0 \rangle} = (\tfrac{4}{5}, 1);$$

and the elements of $\mathcal{P}_4$ are

$$P_{\langle 0,1,2,3 \rangle} = (0, \tfrac{1}{6}), \quad P_{\langle 0,1,3,2 \rangle} = (\tfrac{1}{6}, \tfrac{1}{5}), \quad P_{\langle 0,3,1,2 \rangle} = (\tfrac{1}{5}, \tfrac{2}{9}) \cup (\tfrac{2}{7}, \tfrac{1}{3}),$$

$$P_{\langle 3,0,1,2 \rangle} = (\tfrac{2}{9}, \tfrac{2}{7}), \quad P_{\langle 0,2,1,3 \rangle} = (\tfrac{1}{3}, \tfrac{2}{5}), \quad P_{\langle 2,0,3,1 \rangle} = (\tfrac{2}{5}, \tfrac{4}{9}) \cup (\tfrac{4}{7}, \tfrac{3}{5}),$$

$$P_{\langle 2,3,0,1 \rangle} = (\tfrac{4}{9}, \tfrac{4}{7}), \quad P_{\langle 2,0,1,3 \rangle} = (\tfrac{3}{5}, \tfrac{2}{3}), \quad P_{\langle 3,1,0,2 \rangle} = (\tfrac{2}{3}, \tfrac{4}{5}),$$

$$P_{\langle 1,3,2,0 \rangle} = (\tfrac{4}{5}, \tfrac{5}{6}), \quad P_{\langle 1,2,0,3 \rangle} = (\tfrac{6}{7}, \tfrac{8}{9}), \quad P_{\langle 1,2,3,0 \rangle} = (\tfrac{5}{6}, \tfrac{6}{7}) \cup (\tfrac{8}{9}, 1).$$

See Fig. 6.1 and compare with Fig. 1.7; owing to the order isomorphy of the symmetric tent map and the logistic map, there is a one-to-one relation between their admissible ordinal *L*-patterns. Computation of the metric permutation entropies of orders 2, 3, and 4 of the symmetric tent map Λ (the invariant measure μ is here the Lebesgue measure) yields the following results:

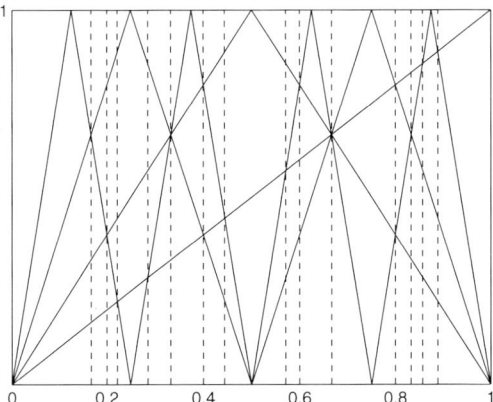

Fig. 6.1 Graphs of the identity, Λ, Λ^2, and Λ^3. The *vertical, dashed lines* separate different P_π, $\pi \in \mathcal{S}_4$

$$h_\mu^{*\mathrm{BKP}}(\Lambda, 2) = \tfrac{2}{3} \log \tfrac{3}{2} + \tfrac{1}{3} \log 3 = 0.9183 \text{ bit/symbol},$$

$$h_\mu^{*\mathrm{BKP}}(\Lambda, 3) = 1.0746 \text{ bit/symbol},$$

$$h_\mu^{*\mathrm{BKP}}(\Lambda, 4) = 1.1807 \text{ bit/symbol}.$$

By Theorem 12,

$$h_\mu^{*\mathrm{BKP}}(\Lambda) = h_\mu(\Lambda) = \log 2 = 1 \text{ bit/symbol}.$$

But in the case of general maps, it seems that only inequality (6.19) below (formally similar to (6.9)) can be proved. Comparing such one-dimensional results with the dimensional generality of Theorem 10, we may conclude that the definition (6.16) of permutation entropy offers some advantages.

Note that the central distinction, which makes formulation (6.16) easier and more natural, is that (6.16) takes the limit of infinite long conditioning ($L \to \infty$) first and the discretization limit ($\|\kappa_n\| \to 0$) last, similar to Kolmogorov–Sinai entropy, and as opposed to (6.18), where an explicit discretization is not taken. Thus we have two limits to take (while $h_\mu^{*\mathrm{BKP}}(f)$ involves only one limit), but the second, $\|\kappa_n\| \to 0$, is harmless and, in principle, can be numerically approximated. We conjecture that for "non-pathological" dynamical systems of the sort one might observe in nature, the two formulations are equivalent, but there are likely to be some non-trivial technicalities involved in a rigorous analysis. More on this, in the next chapter.

Transformations with an infinite number of monotonicity segments are not unusual in ergodic theory.

Example 14 The *Gauss transformation,* $f:[0, 1) \to [0, 1)$ with

$$f(x) = \begin{cases} 0 & \text{if } x = 0 \\ \frac{1}{x} \ (\mathrm{mod}\ 1) & \text{if } x \neq 0 \end{cases},$$

is an ergodic map [52, Chap. 5] with infinitely many monotonicity segments, see Fig. 6.2.

The next theorem shows that, in general, $h_\mu^{*\mathrm{BKP}}(f)$ can only be expected to be an upper bound of $h_\mu(f)$.

Theorem 13 [29] *If $f:I \to I$ is a μ-preserving map with $h_\mu(f) < \infty$, then*

$$\liminf_{L \to \infty} h_\mu^{*BKP}(f, L) \geq h_\mu(f). \tag{6.19}$$

It follows $h_\mu^{*BKP}(f) \geq h_\mu(f)$, provided $h_\mu^{*BKP}(f)$ exists.

Proof Let $\iota = \{I_j, 1 \leq j \leq |\iota|\}$ be a partition of $(I, \mathcal{B} \cap I, \mu)$, where $I_j \subset I$ are intervals. This being the case, let $c_1 < c_2 < \cdots < c_{|\iota|-1}$ be the points that subdivide the interval $I = [a, b]$ into the $|\iota|$ intervals I_j of the partition ι. We consider a fixed

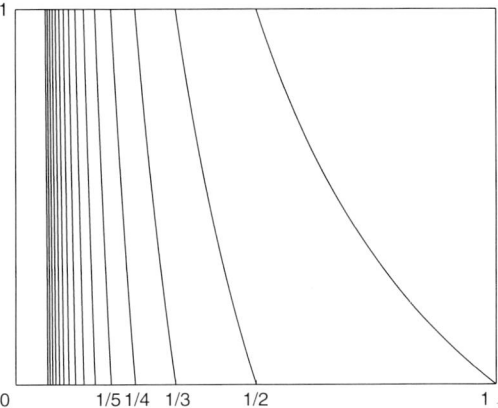

Fig. 6.2 Some monotony intervals of the Gauss transformation

$P_\pi \in \mathcal{P}_L$ and show that it can intersect at most $(L+1)^{|\iota|-1}$ sets of the partition $\iota_0^{L-1} := \vee_{i=0}^{L-1} f^{-i}(I_{j_i})$ with $I_{j_0}, \ldots, I_{j_{L-1}} \in \iota$. For $x \in P_\pi$, let $\Delta_L[x]$ denote the set in ι_0^{L-1} that contains x. Thus, $\Delta_L[x]$ can be written as $I_{j_0} \cap f^{-1}(I_{j_1}) \cap \cdots \cap f^{-(L-1)}(I_{j_{L-1}})$ with $I_{j_0}, \ldots, I_{j_{L-1}} \in \iota$, so that it can be specified by the n-tupel $j[x] = (j_0, \ldots, j_{L-1}) \in \{1, \ldots, |\iota|\}^L$.

Now, π is given by inequalities $x_{k_1} < \cdots < x_{k_L}$ with $\{k_1, \ldots, k_L\} = \{0, \ldots, L-1\}$ and $x_k = f^k(x)$. For each $x \in P_\pi$ we can extend these inequalities so that they give the common order of the c_r and the x_{k_l}, where $1 \leq r \leq |\iota| - 1$ and $1 \leq l \leq L$. It follows that there are at most $(L+1)^{|\iota|-1}$ possible extended orders since each c_r has $L+1$ possible bins to go among the x_{k_l}. Moreover, when we know the common order of the c_r and x_{k_l}, then $j[x]$ is uniquely determined (since $c_{j-1} < x_k < c_j$ implies $x_k \in I_j$ and thus $x \in f^{-k}(I_j)$, with $1 \leq j \leq |\iota|$, $c_0 = a$, and $c_{|\iota|} = b$).

Each $P_\pi \in \mathcal{P}_L$ is then the union of at most $(L+1)^{|\iota|-1}$ sets $V_k \in \iota_0^{L-1} \vee \mathcal{P}_L$ with total measure $\mu(P_\pi)$. Hence,

$$
\begin{aligned}
&- \sum_{k=1}^{(L+1)^{|\iota|-1}} \mu(V_k) \log \mu(V_k) \\
&\leq - \sum_{k=1}^{(L+1)^{|\iota|-1}} \frac{\mu(P_\pi)}{(L+1)^{|\iota|-1}} \log \frac{\mu(P_\pi)}{(L+1)^{|\iota|-1}} \\
&= -\mu(P_\pi) \log \mu(P_\pi) + (|\iota| - 1)\mu(P_\pi) \log (L+1)
\end{aligned}
$$

and summing over all $\pi \in \mathcal{S}_L$,

$$
H_\mu(\iota_0^{L-1}) \leq H_\mu(\iota_0^{L-1} \vee \mathcal{P}_L) \leq H_\mu(\mathcal{P}_L) + (|\iota| - 1) \log (L+1). \tag{6.20}
$$

It follows that

$$\frac{1}{L-1}H_\mu(\mathcal{P}_L) \geq \frac{1}{L-1}\left[H_\mu(\iota_0^{L-1}) - (|\iota|-1)\log(L+1)\right]$$

and

$$\liminf_{L\to\infty}\frac{1}{L-1}H_\mu(\mathcal{P}_L) \geq \liminf_{L\to\infty}\frac{1}{L-1}H_\mu(\iota_0^{L-1}), \qquad (6.21)$$

since $\frac{1}{L-1}\log(L+1) \to 0$ as $L \to \infty$.

On the other hand, the sequence $\frac{1}{L-1}H_\mu(\iota_0^{L-1})$ converges to $h_\mu(f,\iota)$ when $L \to \infty$, hence

$$\liminf_{L\to\infty} h_\mu^{*\mathrm{BKP}}(f,L) = \lim_{L\to\infty}\inf\frac{1}{L-1}H_\mu(\mathcal{P}_L) \geq h_\mu(f,\iota),$$

for any partition ι. Finally,

$$\liminf_{L\to\infty} h_\mu^{*\mathrm{BKP}}(f,L) \geq \sup_\iota h_\mu(f,\iota) = h_\mu(f).$$

□

6.4 Numerical Issues

Our way to the metric permutation entropy of maps was paved by partitions of the state space and the corresponding symbolic dynamics, very much the same way as it happens with the Kolmogorov–Sinai entropy. Therefore, calculating the metric permutation entropy of maps and information sources turns out to be essentially the same task, except for the fact that in the first case this calculation has, in principle, to be repeated with ever finer partitions. In practice, one estimates the true value of the permutation entropy by taking a "sufficiently" fine partition once and for all. This corresponds, by the way, to the numerical practice, as we shall presently explain. If, furthermore, the map (and hence the ensuing source) is ergodic, then it suffices to consider one or a small sample of coarse-grained orbits.

As a by-product of the previous results on metric permutation entropy, the practitioner of time-series analysis will find an alternative way to envision or, eventually, numerically estimate the Kolmogorov–Sinai entropy of real sources. It is worth reminding (see Chap. 1) that the entropy of information sources can be measured by a variety of techniques that go beyond counting word statistics and comprise different definitions of "complexities" such as, for example, counting the patterns along a digital (or digitalized) data sequence [137, 211, 6]. Bandt and Pompe refer in [28] to the permutation entropy of time series as complexity. That the entropy can also be computed by counting ordinal patterns shows once again that it is a so general concept that can be captured with different and seemingly blunt approaches.

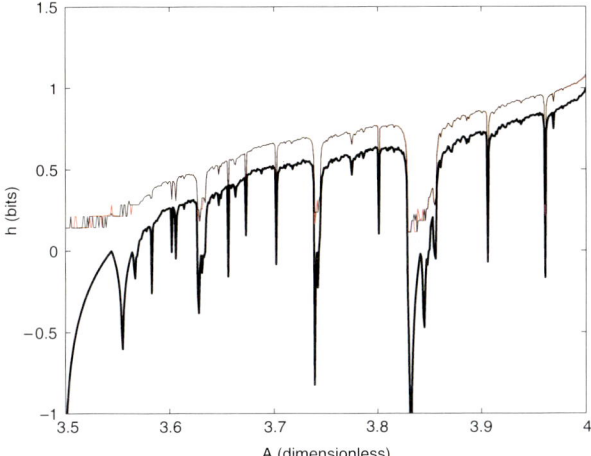

Fig. 6.3 Lyapunov exponent (*black thick line*) of the logistic map g_A, $3.5 \leq A \leq 4$, and metric permutation entropy (rate) estimates $\hat{h} = h^*(X_0^{13})$ in bits/symbol for $N = 10^6$ length time series from the map (*black thin lines*). The metric permutation entropy estimate tracks changes in the Lyapunov exponent well, with a nearly constant bias. Periodic orbits give a finite permutation entropy, but the rate estimate would tend to zero given a sufficiently long word

We demonstrate numerical results on time series $x_{n+1} = g_A(x_n)$ from the logistic map $g_A(x) = Ax(1 - x)$, where $0 \leq A \leq 4$ and $0 \leq x \leq 1$. Figure 6.3 shows an estimate of the metric permutation entropy on noise-free data as a function of A, comparing the Lyapunov exponent $L_\mu(g_A)$ (computed from the orbit knowing the equation of motion) to the metric permutation entropy of g_A for $3.5 \leq A \leq 4$. To be precise, we are estimating $h_\mu^*(\mathbf{X})$ with $\mathbf{X}$ discretized from the logistic map iterated at the discretization of double-precision numerical representation, i.e., $\mathbf{X}$ is the output of a standard numerical iteration and μ is the natural invariant measure with density $d\mu/dx = \frac{1}{\pi \sqrt{x(1-x)}}$. The entropy estimator of the block ranks was the plug-in estimator (substituting observed frequencies for probabilities) plus the classical bias correction, first order in $1/N$, N being here the number of samples (which can be taken, for instance, from sliding windows of fixed length L along the orbit/orbits considered) [167]. Let us remind that

$$h_\mu(g_4) = L_\mu(g_4) = \int_0^1 \log \left| g_4'(x) \right| d\mu(x) = \log 2.$$

Thus, in practice the BKP approach (Sect. 6.3) and our approach (Sect. 6.2) boil down to the same recipe: generate orbits and count ordinal patterns in sliding windows of increasing sizes; for more details, see Chap. 9. The most intriguing characteristic of order relations is that they define, on their own, partitions $\mathcal{P}_L$ for the mapping from continuous values (as the discretization level $\|\kappa_n\|$ goes to zero) to a lower precision symbolic representation which has the natural structure for entropy. When estimating entropy from the discrete information source induced from a *fixed* discretization, the entropy of the symbol stream will not generally equal the Kolmogorov–Sinai entropy unless a generating partition is used, and that can be difficult to find, especially for observed data alone, although some recent works show progress in this direction (e.g., [40] and references therein). The "magic" in using ordinal patterns is that the self-defined partitions $\mathcal{P}_L$ give the Kolmogorov–Sinai entropy, at least asymptotically. Permutation entropy may offer a significant opportunity to advance analytical computations of entropies for various dynamical systems, where generating partitions might be too difficult to find rigorously.

It turns out that using metric permutation entropy to accurately estimate the Kolmogorov–Sinai entropy is more difficult than using it as a very rapid and easy-to-compute *relative* quantification of entropy or complexity which can be computed without requiring a fixed partition (see, e.g., [45]). The key issue in using permutation entropy for empirical data analysis as an entropy estimator is the same as with standard Shannon entropy estimation: balancing the tension between larger word lengths L, to capture more dependencies, and the loss of sufficient sampling for good statistics in the ever larger discrete space. Extracting permutation entropies is rapid and easy—but taking the limits is not at all simple numerically. The finite L performance and convergence rate and bias of any specific computational method are major issues when it comes to accurately estimating the entropy of a source from observed data. It is now appreciated that numerically estimating the Shannon block entropy from finite data and, especially, the asymptotic entropy can be surprisingly tricky [195, 127, 6, 121, 122]. The theoretical definitions of entropy do not necessarily lead to good statistical methods, and superior alternatives have been developed over the many years since Shannon. We believe that some of these ideas may similarly be applicable to the permutation entropy situation, either in terms of using some of the superior entropy estimation methods for block entropies or developing algorithms based on more sophisticated data compression principles to extract the entropy itself.

Also important for practical time-series analysis is the usual situation where observations of a predominantly deterministic source is contaminated with a small level of observational noise. Here, we recommend that the user *fix* some discretization level $\|\kappa_n\|$ characteristic of the noise and evaluate the permutation entropies via entropies of rank words evaluated from the discretized observables.

In regard to vector-valued sources, we used (without restriction) lexicographic ordering in the theoretical part because of definiteness and simplicity. For analyzing chaotic observed data, however, it may be acceptable to still use but one scalar projection subject to the traditional caveats of time-delay embedology. We would expect that for appropriately mixing sources and generic observation functions, the Kolmogorov–Sinai entropy estimated through that scalar still equals the true value, and likewise so might permutation entropy. We have found that numerically this appears to work in practice. Moreover, the lexicographic ordering will effectively reduce to this case anyway except for the few cases where the symbols on the dominant coordinate match, which will be less frequent as L increases. More on this in Chaps. 7 and 9.

Chapter 7
Topological Permutation Entropy

Permutation entropy, as conventional entropy, comes in the metric version (Chap. 6) and in the topological version (this chapter). Topological permutation entropy was also introduced by Bandt et al. [29], together with metric permutation entropy. Let us stress once more that the concept of metric permutation entropy of a map introduced in the last chapter differs from the original one, the difference consisting basically in the order of an iterated limit (first the length of the orbit, then the precision of the measurement, as in the definition of the Kolmogorov–Sinai entropy). This technical change made possible to generalize one of the main results of [29], namely, the equality of metric entropy and metric permutation entropy for piecewise monotone maps on one-dimensional intervals to higher dimensions at the expense of requiring ergodicity (Theorem 10).

   In this chapter we will apply the same approach to topological entropy with the parallel result that the equality of topological entropy and topological permutation entropy for piecewise monotone maps on one-dimensional intervals (the other main result of [29]) can also be generalized to higher dimensions, this time requiring the map to be expansive (Theorem 15). The possibility of going higher dimensional is an advantage of the definitions of metric and topological permutation entropies used in this book.

7.1 Topological Permutation Entropy of Sources

Let $\mathbf{X} = \{X_n\}_{n \in \mathbb{N}_0}$ be an information source with finite alphabet S. We define the *topological entropy of order L of $\mathbf{X}$* as

$$h_{\text{top}}(X_0^{L-1}) = \frac{1}{L} \log N(\mathbf{X}, L), \tag{7.1}$$

where X_0^{L-1} is shorthand for the block of random variables $X_0, \ldots, X_{L-1}$ and $N(\mathbf{X}, L)$ is the number of sequences (words, blocks, etc.) of length L, $x_0^{L-1} = x_0, \ldots, x_{L-1}$, that $\mathbf{X}$ can output. Put in a different way, $N(\mathbf{X}, L)$ is the number of words of length L, built by consecutive letters, that are *allowed* or *admissible* in

J.M. Amigó, *Permutation Complexity in Dynamical Systems,* Springer Series in Synergetics, DOI 10.1007/978-3-642-04084-9_7, © Springer-Verlag Berlin Heidelberg 2010

the messages of $\mathbf{X}$ (since $\mathbf{X}$ is stationary, we may restrict ourselves to an initial segment). The *topological entropy* of $\mathbf{X}$ is then defined as

$$h_{\text{top}}(\mathbf{X}) = \lim_{L \to \infty} h_{\text{top}}(X_0^{L-1}), \tag{7.2}$$

provided the limit exists. In an information-theoretical framework, $h_{\text{top}}(\mathbf{X})$ is called the *capacity* of $\mathbf{X}$ [186]. If, furthermore,

$$h_\mu(X_0^{L-1}) = -\frac{1}{L} \sum_{x_0,\dots,x_{L-1} \in S} p(x_0,\dots,x_{L-1}) \log p(x_0,\dots,x_{L-1}) \tag{7.3}$$

is the *Shannon* (or *metric*) *entropy of order L* of $\mathbf{X}$, then clearly $h_\mu(X_0^{L-1}) \leq h_{\text{top}}(X_0^{L-1})$ (for any logarithm base > 1). Therefore

$$h_\mu(\mathbf{X}) = \lim_{L \to \infty} h_\mu(X_0^{L-1}) \leq h_{\text{top}}(\mathbf{X}), \tag{7.4}$$

where $h_\mu(\mathbf{X})$ is the *Shannon* (or *metric*) *entropy* of $\mathbf{X}$. Also

$$h_\mu(\mathbf{X}) = h_{\text{top}}(\mathbf{X}) \iff p(x_0,\dots,x_{L-1}) = \frac{1}{N(\mathbf{X},L)} \quad \forall L \geq 1.$$

Suppose now that the alphabet S of the source $\mathbf{X}$ is endowed with a total ordering $\leq$, so that one can also define the corresponding *permutation entropies* of order L via the ordinal patterns realized by the words of finite lengths $L \geq 2$. Then the topological *permutation* entropy of an information source is defined analogous to the topological entropy, using *rank variables*.

Thus, the *topological permutation entropy* of $\mathbf{X}$, $h_{\text{top}}^*(\mathbf{X})$, is defined as

$$h_{\text{top}}^*(\mathbf{X}) = \lim_{L \to \infty} h_{\text{top}}^*(X_0^{L-1}), \tag{7.5}$$

provided the limit exists, with

$$h_{\text{top}}^*(X_0^{L-1}) \equiv h_{\text{top}}(R_0^{L-1}) = \frac{1}{L} \log N(\mathbf{R}, L). \tag{7.6}$$

Analogous to (7.1), $N(\mathbf{R}, L)$ stands for the number of allowed words of length L of the process $\mathbf{R} = \{R_n\}_{n \in \mathbb{N}_0}$ (see Sect. 6.1). Note that

$$N(\mathbf{R}, L) \leq N(\mathbf{X}, L), \tag{7.7}$$

since several finite symbol sequences may produce the same sequence of rank variables (i.e., $x_0^{L-1} \mapsto r_0^{L-1} = \text{rank}\,(x_0^{L-1})$ is many-to-one).

As in (7.4), the metric permutation entropy,

$$h_\mu^*(\mathbf{X}) = - \lim_{L\to\infty} \frac{1}{L} \sum p(r_0,\dots,r_{L-1}) \log p(r_0,\dots,r_{L-1}),$$

is upper bounded by the topological permutation entropy,

$$h_\mu^*(\mathbf{X}) \le h_{top}^*(\mathbf{X}) \tag{7.8}$$

and, moreover,

$$h_\mu^*(\mathbf{X}) = h_{top}^*(\mathbf{X}) \iff p(r_0,\dots,r_{L-1}) = \frac{1}{N(\mathbf{R},L)} \quad \forall L \ge 2.$$

From these definitions and (7.7), it follows that

$$h_{top}^*(\mathbf{X}) \le h_{top}(\mathbf{X}). \tag{7.9}$$

Therefore, the topological permutation entropy is always a lower bound of the topological entropy for information sources.

Remark 2 The topological permutation entropy of sources can also be introduced using ordinal patterns instead of rank variables:

$$h_{top}^*(X_0^{L-1}) = \frac{1}{L} \log N^*(\mathbf{X},L), \tag{7.10}$$

where $N^*(\mathbf{X},L)$ is the number of admissible ordinal L-patterns in the messages produced by $\mathbf{X}$.

7.2 Constrained Sequences

Let $N(\mathbf{X},L)$ be as before the number of allowed sequences of length L of a source $\mathbf{X}$ with finite alphabet. If all possible sequences of length L are allowed, i.e., $N(\mathbf{X},L) = |S|^L$, then

$$h_{top}(\mathbf{X}) = \lim_{L\to\infty} \frac{1}{L} \log |S|^L = \log |S|.$$

To calculate $h_{top}^*(\mathbf{X})$ for an unconstrained source $\mathbf{X}$, we assume for simplicity a binary alphabet. Remember from Example 11, that, given the length-L word x_0^{L-1}, $L \ge 1$, then

$$\begin{aligned} x_L = 0 &\implies r_L = N_0 + 1, \\ x_L = 1 &\implies r_L = L + 1, \end{aligned}$$

where N_0 is the number of 0's in x_0^{L-1} (remember also that $1 \leq r_L \leq L+1$). How many distinct ranks of length $L+1$, r_0^L, can produce a word x_0^L?

The case $r_L = 1$ is only possible if $x_0 = x_1 = \cdots = x_{L-1} = 1$ (i.e., $N_0 = 0$) and $x_L = 0$.

The case $r_L = 2$ requires $N_0 = 1$ and $x_L = 0$. If $x_i = 0$, $0 \leq i \leq L-1$, (otherwise 1), then

$$r_0^L = 1, 2, \ldots, i, 1, i+2, \ldots, L, 2.$$

This case contributes $L = \binom{L}{1}$ distinct rank blocks of length $L+1$.

The case $r_L = 3$ requires $N_0 = 2$ and $x_L = 0$. If $x_i = x_j = 0$, $0 \leq i < j \leq L-1$, (otherwise 1), then

$$r_0^L = 1, 2, \ldots, i, 1, i+2, \ldots, j, 2, j+2, \ldots, L, 3.$$

This case contributes $\binom{L}{2}$ distinct rank blocks of length $L+1$.

Proceeding further in this way, we conclude that the case $r_L = k$, $1 \leq k \leq L$ contributes $\binom{L}{k-1}$ distinct rank blocks of length $L+1$.

Finally, the case $r_L = L+1$ requires $N_0 = L$ and $x_L = 0$, or $0 \leq N_0 \leq L$ and $x_L = 1$. There are $1 + 2^L$ such cases. Therefore, for $L \geq 1$,

$$N(\mathbf{R}, L+1) = 1 + \binom{L}{1} + \cdots + \binom{L}{L-1} + 1 + 2^L = 2^{L+1}$$

and

$$h_{\text{top}}^*(\mathbf{X}) = \lim_{L \to \infty} \frac{1}{L+1} \log N(\mathbf{R}, L+1) = \lim_{L \to \infty} \frac{1}{L+1} \log 2^{L+1}$$
$$= \log |S|.$$

In general, the information source $\mathbf{X}$ has forbidden words. In this case, one speaks also of *constrained sequences* or *constrained sources* [186]. Constrained sequences are very important in information theory, where the constrains are imposed by technological feasibility or convenience. For example, to ensure proper synchronization in magnetic recording, it is often necessary to limit the length of runs of 0's between two 1's when reading and recording bits. Also to reduce intersymbol interference, it may be required at least one 0 between any two 1's [59].

Alternatively, a constrained source can be defined as the set of sequences generated by walks on a labeled, oriented graph G. Formally, an oriented graph G is an ordered pair of sets, $G = (V, E)$, where E is a subset of *ordered* pairs of V. The elements of V are called vertices, and will be denoted as i, j, etc.; the elements $(i, j) \in E$ are called (oriented or directed) edges, with initial vertex i and terminal vertex j, and will denoted by e_{ij}. Without restriction we take $V = \{1, 2, \ldots, |V|\}$. The vertices i of the graph represent "states" and the directed edges e_{ij} show the state transitions allowed to the system. The system outputs the letter attached to each oriented edge when performing the corresponding transition. Depending on how the transition probabilities p_{ij} are defined, we have different kinds of stochastic processes: Markovian, finite type, etc.

Example 15 [59] Suppose that in the example mentioned above, borrowed from magnetic recording, we are required to have at least one 0 and at most two 0's between any pair of 1's in a sequence. The forbidden words are 11 and any word of the form 10...01 containing more than two 0's. Show that the set of constrained sequences is the same as the set of allowed paths on the state diagram in Fig. 7.1.

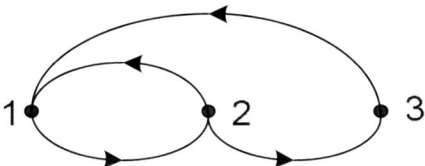

Fig. 7.1 Allowed paths between the nodes 1, 2, and 3

Given an oriented graph G, the *connection matrix* of G is a $|V| \times |V|$ matrix A_G whose entries $(A_G)_{i,j}$, $1 \leq i, j \leq |V|$, are defined as follows:

$$(A_G)_{i,j} = \begin{cases} 1 & \text{if } (j, i) \in E, \\ 0 & \text{otherwise.} \end{cases}$$

A *path* P of length l is a graph of the form

$$V(P) = \{i_0, i_1, \ldots, i_l\}, \quad E(P) = \{e_{i_0 i_1}, e_{i_1 i_2}, \ldots, e_{i_{l-1} i_l}\}.$$

An oriented graph is irreducible if, given any two vertices, there exists a path from the first vertex to the second. If $N_i(L)$ is the number of valid paths of lengths L ending at node (or state) i and $\mathbf{N}(L)$ is the column vector

$$\mathbf{N}(L) = (N_1(L), N_2(L), \ldots, N_{|V|}(L))^\top,$$

where the upper index $\top$ stands for "transposed," then

$$\mathbf{N}(L) = A_G \mathbf{N}(L - 1),$$

and by induction,

$$\mathbf{N}(L) = A_G^{L-1} \mathbf{N}(1),$$

where the entries in A_G^L correspond to paths in G of length L.

By the *Perron–Frobenius theorem* [202] for non-negative matrices, there is an eigenvalue $\lambda \geq 0$ such that no other eigenvalue of A_G has absolute value greater than λ. Corresponding to λ there is a non-negative left (row) eigenvector $\mathbf{u} = (u_1, \ldots, u_{|V|})$ and a non-negative right (column) eigenvector $\mathbf{v} = (u_1, \ldots, u_{|V|})^\top$. Moreover, if A_G is *irreducible* (i.e., for any pair i, j there is some $n > 0$ such that $(A_G^n)_{i,j} > 0$), then $\lambda > 0$ (in fact, $\min_i \sum_{j=1}^{|V|} (A_G)_{i,j} \leq \lambda \leq \max_i \sum_{j=1}^{|V|} (A_G)_{i,j}$), λ is a simple eigenvalue, and the corresponding eigenvectors are strictly positive (i.e., $u_i > 0$, $v_i > 0$ for all i).

The connection matrix A_G is *irreducible* and *aperiodic* if there exists $n \geq 1$ such that $(A_G^n)_{i,j} > 0$ for all i, j. In this case [202],

$$\lim_{n \to \infty} \frac{1}{\lambda^n} (A_G^n)_{i,j} = u_j v_i = (\mathbf{v} \otimes \mathbf{u})_{i,j},$$

where $\mathbf{v} \otimes \mathbf{u}$ denotes the tensor product of the vectors $\mathbf{v}$ and $\mathbf{u}$. This means that the matrices A_G^n and $\lambda^n (\mathbf{v} \otimes \mathbf{u})$ have the same limit when $n \to \infty$.

Lastly,

$$\lim_{L \to \infty} \frac{1}{L} \log N_i(L)$$

$$= \lim_{L \to \infty} \frac{1}{L} \log \sum_{j=1}^{|V|} (A_G^{L-1})_{i,j} N_j(1)$$

$$= \lim_{L \to \infty} \frac{1}{L} \log \lambda^{L-1} \sum_{j=1}^{|V|} (\mathbf{v} \otimes \mathbf{u})_{i,j} N_j(1)$$

$$= \lim_{L \to \infty} \frac{1}{L} \log \lambda^{L-1} + \lim_{L \to \infty} \frac{1}{L} \log \sum_{j=1}^{|V|} (\mathbf{v} \otimes \mathbf{u})_{i,j} N_j(1)$$

$$= \log \lambda.$$

This shows that the number of allowed sequences of length L grows as λ^L for large L and provides sufficient conditions for the limit $h_{\text{top}}(\mathbf{X})$ to exist.

Proposition 7 [186] *If $\mathbf{X}$ is a constrained source such that the connection matrix A_G of its oriented graph is irreducible and aperiodic, then*

$$h_{top}(\mathbf{X}) = \log \rho(A_G),$$

where $\rho(A_G)$ is the spectral radius of the matrix A_G,

$$\rho(A_G) = \max\{|\lambda| : \lambda \text{ is an eigenvalue of } A_G\}.$$

7.3 Topological Permutation Entropy of Maps

Once more let D be a simple domain of $\mathbb{R}^q$ endowed with the Borel sigma-algebra $\mathcal{B}$, and let f be a map from D to itself. Furthermore, consider a quasi-box partition

$$\kappa_0 = \{K_j : 1 \le i \le |\kappa_0|\}, \quad K_j < K_{j+1},$$

of D and an increasing sequence $(\kappa_n)_{n \in \mathbb{N}}$ of refinements of κ_0 with $\|\kappa_n\| \to 0$ (see Sect. 6.2).

Analogous to the definition of the metric permutation entropy of f with respect to an f-invariant measure μ on $(D, \mathcal{B} \cap D)$ (6.16),

$$h_\mu^*(f) = \lim_{n \to \infty} h_\mu^*(\mathbf{X}^{\kappa_n}),$$

where $\mathbf{X}^{\kappa_n}$ is the symbolic dynamics of f with respect to the partition κ_n, we define now the *topological permutation entropy of f* as

$$h_{\text{top}}^*(f) = \lim_{n \to \infty} h_{\text{top}}^*(\mathbf{X}^{\kappa_n}). \tag{7.11}$$

Note that limit (7.11) exists or diverges to $+\infty$, since $h_{\text{top}}^*(\mathbf{X}^{\kappa_n})$ is non-decreasing with ever finer partitions κ_n. Moreover, as shown in the proof of Lemma 9, this limit does not depend on the particular initial partition α_0 and its successive refinements α_n as long as $(\alpha_n)_{n \in \mathbb{N}}$ converges to the partition of D into separated points, and the order relations are preserved when going from α_n to α_{n+1}. This implies the following result.

Theorem 14 *Let D_1, D_2 be two simple domains of $\mathbb{R}^q$, and suppose that the maps $f_i : D_i \to D_i$, $i = 1, 2$, are order isomorphic by means of a homeomorphism $\phi : D_1 \to D_2$. If the topological permutation entropy exists for one of the maps, then it also exists for the other map, and in this case*

$$h_{top}^*(f_1) = h_{top}^*(f_2).$$

Proof Let κ be a quasi-box partition of D_1. Then $\phi(\kappa)$ is a partition of D_2 which, furthermore, generates an increasing sequence of partitions preserving the order relations and converging to the partition of D_2 into separate points as $\|\kappa\| \to 0$.

Let $\mathbf{X}^\kappa$ be the symbolic dynamics of f_1 with respect to the partition $\kappa = \{K_j : 1 \le j \le |\kappa|\}$ and $\mathbf{Y}^{\phi(\kappa)}$ be the symbolic dynamics of f_2 with respect to the partition $\phi(\kappa) = \{\phi(K_j) : 1 \le j \le |\kappa|\}$. Then

$$
\begin{aligned}
X_n^\kappa(x) = j &\Leftrightarrow f_1^n(x) \in K_j \\
&\Leftrightarrow \phi^{-1} \circ f_2^n \circ \phi(x) \in K_j \\
&\Leftrightarrow f_2^n \circ \phi(x) \in \phi(K_j) \\
&\Leftrightarrow Y^{\phi(\kappa)}(\phi(x)) = j.
\end{aligned}
$$

It follows that $\mathbf{X}^\kappa$ and $\mathbf{Y}^{\phi(\kappa)}$ have the same admissible ordinal patterns of any length, hence

$$h^*_{top}(f_1) = \lim_{\|\kappa\| \to 0} h^*_{top}(\mathbf{X}^\kappa) = \lim_{\|\phi(\kappa)\| \to 0} h^*_{top}(\mathbf{Y}^{\phi(\kappa)}) = h^*_{top}(f_2). \square$$

Note for further reference that (7.8) implies

$$h^*_\mu(f) \le h^*_{top}(f). \tag{7.12}$$

Therefore, the topological permutation entropy is always an upper bound of the topological entropy for maps, as it happens with the conventional metric and topological entropies.

Since the (conventional) topological entropy is usually defined for continuous maps (see Sect. B.3.1), we shall assume continuity in the following propositions. In dimension 1, continuity may be replaced by piecewise monotonicity.

Lemma 10 *Let $f : D \to D$ be a continuous map. Then*

$$h_{top}(f) \le h^*_{top}(f). \tag{7.13}$$

Proof From Theorem 10, $h_\mu(f) = h^*_\mu(f)$ holds for all $\mu \in E(D, f)$, the set of f-invariant, ergodic measures on $(D, \mathcal{B} \cap D)$. Thus, in virtue of the variational principle (B.27),

$$h_{top}(f) = \sup_{\mu \in E(D,f)} h^*_\mu(f) \le h^*_{top}(f), \tag{7.14}$$

where the last inequality follows from (7.12). $\square$

Observe from (7.14) that if a variational principle like (B.27) would also hold for the metric and topological permutation entropies, that is,

$$\sup_{\mu \in E(D,f)} h^*_\mu(f) = h^*_{top}(f), \tag{7.15}$$

then $h_{top}(f) = h^*_{top}(f)$ would follow.

Proposition 8 *Let $f : D \to D$ be a continuous map. Then the variational principle (7.15) holds if and only if $h_{top}(f) = h^*_{top}(f)$.*

Another equivalent condition for the variational principle (7.15) to hold follows from the inequality (7.9) applied to the sources $\mathbf{X}^{\kappa_n}$ in (7.11):

$$h^*_{top}(\mathbf{X}^{\kappa_n}) \le h_{top}(\mathbf{X}^{\kappa_n}).$$

Letting $n \to \infty$, we conclude

$$h_{top}^*(f) \leq \lim_{n \to \infty} h_{top}(\mathbf{X}^{\kappa_n}), \qquad (7.16)$$

provided $\lim_{n \to \infty} h_{top}^*(\mathbf{X}^{\kappa_n})$ converges.

Proposition 9 *Let $f : D \to D$ be a continuous map. Then the variational principle (7.15) holds if and only if $\lim_{n \to \infty} h_{top}(\mathbf{X}^{\kappa_n}) = h_{top}(f)$.*

Proof If $\lim_{n \to \infty} h_{top}(\mathbf{X}^{\kappa_n}) = h_{top}(f)$, then (7.16) implies $h_{top}^*(f) \leq h_{top}(f)$. On the other hand, $h_{top}(f) \leq h_{top}^*(f)$ holds true in general (Lemma 10). Apply now Proposition 8.   □

7.4 Relation Between Topological Entropy and Topological Permutation Entropy

One of the main interests of $h_{top}^*(f)$ is that, under some assumptions on f, it coincides with $h_{top}(f)$, the topological entropy of f, thus eventually providing an estimator of it.

Lemma 11 *Let $D \subset \mathbb{R}^q$, $q \geq 2$, be a simple domain and $f : D \to D$ a positively expansive map. Then*

$$\lim_{n \to \infty} h_{top}(\mathbf{X}^{\kappa_n}) = h_{top}(f), \qquad (7.17)$$

where $(\kappa_n)_{n \in \mathbb{N}}$ is an increasing sequence of quasi-box partitions of D and $\mathbf{X}^{\kappa_n}$ is the symbolic dynamics of f with respect to κ_n.

Intuitively speaking, a self-map is positively expansive if every pair of sufficiently close points eventually separate by a finite distance under iteration of the map. Expansive and positively expansive maps are defined in Sect. B.3.1, Definition 26. Typical examples of positively expansive maps are the one- and two-sided shifts. The condition $q \geq 2$ recalls that one-dimensional closed intervals do not admit expansive maps. To establish a connection between $h_{top}(\mathbf{X}^{\kappa_n})$ and $h_{top}(f)$, we will use (n, ε)-separated sets (Definition 23).

Proof For definiteness we will take the metric d in $\mathbb{R}^q$ to be the Euclidean distance (any other equivalent distance would do as well). Let $A \subset D$ be (n, ε)-separated with respect to f, i.e., $x, y \in A$, $x \neq y$, implies $d_n(x, y) > \varepsilon$, where

$$d_n(x, y) = \max_{0 \leq i \leq n-1} d(f^i(x), f^i(y)).$$

Lay on D a quasi-box partition $\kappa = \{K_j : 1 \leq j \leq |\kappa|\}$ such that

$$\|\kappa\| < \varepsilon,$$

so as points lying at a distance greater than ε belong necessarily to different bins of κ. Then,

$$d_n(x, y) > \varepsilon \quad \Leftrightarrow \quad d(f^i(x), f^i(y)) > \varepsilon \quad \text{for some } 0 \leq i \leq n-1$$
$$\Rightarrow \quad (X^\kappa)_0^{n-1}(x) \neq (X^\kappa)_0^{n-1}(y).$$

Thus, every point $x \in A \cap K_{j_0}$, $1 \leq j_0 \leq |\kappa|$, generates a different sequence $(X^\kappa)_0^{n-1}(x) = j_0, \ldots$ of length n. Of course, there can be points $x' \in K_{j_0}$, $x' \notin A$, such that $(X^\kappa)_0^{n-1}(x') = j_0, \ldots \neq (X^\kappa)_0^{n-1}(x)$ for all $x \in A \cap K_{j_0}$, but the number of such points will vanish when $n \to \infty$ if $\varepsilon \leq \delta$, δ being an expansiveness constant for f (see Definition 26). In this limit (and $\varepsilon \leq \delta$) we also have $A \cap K_j \neq \emptyset$ for $\forall j$, $1 \leq j \leq |\kappa|$, hence there is a one-to-one relation between points in A and outputs $(x^\kappa)_0^\infty$ of X^κ. If, as in Definition 23, $s_n(\varepsilon, D)$ denotes the largest cardinality of any (n, ε)-separated subset of D with respect to f and $N(X^\kappa, n)$ denotes the number of distinct symbolic sequences of length n, it follows that

$$\limsup_{n \to \infty} \frac{1}{n} \log N(X^\kappa, n) = \limsup_{n \to \infty} \frac{1}{n} \log s_n(\varepsilon, D),$$

for $\varepsilon \leq \delta$, and thus (see (7.11) and (B.25))

$$\lim_{\|\kappa\| \to 0} h_{\text{top}}(X^\kappa) = \lim_{\|\kappa\| \to 0} \limsup_{n \to \infty} \frac{1}{n} \log N(X^\kappa, n)$$
$$= \lim_{\varepsilon \to 0} \limsup_{n \to \infty} \log s_n(\varepsilon, I)$$
$$= h_{\text{top}}(f). \qquad \Box$$

Theorem 15 *Let D be a q-dimensional simple domain, $q \geq 2$, and $f : D \to D$ a positively expansive map. Then*

$$h_{\text{top}}^*(f) = h_{\text{top}}(f) \qquad (7.18)$$

and

$$\sup_{\mu \in E(D, f)} h_\mu^*(f) = h_{\text{top}}^*(f).$$

Proof Apply Lemma 11 and Propositions 8 and 9. $\Box$

From the proof of Lemma 11, it should be clear where the need for expansiveness comes from: it can otherwise happen that points x of the (n, ε)-separated subset $A \subset D$ have neighboring points x_ε' that shadow their trajectories at arbitrarily close distance (hence $x_\varepsilon' \notin A$) but define symbolic sequences $X^\kappa(x_\varepsilon') \neq X^\kappa(x)$. This will be certainly the case when, for instance, x belongs to the stable manifold of a hyperbolic fixed point $p \in D$ or, more generally, whenever the state space have lower dimensional manifolds whose points are not sensitive to initial conditions. The good news for the practitioner is that, since such local manifolds have Lebesgue measure zero, at least for sufficiently smooth dynamics, equality (7.18) will hold in numerical calculations for smooth maps with sensitivity to initial conditions almost everywhere (with respect to the Lebesgue measure). The bad news is that expansive maps are difficult to approximate numerically: small errors in computations (like those due to round-off) get magnified upon iteration.

From Theorems 14 and 27 (Sect. B.3) it follows:

Corollary 6 *Let D_1, D_2 be simple domains of $\mathbb{R}^q$, and $f_i : D_i \rightarrow D_i$, $i = 1, 2$, positively expansive maps. Suppose that $\phi : D_1 \rightarrow D_2$ is a homeomorphism such that $\phi \circ f_1 = f_2 \circ \phi$. Then*

$$h^*_{top}(f_1) = h^*_{top}(f_2).$$

Thus topological conjugacy is a sufficient condition for two positively expansive self-maps of simple domains to have the same topological permutation entropy.

Let us remark at this point that the original definition of the topological permutation entropy of a self-map f of a closed one-dimensional interval I, given by Bandt, Keller, and Pompe in [29], is

$$h^{*BKP}_{top}(f) = \lim_{n \to \infty} h^{*BKP}_{top}(f, L), \tag{7.19}$$

where

$$h^{*BKP}_{top}(f, L) = \frac{1}{L-1} \log |\mathcal{P}_L| \tag{7.20}$$

is the topological permutation entropy of f of order L, and remember from (3.4) and (3.5),

$$|\mathcal{P}_L| = |\{P_\pi \neq \emptyset : \pi \in \mathcal{S}_L\}| \tag{7.21}$$

gives the number of ordinal patterns realized by the orbits of length L, $(f^n(x))_{n=0}^{L-1}$ with $x \in I$. The following result holds.

Theorem 16 [29] *If I is a closed one-dimensional interval and $f : I \rightarrow I$ is piecewise monotone, then $h^{*BKP}_{top}(f) = h_{top}(f)$, where $h_{top}(f)$ is the topological entropy of f.*

On the other hand, Misiurewicz proved that this result is not true if the map is not piecewise monotone [157]. His counterexample is a continuous map with infinite monotonicity segments that has zero topological entropy but positive topological permutation entropy. He also shows in [157] that for piecewise monotone interval maps, the topological entropy can be computed by counting the permutations exhibited by the periodic orbits.

Example 16 For the symmetric tent map Λ, the partitions $\mathcal{P}_2$, $\mathcal{P}_3$, and $\mathcal{P}_4$ have cardinalities 2, 5, and 12 (Example 13), respectively. Hence, the topological permutation entropies of orders 2, 3, and 4 are the following:

$$h_{top}^{*BKP}(\Lambda, 2) = \log |\mathcal{P}_2| = \log 2 = 1 \; bit/symbol,$$

$$h_{top}^{*BKP}(\Lambda, 3) = \frac{1}{2} \log |\mathcal{P}_3| = \frac{1}{2} \log 5 = 1.1610 \; bit/symbol,$$

$$h_{top}^{*BKP}(\Lambda, 4) = \frac{1}{3} \log |\mathcal{P}_4| = \frac{1}{3} \log 12 = 1.1950 \; bit/symbol.$$

By Theorem 16,

$$h_{top}^{*BKP}(\Lambda) = h_{top}(\Lambda) = \log 2 = 1 \; bit/symbol.$$

To conclude, it was pointed out in Sect. 3.4.1 that order-isomorphic maps have the same admissible and forbidden ordinal patterns of any length. This fact together with Theorem 16 lead to the following results.

Corollary 7 *Let I_1, I_2 be two closed intervals of $\mathbb{R}$, and suppose that the maps $f_i : I_i \to I_i$, $i = 1, 2$, are order isomorphic. Then,*

(1) $h_{top}^{*BKP}(f_1) = h_{top}^{*BKP}(f_2)$, provided one of them exists.

(2) Furthermore, if f_1 and f_2 are piecewise monotone, then $h_{top}(f_1) = h_{top}(f_2)$.

7.5 Estimating Topological Entropy

Estimation of topological entropies from naive numerical simulation of long orbits is notoriously difficult. Metric entropy by itself can be quite tricky and difficult, requiring very long data sets for increasing L, but topological entropy is worse yet, because it weighs each pattern equally. This means that patterns which are exceptionally infrequent on the natural measure of the attractor can still have a significant influence on the result. Attempting to estimate the same quantities using empirical occurrences of ordinal patterns is even more difficult, requiring more data than would a good, low-alphabet generating partition for ordinary symbolic dynamics.

For the present purpose, we consider a continuous system in greater than one dimension, with a chaotic attractor, and whose topological entropy can be found by independent rigorous means. The *Lozi map*,

$$x_{i+1} = y_i,$$
$$y_{i+1} = 1 + bx_i - a|y_i|,$$

with parameters $a, b \in \mathbb{R}$, $b \neq 0$, satisfies all these criteria. A mathematical proof for the existence of an attractor for the Lozi map was given by Misiurewicz [156]. In particular, $a = 6/5, b = -2/15$ yield a low-entropy chaotic attractor (roughly 0.3 bits/iteration) and for those parameters, the topological entropy has been bounded rigorously with computer-assisted analytical computations [102, 178], and we use their results.

We found that the best numerical procedure was to look at the "outgrowth ratio" of ordinal patterns of a given length L. The outgrowth ratio for some pattern of length L is the cardinality of the set of distinct ordinal patterns of length $L + 1$ which have the given length-L pattern as a prefix. More concretely, we find vectors of length $L + 1$ from an orbit of the map. The ordinal pattern on the first L points is the prefix pattern. Regardless of the dynamics, there can be at most $L + 1$ ordinal patterns of length $L + 1$ conditioned on the length-L ordinal pattern, since the single new element belongs to the alphabet $\{1, \ldots, L + 1\}$.

Indeed, according to definitions (7.11), (7.5), and (7.6), the topological permutation entropy $h_{\text{top}}^*(f)$ is the scaling rate of the logarithm of the number of patterns with L of the "coarse-grained" dynamics $\mathbf{X} \equiv \mathbf{X}^\kappa$ for κ sufficiently fine, i.e.,

$$\log N(\mathbf{R}, L) \approx L h_{\text{top}}^*(\mathbf{X}),$$

(R_0^{L-1} are the rank variables defined by X_0^{L-1}), so

$$\log \frac{N(\mathbf{R}, L + 1)}{N(\mathbf{R}, L)} \approx h_{\text{top}}^*(\mathbf{X}).$$

Therefore, a reasonable estimator for $h_{\text{top}}^*(f)$ is the logarithm of the outgrowth ratio averaged uniformly over all extant prefix patterns. This value, for sufficiently large L and sufficiently large simulation sets, ought to be $h_{\text{top}}^*(f)$ on average. Note that independent white noise would give an estimate of $\log(L + 1)$, i.e., not converging with L.

Figure 7.2 shows the numerical result of estimating $h_{\text{top}}^*(f)$ on long orbits of the Lozi map with $a = 6/5, b = -2/15$, using two specific instantiations of the outgrowth method. The dotted lines are the bounds on the true topological entropy.

The first strategy involves computing $N_1 = 50 \times 10^6$ ordinal patterns of length $L + 1$ and their length L prefix. For every element in the prefix set we accumulate the number of distinct elements in the conditioning set and average the logarithm of the number of distinct occurrences over the observed length-L ordinal patterns—as long as each of those ordinal patterns had at least two successors. This method will typically have a bias downward for large L on account of undersampling the space.

The second strategy starts by computing $N_2 = 10^6$ ordinal patterns of length $L+1$ from orbits of the map. The set of distinct order-L prefixes forms the "conditioning" set. The N_2 length $L + 1$ ordinal patterns from these are accumulated, and then the map is iterated and ordinal patterns computed, until there have been $(K - 1)N_2$ more observations of length-L ordinal patterns which were in the prefix, so that there are $KN_2 = N_1$ with $K = 50$ observations, all of whose order L prefixes are in the conditioning set. Then similarly the logarithm of the outgrowth ratio is estimated over the conditioning set for all conditioning patterns with at least two observations. This method has positive and negative biases due to finiteness of observations. First, because of finite K there is a downward bias, as the number of observed outgrowths is a strict lower bound on the number of allowed outgrowths in the dynamical

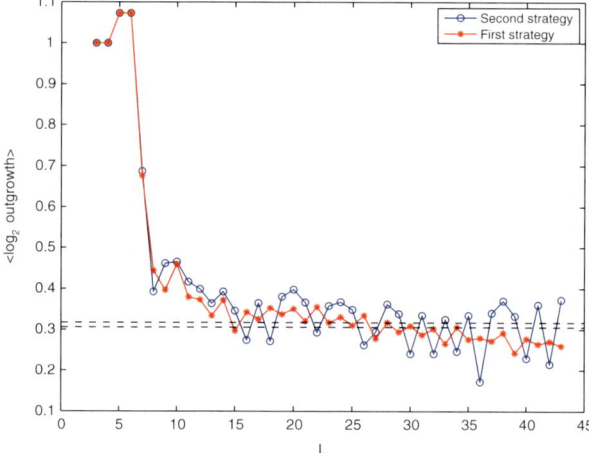

Fig. 7.2 Logarithmic outgrowth ratios for the Lozi map vs L. The *dotted lines* represent rigorous bounds on the topological entropy rate, computed by computer-assisted analytical methods. The outgrowth ratio approximates the topological permutation entropy rate, is practical for computing, and can scale to significant L

system. There is a more subtle upward bias, which changes with L as well. It is because the ordinal patterns which were selected as conditioning states came from an ergodic sample on the natural measure which does not sample the support uniformly. More frequently occurring patterns are more likely to occur in the conditioning set—and we have observed heuristically that in chaotic systems the outgrowth ratio tends to be roughly correlated in the same direction as the frequency of the conditioning pattern. The measure on the allowable patterns does vary very widely hence it can take very long simulations to find more of the allowable conditioning patterns even though their total number is far smaller than the number of samples from the map. This effect is also present in the first method as well, but appears to be dominated by the downward bias.

7.6 Existence of Forbidden Ordinal Patterns

We turn to the study of forbidden patterns for self-maps of q-dimensional simple domains and, more specifically, to the issue of finding sufficient conditions that guarantee forbidden patterns of any length. The existence of one-dimensional interval maps with no forbidden patterns (Fig. 4.2) shows that this question is pertinent.

Let D be a q-dimensional simple domain and $f : D \to D$ a map with $h^*_{\text{top}}(f) < \infty$. According to the definition of $h^*_{\text{top}}(f)$, (7.11), given $\varepsilon > 0$ arbitrarily small there exists a quasi-box partition κ_0 of D such that

$$\left| h^*_{\text{top}}(f) - h^*_{\text{top}}(\mathbf{X}^\kappa) \right| < \frac{\varepsilon}{2}$$

whenever the quasi-box partition κ is a refinement of κ_0. Furthermore, according to the definition of $h^*_{\text{top}}(\mathbf{X}^\kappa)$ ((7.5) and (7.10) with $\mathbf{X}^\kappa$ instead of $\mathbf{X}$), there exists a length L_0 such that

$$\left| h^*_{\text{top}}(\mathbf{X}^\kappa) - \frac{1}{L} \log N^*(\mathbf{X}^\kappa, L) \right| < \frac{\varepsilon}{2}$$

whenever $L \geq L_0$, where $N^*(\mathbf{X}^\kappa, L)$ is the number of *admissible* ordinal L-patterns of the symbolic dynamics $\mathbf{X}^\kappa$ with respect to κ. Therefore, with κ sufficiently fine and L sufficiently large we have

$$\left| h^*_{\text{top}}(f) - \frac{1}{L} \log N^*(\mathbf{X}^\kappa, L) \right| < \varepsilon,$$

hence,

$$N^*(\mathbf{X}^\kappa, L) = e^{L h^*_{\text{top}}(f)} + \mathcal{O}_L(\varepsilon), \tag{7.22}$$

where the term $\mathcal{O}_L(\varepsilon)$ depends also on L, as indicated by the subindex, and $\mathcal{O}_L(\varepsilon) \to 0$ when $\varepsilon \to 0$ (or $\|\kappa\| \to 0$).

On the other hand, we already know that the number of *possible* ordinal L-patterns, $|\mathcal{S}_L| = L!$, grows superexponentially with L, (3.8). We conclude from (7.22) that the symbolic dynamics $\mathbf{X}^\kappa$ has forbidden patterns whenever $h^*_{\text{top}}(f)$ exists and is finite. Then, the same must happen with maps, since their dynamic can be approximated by symbolic dynamics.

Theorem 17 *Let $D \subset \mathbb{R}^q$ be a simple domain and $f : D \to D$ a map. Then*

$$\lim_{\|\kappa\| \to 0} N^*(\mathbf{X}^\kappa, L) = |\mathcal{P}_L|,$$

where we use the notation $|\mathcal{P}_L|$ as in (7.21) for the number of admissible ordinal L-patterns for f.

Proof We claim that the admissible L-patterns for f will coincide with the admissible L-patterns for the corresponding symbolic dynamics $\mathbf{X}^\kappa$ with respect to a quasi-box partition $\kappa = \{K_j\}_{1 \leq j \leq |\kappa|}$ in the limit $\|\kappa\| \to 0$. Indeed, if $x \in D$ is of type $\pi \in \mathcal{S}_L$, the only way that the length-L word $x, f(x), \ldots, f^{L-1}(x)$ does not define π when observed with the precision set by κ is that at least two letters, say $f^{i_1}(x)$ and $f^{i_2}(x)$, $0 \leq i_1 < i_2 \leq L - 1$, fall in the same bin $K_{j_0} \in \kappa$, since then we cannot discern the order relation between both letters. But this will not happen when κ is so fine that $x, f(x), \ldots, f^{L-1}(x)$ fall in different bins. We conclude that the number of such discrepancies will diminish as the partition κ gets finer, and finally vanish in the limit $\|\kappa\| \to 0$. $\square$

Theorem 17 and (7.22) imply the following result.

Corollary 8 *The number of allowed L-patterns of self-maps f of q-dimensional simple domains grows asymptotically with L as*

$$|\mathcal{P}_L| \sim e^{Lh^*_{top}(f)}, \tag{7.23}$$

provided $h^*_{top}(f)$ exists and is finite.

The same conclusion follows directly from (7.19) when $h^*_{top}(f)$ is replaced by $h^{*BKP}_{top}(f)$ in (7.23) for one-dimensional interval maps. Since calculating $h^*_{top}(f)$ requires in practice the calculation of the growth rate with L of the allowed L-patterns for f, we use Theorems 15 and 16 to provide more natural conditions for (7.23).

Corollary 9 *Let $D \subset \mathbb{R}^q$ be a simple domain and $f : D \rightarrow D$ a map with $h_{top}(f) < \infty$. (i) If $q = 1$ and f is piecewise monotone or (ii) $q \geq 2$ and f is positively expansive, then*

$$|\mathcal{P}_L| \sim e^{Lh_{top}(f)}.$$

Corollary 9 provides sufficient conditions for the existence of forbidden ordinal patterns since, as already pointed out in some previous passages, the number of possible ordinal L-patterns grows superexponentially with L: $|\mathcal{S}_L| = L!$. In more quantitative terms, forbidden patterns proliferate in these two cases as (see (3.8))

$$|\{P_\pi = \varnothing : \pi \in \mathcal{S}_L\}| \sim L! - e^{Lh_{top}(f)} = e^{L \ln L}\left(1 - e^{-L(\ln L - h_{top}(f))}\right).$$

It is an open problem to find a more general condition than expansiveness in higher dimensional dynamics for the existence of forbidden pattern. Numerical simulations support the existence of forbidden patterns also for non-expansive multidimensional maps (see next section).

Apart from the superexponential scaling law with L, it is quite difficult to make more specific statements on the forbidden patterns for a map like, for instance, the minimal length of its forbidden patterns or the lengths of its root forbidden patterns. One important exception is the shift and signed shift transformations (and all order-isomorphic maps) we studied in Chaps. 4 and 5.

Last but not the least, forbidden patterns, be in one-dimensional dynamics or in higher dimensional dynamics, have the properties discussed in Sect. 3.4.

7.7 Numerical Simulations

We demonstrate numerical evidence for the existence of forbidden ordinal patterns in multi-dimensional maps. Of course, direct simulation of dynamical systems directly yields only *allowed* ordinal patterns. The failure to observe any given ordinal pattern in any finite time series does not mean of course that it is forbidden (probability zero) but only that its probability is sufficiently low in the natural measure induced by the dynamics that it has not yet been seen.

However, with reasonable L (as effort and memory increases radically with L) and robust computational ability we can infer in many cases, the existence of forbidden patterns by examining the convergence of allowed patterns with N, the number of data emitted by the source. In particular, we suggest examining the logarithmic ratio of the cardinality of all L-patterns to the number of observed L-patterns $\log(L!/P_{obs})$ vs $\log N$. If a system has a "core" of forbidden patterns, as with deterministic systems, then we expect that this ratio will decline with N and eventually level off with increasing N, assuming the asymptotic behavior can be observed. Here, P_{obs} is the naive, biased-downward, estimator of the unknown $P_{allowed}$, the number of allowed L-patterns.

When N is much larger than $P_{allowed}$, P_{obs} is likely to be a good estimator, assuming most patterns have a reasonable probability of occurring. With increasing L, however, this is difficult to achieve practically because of memory limitations, as the identities and counts of each observed patterns (a subset of the allowed patterns) must be retained. The number of allowed patterns increases exponentially with L in deterministic chaos and faster than exponentially with noise, and therefore one must increase N, the number of iterates, substantially to permit a commensurately large number of distinct patterns to be actually observed.

This motivates using a superior statistical estimator of $P_{allowed}$. This equivalent problem has a significant history, motivated especially from the ecology community. Consider a situation where one can observe a finite sample of individual organisms, from a presumably large population. What is the estimated number of distinct species, the biodiversity, and how can we estimate this given the individual counts of observed species? (For reviews of approaches to this problem, see [41, 100].) This is analogous to our situation where we can distinguish individual ordinal patterns but each observation is drawn from the natural distribution induced by typical orbits of the dynamical system. For our needs we wish to go reasonably deep into the undersampled regime and impose few probabilistic priors. We adopt the non-parametric estimator of Chao [49], motivated by comments in the reviews and our experience, as a simple but reasonably effective improvement:

$$P_{Chao} = P_{obs} + \frac{c_1^2}{2c_2^2}, \tag{7.24}$$

where c_k are the "meta-counts" of observations, i.e., c_1 is the number of distinct ordinal patterns which were observed exactly once in the sample, c_2 the number which were observed exactly twice, etc. In practice this is accomplished by counting frequencies of observed patterns through a hash table and in a second phase, counting the frequencies of such frequencies with a similar hash table. Note that if the sample size is particularly small (relative to what is necessary to see a substantial fraction of allowed patterns), P_{Chao} will still be an underestimate. Consider that its maximum value is obtained with $c_1 = N - 1$ and $c_2 = 1$, i.e., one doubleton and all remaining observations being unique (all unique naturally leads to an undefined estimate), and so P_{Chao} is bounded by $(P_{\text{obs}}^2 + 1)/2$. Bunge and Fitzpatrick [41] call P_{Chao} to be an "estimated lower bound." We believe that no statistical estimator can perform well in the extremely undersampled regime and there is no substitute for substantial computational effort when L becomes sufficiently large; however, we will see an improvement over the naive estimator.

Our first numerical example is Arnold's *cat map*:

$$x_{i+1} = x_i + y_i \qquad \mod 1,$$
$$y_{i+1} = x_i + 2y_i \qquad \mod 1.$$

As a hyperbolic toral automorphism, this is an expansive transformation [115]. We start with initial conditions drawn uniformly in $[0, 1) \times [0, 1)$ and iterate. Ordinal patterns are computed using order relations on the x-coordinate only; since coincidences in the x-coordinate are unlikely, this amounts in practice to using lexicographic order in $[0, 1) \times [0, 1)$. Figure 7.3 shows the strong numerical evidence for forbidden patterns characteristic of deterministic systems. As a demonstration of the genericity of the results, Fig. 7.4 shows the equivalent except that the observable upon which ordinal patterns were computed is $3x^3 - y$. Results are nearly identical, as one expects.

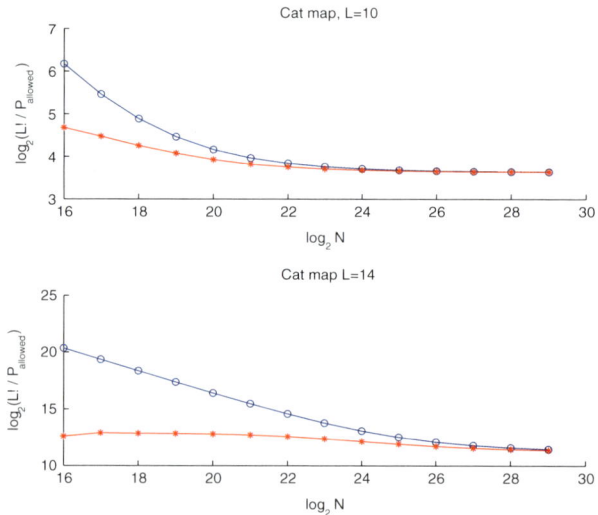

Fig. 7.3 Convergence of estimated forbidden patterns with N, cat map. *Circles* (o) are for P_{allowed} estimated by P_{obs}, *asterisks* (*) have P_{allowed} estimated by P_{Chao}. *Top*, $L = 10$, *bottom* $L = 14$. Both figures show clear evidence of convergence to a constant, evidence of true forbidden patterns as $N \to \infty$. In the *lower figure* especially, the improved estimator P_{Chao} "senses" the approach to a convergence earlier than the naive counting estimator. Note the differing scales on the y-axes

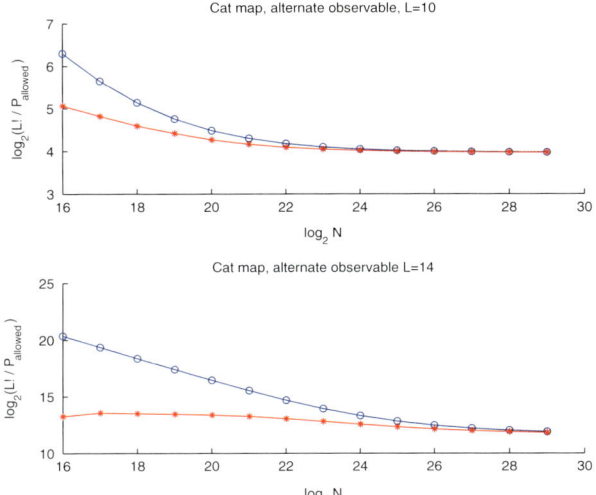

Fig. 7.4 Convergence of estimated forbidden patterns with N, cat map, alternative observable. *Circles* (o) are for $P_{allowed} = P_{obs}$, *asterisks* (*) have $P_{allowed} = P_{Chao}$. *Top*, $L = 10$, *bottom* $L = 14$. Both figures show clear evidence of convergence to a constant, evidence of true forbidden patterns as $N \to \infty$. In the *lower figure* especially, the improved estimator P_{Chao} "senses" the approach to a convergence earlier than the naive counting estimator. Note the differing scales on the y-axes

By comparison, consider Fig. 7.5, generated by an i.i.d. noise source (ordinal patterns are insensitive to changes in distribution). Here, the observed patterns imply convergence to zero forbidden patterns with increasing N. More remarkably, the estimator P_{Chao} senses this long before and predicts zero forbidden patterns with orders of magnitude lower than N, apparently because the assumptions made by the estimator of equiprobable patterns for both observed and unobserved are exactly fulfilled.

As an example of a non-expansive map, we turn to a chaotic system, the *Hénon map*,

$$x_{i+1} = 1 - ax_i^2 + by_i,$$
$$y_{i+1} = x_i,$$

with $a = 1.4$, $b = 0.3$, observable being the x-coordinate. This map is not uniformly hyperbolic (it has two fixed points, one attractive and one repellent), more characteristic of real dynamics seen in nature. (The Hénon map is non-expansive for "almost all" values of the parameter a [154].) In Fig. 7.6, we see convergence to a finite core of forbidden patterns with larger N. Note that the performance of P_{Chao} is still improved over the naive estimator but it is not as good as with noise, because with real dynamics there is a wide variation in the probability of the various allowed patterns, and so larger N feels the "tail" of the distribution of rare patterns. By comparison consider Fig. 7.7, which shows results from the same dynamics but

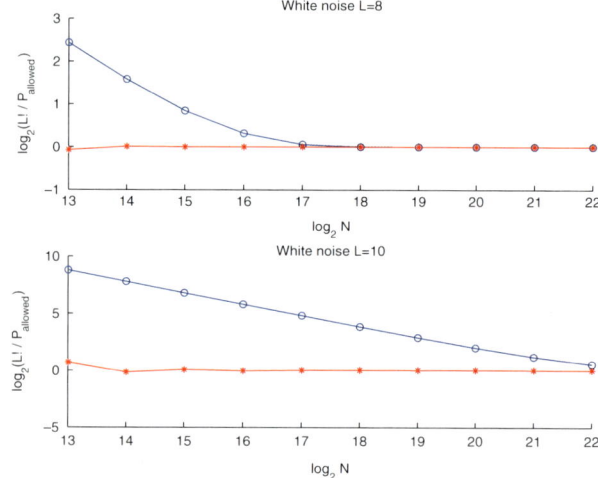

Fig. 7.5 Convergence of estimated forbidden patterns with N, i.i.d. noise. *Circles* (o) are for $P_{\text{allowed}} = P_{\text{obs}}$, *asterisks* (*) have $P_{\text{allowed}} = P_{\text{Chao}}$. *Top*, $L = 8$, *bottom* $L = 10$. P_{obs} shows convergence to zero forbidden patterns; P_{Chao} estimates zero forbidden patterns well before convergence of naive estimator

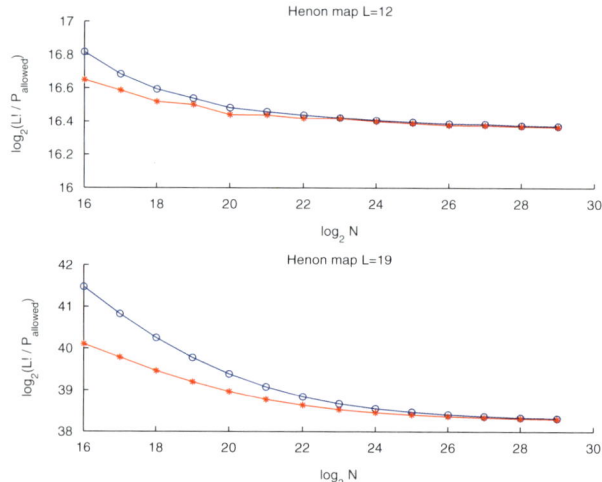

Fig. 7.6 Convergence of estimated forbidden patterns with N, Hénon map. *Circles* (o) are for P_{allowed} estimated by P_{obs}, *asterisks* (*) have P_{allowed} estimated by P_{Chao}. *Top*, $L = 12$, *bottom* $L = 19$. Both naive and improved estimators show convergence to a finite number

each observable was contaminated with uniform i.i.d. noise $\eta \in [0, 0.2)$. This time, increasing N clearly shows increasing allowed/decreasing forbidden patterns, proportional to N as expected with noise. As a matter of fact, arbitrarily small noise will eventually lead to noise-like scaling, but the size of the word necessary to

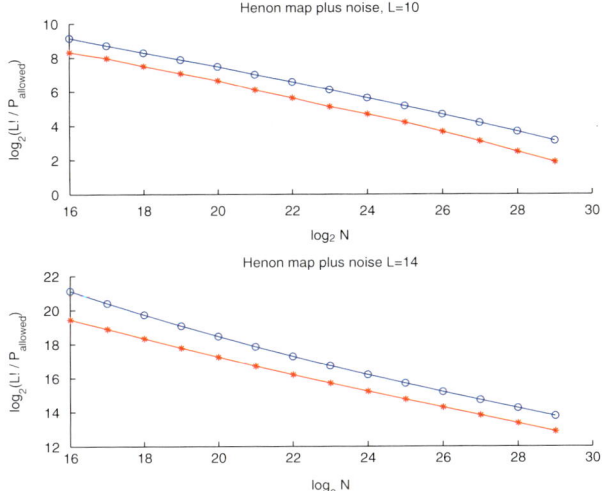

Fig. 7.7 Lack of convergence of estimated forbidden patterns with N, Hénon map with additive i.i.d. noise. *Circles* (o) are for $P_{\text{allowed}} = P_{\text{obs}}$, *asterisks* (*) have $P_{\text{allowed}} = P_{\text{Chao}}$. *Top*, $L = 12$, *bottom* $L = 14$. Both naive and improved estimators show continued increase in allowed patterns (decrease in forbidden patterns) with increasing N

see this (and consequently the size of the data set necessary to see the effect) will grow astronomically. If the noise support is bounded (or, we conjecture, thin-tailed), then fairly small noise levels will not be visible in the ordinal patterns if they are substantially smaller than typical sizes of $x_i - x_j$ for $1 \leq j \leq L$, and hence will not change the patterns. The behavior with N clearly distinguishes low-dimensional dynamics from noise.

As a philosophical point it is true that the "noise" generator in a computer software is but a deterministic dynamical system on its own, but in practice it has an extremely long period and virtually no correlation, and hence if one wanted to see ordinal pattern scaling different from true noise, one would need exceptionally long L and impractically astronomical memory requirements. We use a validated high-quality random number generator [148] from the Boost C++ library.

Chapter 8
Discrete Entropy

From a mathematical point of view, entropy made its first appearance in continuous-time dynamical systems (more exactly, in Hamiltonian flows), and from there it was extended to quantum mechanics by von Neumann, to information theory by Shannon, and to discrete-time dynamical systems by Kolmogorov and Sinai. In all these cases we observe that (i) if the state space is discrete and/or finite (like in quantum mechanics and finite-alphabet information sources), then the evolution is random and (ii) if the evolution is deterministic (like in continuous- and discrete-time dynamical systems), then the state space is infinite. Still today one speaks of random dynamical systems in the first case and of deterministic dynamical systems in the second case. But not all dynamical systems of interest fall under one of the previous categories. An important example of a deterministic physical system where both state space and dynamics are discrete is a digital computer; this entails that any dynamical trajectory in computer becomes eventually periodic—a well-known effect in the theory and practice of pseudo-random number generation. Dynamical systems with discrete and even with a finite number of states have been considered by a number of authors, in particular in the development of *discrete chaos* [125]—an attempt to formalize the idea that maps on finite sets may have different diffusion and mixing properties. From this perspective, it seems desirable to export some concepts and tools from the general theory to this new setting. This is the rationale behind, e.g., the discrete Lyapunov exponent [124, 125].

The topic of this chapter is precisely the extension of entropy to maps on finite sets—a concept we call *discrete entropy*. When going from the conventional framework to considering maps F on sets S with cardinality $|S| < \infty$ (and, eventually, an atomic measure), one main difficulty arises at the very beginning: the entropy of F with respect to any partition of S vanishes, rendering null entropy. It is not clear how to modify the concept of entropy while still gauging the "randomness" generated by F on S in the limit $|S| \to \infty$. Thanks to its combinatorial nature, permutation entropy lends itself especially well to the methods of discrete mathematics, allowing in fact to define a *discrete entropy* concept (Sect. 8.1). The definition of discrete entropy can then be justified by showing that, for a large class of maps, the discrete entropy converges to the measure-theoretic entropy in the "infinite" limit (Sect. 8.2). More precisely, let $f:I^d \to I^d$ be a d-dimensional interval map, which is ergodic with respect to a measure μ, and let F_M be a permutation on M elements

J.M. Amigó, *Permutation Complexity in Dynamical Systems,*
Springer Series in Synergetics, DOI 10.1007/978-3-642-04084-9_8,
© Springer-Verlag Berlin Heidelberg 2010

obtained from f via discretization and orbit truncation (see Sect. 8.2 for details). Then $\lim_{M\to\infty} h_\delta(F_M) = h_\mu(f)$, where $h_\delta(F_M)$ is the *discrete entropy* of F_M, and $h_\mu(f)$ is the metric entropy of f with respect to μ. An alternative approach using *topological* entropy is also possible and will be discussed in Sect. 8.3.

Apart from their role as entropy-like tools of discrete chaos, metric and topological discrete entropy can be viewed also as estimators of the corresponding "continuous" counterpart, thanks to the infinite limit mentioned above. This more practical side of discrete entropy is somewhat hampered by the fact that discrete entropy requires large amounts of data to converge—albeit a property shared with most of the entropy estimators.

8.1 Discrete Entropy

Let $A = \{a_1, \ldots, a_L\}$ be a finite set endowed with a linear ordering $\leq$, $F: A \to A$ a bijection, and $\pi = \langle \pi_0, \ldots, \pi_{n-1} \rangle \in S_n$, $2 \leq n \leq L$. Define

$$Q_\pi(n) = \left\{ a \in A : F^{\pi_0}(a) < \cdots < F^{\pi_{n-1}}(a) \right\} \tag{8.1}$$

and

$$q_\pi(n) = \frac{|Q_\pi(n)|}{\sum_{\tau \in S_n} |Q_\tau(n)|} \tag{8.2}$$

if $\sum_{\tau \in S_n} |Q_\tau(n)| \neq 0$ (in which case, $\sum_{\pi \in S_n} q_\pi(n) = 1$) and $q_\pi(n) = 0$ otherwise. We shall drop the argument n of Q_π and q_π when it is clear from the context that $\pi \in S_n$. We say that $a \in A$ defines the ordinal pattern $\pi \in S_n$ if $s \in Q_\pi$.

Without restriction we take $A = \{0, \ldots, L-1\}$ with the natural order inherited from $\mathbb{N}_0$. Then we write the permutation F as in (1.22),

$$F = [F(0), F(1), \ldots, F(L-1)].$$

On the other hand, F can also be written as a product of cycles. As in Sect. 2.4, we denote by $(i_1, i_2, \ldots, i_n)$ the *cycle* $i_1 \mapsto F(i_1) = i_2 \mapsto \cdots \mapsto F(i_{n-1}) = i_n \mapsto F(i_n) = i_1$ of length n. If $F = (i_1, \ldots, i_L)$ (i.e., a cycle of maximal length), we say that F is *irreducible*, otherwise it is *reducible*.

In view of (6.17), we introduce the following concept.

Definition 4 The discrete entropy of F of order $n \geq 2$, is

$$h_\delta^{(n)}(F) = -\frac{1}{n-1} \sum_{\pi \in S_n} q_\pi \log q_\pi.$$

The subscript δ stands for "discrete" but also for "Dirac measure" on A. Observe that (i) $a \in Q_\pi(n)$ as long as $n \leq \text{Per}(a)$, the period of a and (ii) alternatively, one can take the truncated orbits $\mathrm{orb}(a) = \{a, F(a), \ldots, F^{p-1}(a)\}$, with $p = \mathrm{Per}(a)$, and normalize the count of a's defining the ordinal pattern $\pi \in \mathcal{S}_n$ for $n = 2, \ldots, p_{\max} = \max_{a \in A}\{\mathrm{Per}(a)\}$.

So to speak, $h_\delta^{(n)}(F)$ senses the mixing properties of F in the short run $1 \leq n \leq p_{\max}$—before periodicity sets in on the whole "phase space" A. This is the timescale we are interested in. This explains also why we do not allow for repetition of symbols and use strict order in (8.1) instead of ranks. In the infinite limit $L \to \infty$ (Sect. 8.2) it makes no difference, but for finite L we want to switch off periodicities.

Let us tackle this discretization of entropy by considering first some special cases.

Case 1. If $F = (i_1, i_2, \ldots, i_L)$, then each $a \in A$ defines permutations $\pi \in \mathcal{S}_n$ for $2 \leq n \leq L$, the corresponding sets $Q_\pi(n)$ building thus partitions of A, and we can write down a whole hierarchy of entropies of orders $2, \ldots, L$. In particular, $q_\pi(L) = 1/L$ if $Q_\pi \neq \emptyset$, so

$$h_\delta^{(L)}(F) = \frac{1}{L-1} \log L.$$

As a result, $h_\delta^{(L)}(F)$ and possibly other entropies of lower orders cannot discriminate two permutations of the same maximal length L.

Example 17 For the right shift modulo $L - 1$, defined as $\theta_L(i) = i + 1$ for $i = 0, 1, \ldots, L - 2$ and $\theta_L(L - 1) = 0$, i.e.,

$$\theta_L = (0, 1, \ldots, L - 1), \tag{8.3}$$

we get

$$h_\delta^{(n)}(\theta_L) = \frac{L - n + 1}{L(n-1)} \log \frac{L}{L - n + 1} + \frac{1}{L} \log L \tag{8.4}$$

for $2 \leq n \leq L$. In particular, for $L = 4$ we have

$$h_\delta^{(2)}(\theta_4) = 0.811, \quad h_\delta^{(3)}(\theta_4) = 0.750, \quad h_\delta^{(4)}(\theta_4) = 0.667,$$

in bit/symbol.

Case 2. On the opposite (non-trivial) end, let $F = (i_1, i_2)(j_1, j_2) \cdots (k_1, k_2)$ with, say, $i_1 < i_2, j_1 < j_2, \ldots, k_1 < k_2$. In this case, every $a \in A$ defines only one ordinal pattern of order 2, the symbols $i_1, \ldots, k_1$ belonging to $Q_{\langle 0,1 \rangle}$ and the symbols $i_2, \ldots, k_2$ to $Q_{\langle 1,0 \rangle}$. Hence, $q_{\langle 0,1 \rangle} = q_{\langle 1,0 \rangle} = 1/2$ and $h_\delta^{(2)}(F) = 1$; entropies of higher order are not defined ($Q_\pi(n) = \emptyset$ for $n \geq 3$).

In general, $F = (i_1, \ldots, i_{p_1})(j_1, \ldots, j_{p_2}) \ldots (k_1, \ldots, k_{p_r})$ with $1 \leq p_1, \ldots, p_r \leq L$ ($p_i = 1$ for the fixed points), $p_1 + \cdots + p_r = L$ and $p_{\max} := \max\{p_1, \ldots, p_r\} \geq 2$ (otherwise, F is the identity). If the symbol $a \in A$ appears in a cycle of length p, then a defines ordinal patterns of order $2, 3, \ldots, p$. Hence, F has entropies of order $2, 3, \ldots, p_{\max}$, although from some order on (depending on F), both the number and cardinality of the sets $Q_\pi(n)$ will decrease with n, rendering their contribution less and less significant.

Let us mention in passing that the normalized expected maximum cycle length of a random permutation of L symbols tends to $0.62432\ldots$ as $L \to \infty$, a result first observed experimentally by Golomb [86] and proved by Shepp and Lloyd [187]. So, we expect on average

$$h_\delta^{p\max}(F) \approx \frac{1}{0.6L - 1} \log 0.6L.$$

Remark 3 By definition, the discrete entropy of order n and, thus, the discrete entropy do not sense the presence of fixed points. For example, $\theta_L = (0, 1, \ldots, L-1)$ (Example 17) and $F_{L+1} = (0, 1, \ldots, L-1)(L)$ or $F_{L+2} = (0, 1, \ldots, L-1)(L)(L+1)$ have the same entropies (8.4).

Example 18 Given a permutation F_L of $\{0, 1, .., L-1\}$, we call

$$\lambda_{F_L} = \frac{1}{L-1} \sum_{i=0}^{L-2} \log |F_L(i+1) - F_L(i)|,$$

the *discrete Lyapunov exponent* [125] of F_L. If $L = 2l$, it can be proved [13, Thm. II.2] that $\lambda_{F_{2l}}$ is maximal for the permutation

$$\Gamma_{2l} = [l, 0, l+1, 1, l+2, 2, \ldots, 2l-1, l-1], \tag{8.5}$$

in which case

$$\lambda_{F_{2l}} \le \lambda_{\Gamma_{2l}} = \frac{l}{2l-1} \ln l + \frac{l-1}{2l-1} \ln (l+1).$$

For $l = 2$ we get

$$h_\delta^{(2)}(\Gamma_4) = 1, \quad h_\delta^{(3)}(\Gamma_4) = 1, \quad h_\delta^{(4)}(\Gamma_4) = 0.667,$$

in bit/symbol. Comparison with Example 17 shows that $h_\delta^{(n)}(\theta_4) \le h_\delta^{(n)}(\Gamma_4)$ for $n = 2, 3, 4$. In particular, the smaller orders $n = 2, 3$ show that Γ_4 is more "random" than θ_4.

The possibly simplest way to encapsulate in a single number the information contained in the whole hierarchy $h_\delta^{(2)}(F), \ldots, h_\delta^{(n_{\max})}(F)$, $n_{\max} = \max\{n:h_\delta^{(n)}(F) \ne 0\}$, without having to dissect F into cycles, is taking the arithmetic mean of it.

Definition 5 We call

$$h_\delta(F) = \frac{1}{n_{\max} - 1} \sum_{n=2}^{n_{\max}} h^{(n)}(F) \tag{8.6}$$

the discrete entropy (or just the entropy) of F.

Hence, $h_\delta(F)$ takes into account both high and, most importantly, low and middle orders on an equal footing. Indeed, although the number of summands in $h_\delta^{(n)}(F)$ grows as $n!$, the sum of the non-zero terms (before getting multiplied by $1/(n-1)$) actually scales linearly in n, rendering the different $h_\delta^{(n)}(F)$ of comparable magnitudes. Moreover, if we let formally $n_{max} \to \infty$ (the limit of ordered sets with arbitrary cardinality), we recover the usual definition of entropy, $h_\delta(F) = \lim_{n\to\infty} h_\delta^{(n)}(F)$, since a convergent sequence and the arithmetic mean of their successive terms (*Césaro mean*) have the same limit.

Example 19 In cryptography, any substitution on n-bit blocks is called an $n \times n$ S-box (for "substitution box"). The cryptographic security of S-boxes can be analyzed with a variety of tools. Consider, for instance, the 4×4 S-boxes defined by the permutations

$$F_1 = [15, 12, 2, 1, 9, 7, 10, 4, 6, 8, 5, 11, 0, 3, 13, 14]$$
$$= (0, 15, 14, 13, 3, 1, 12)(2)(4, 9, 8, 6, 10, 5, 7)(11)$$

and

$$F_2 = [8, 2, 4, 13, 7, 14, 11, 1, 9, 15, 6, 3, 5, 0, 10, 12]$$
$$= (0, 8, 9, 15, 12, 5, 14, 10, 6, 11, 3, 13)(1, 2, 4, 7).$$

The action of the corresponding S-box on the binary block $b_1b_2b_3b_4$ is identified with the action of F_1 or F_2 on the number $b_1 2^3 + b_2 2^2 + b_3 2^1 + b_4 \in \mathbb{Z}_{16}$. F_1 and F_2 share some standard properties of secure S-boxes, like being 0/1 balanced, nonlinear, and fulfilling the maximum entropy criterion [172]. But from the discrete entropy point of view, they are quite different. The discrete entropies of F_1 in bit/symbol are

$$h_\delta^{(2)}(F_1) = 0.99, \quad h_\delta^{(3)}(F_1) = 1.04, \quad h_\delta^{(4)}(F_1) = 0.96,$$
$$h_\delta^{(5)}(F_1) = 0.84, \quad h_\delta^{(6)}(F_1) = 0.70, \quad h_\delta^{(7)}(F_1) = 0.58,$$

and $h_\delta(F_1) = 0.85$. The discrete entropies of F_2 in bits/symbol are

$$h_\delta^{(2)}(F_2) = 0.99, \quad h_\delta^{(3)}(F_2) = 1.08, \quad h_\delta^{(4)}(F_2) = 1.17,$$
$$h_\delta^{(n)}(F_2) = 3.59/(n-1) \quad \text{for } n = 5, \ldots, 12$$

and $h_\delta(F_2) = 0.68$. Thus F_1, with a more even cycle decomposition than F_2, has a higher discrete entropy. Whether discrete entropy is useful for S-box design is an open problem in discrete chaos [125].

Exercise 12 A primitive root for a modulo m is a cyclic generator of $\mathbb{Z}_m^*$, the multiplicative group built by the residues modulo m coprime to m. Prove that the permutation Γ_{2l}, (8.5), is irreducible if and only if $2l + 1$ is a prime with primitive root 2 (i.e., $\mathbb{Z}_{2l+1}^*$ is cyclic and generated by 2). The primes under 100 with primitive root 2 are the following [2, Table 24.8]:

$$3, 5, 11, 13, 19, 29, 37, 53, 59, 61, 67, \text{ and } 83.$$

(Hint: consider the permutation $\tilde{\Gamma}_{2l}:\{1, 2, \ldots, 2l\} \to \{1, 2, \ldots, 2l\}$ defined as $\tilde{\Gamma}_{2l}(i) = \Gamma_{2l}(i - 1) + 1$ and show that

$$orb(1) = \{2^k \mod (2l + 1):0 \le k \le 2l - 1\}$$

under the permutation $(\tilde{\Gamma}_{2l})^{-1}$).

8.2 The Infinite Limit

Next we want to establish a more quantitative link between "continuous" and discrete entropies. The transition from the former to the latter proceeds over the discretization and truncation of orbits.

For simplicity, we will consider an *ergodic* map f on the unit interval $I = [0, 1]$ endowed with the Borel sigma-algebra, preserving a measure μ. Without loss of generality, let $\iota = \{I_i:0 \le i \le 10^k - 1\}$, with $I_i = [i10^{-k}, (i + 1)10^{-k})$ for $0 \le i \le 10^k - 2$ and $I_{10^k-1} = [1 - 10^{-k}, 1]$, be a box partition of I with norm $\|\iota\| = 10^{-k}$. Therefore, the *alphabet* of the ensuing ergodic symbolic dynamic $\mathbf{X}^\iota$ of f with respect to the partition ι is $S = \{0, 1, \ldots, 10^k - 1\}$. Furthermore, let $\{x_j = f^j(x_0):j \ge 0\}$ be a generic trajectory. Given x_0 and $\|\iota\|$, there is a maximal $M \le |S| = 10^k$ such that all points in the initial segment $\{x_j = f^j(x_0):0 \le j \le M - 1\}$ fall in different bins I_i of the partition ι, hence $S_i^\iota(x_0) \ne S_j^\iota(x_0)$ for all $0 \le i, j \le M - 1$ and $i \ne j$. This allows us to define a permutation (actually, a cycle) F_M on $S_M = \{0, 1, \ldots, M - 1\}$ in the following way. First, arrange the symbols $s_n = X_n^\iota(x_0) \in S, 0 \le n \le M - 1$, according to their sizes,

$$s_{n_0} < s_{n_1} < \cdots < s_{n_{M-1}}. \tag{8.7}$$

Then define

$$F_M(i) = j \Leftrightarrow \begin{cases} \text{(i) } n_i \ne M - 1 \text{ and } s_{n_i+1} = s_{n_j} \text{ or} \\ \text{(ii) } n_i = M - 1 \text{ and } n = M - 1. \end{cases}$$

By construction, F_M is order isomorphic (Definition 1) to the permutation $\tilde{F}_M:S \to S$ defined as

$$\tilde{F}_M(s_n) = \begin{cases} s_{n+1} & \text{for } n = 0, \ldots, M - 2, \\ s_0 & \text{for } n = M - 1. \end{cases}$$

Note that $\tilde{F}_M$ is a coarse-grained version of f, conveniently "short circuited" at the last orbit point by sending it back to the first one. Let $\phi:(S, <) \to (S_M, <)$ be the order isomorphism $s_{n_i} \mapsto i$ (so, $\tilde{F}_M = \phi^{-1} \circ F_M \circ \phi$). In particular, if $s_i = s_{n_{k(i)}}$ then s_i and $n_{k(i)}$ define the same ordinal patterns of lengths $l = 2, \ldots, M$ under $\tilde{f}_M$ and

F_M, respectively. With $\|\iota\| \to 0$, it follows that $x_i \in I$, $s_i \in S$, and $n_{k(i)} \in S_M$ define the same ordinal patterns $\pi \in S_l$ for arbitrarily long l.

On the other hand, since the map $f : I \to I$ is ergodic with respect to the measure μ, its entropy and permutation entropy can be determined from a typical trajectory, i.e., except for a set of initial conditions of measure zero. To be specific, let (i) $S^{\mathbb{N}_0}$ be the sample path space of the ergodic process $\mathbf{X}^\iota$, (ii) m_ι the measure induced by μ on $S^{\mathbb{N}_0}$

$$m = \mu \circ \Phi^{-1}$$

(see (B.22)), where $\Phi : I \to S^{\mathbb{N}_0}$ is the coding map (1.6) with respect to the partition ι, and (iii) Σ the shift transformation (1.8). Furthermore, for $L \le M$ and $\pi \in S_L$ set

$$P_\pi = \{s_0^\infty \in S^{\mathbb{N}_0} : s_{\pi_0} < s_{\pi_1} < \cdots < s_{\pi_{L-1}}\} \in \mathcal{P}_L$$

(notation as in (3.4) and (3.5)), and

$$Q_\pi = \{i \in S_M : F_M^{\pi_0}(i) < \cdots < F_M^{\pi_{L-1}}(i)\}$$

(notation as in (8.1)). Observe that in virtue of the order isomorphy ϕ between the permutations $\tilde{F}_M$ and F_M, $s_0^\infty \in P_\pi$ if and only if $\phi(s_0) \in Q_\pi$.

More generally, consider the shift Σ on the sequences $s_0^\infty \in S^{\mathbb{N}_0}$. Then $\Sigma^n(s_0^\infty) = (s_n, s_{n+1}, \dots) \in P_\pi$, $0 \le n \le M - L$, if and only if $\phi(s_n) \in Q_\pi$, $\pi \in S_L$. Apply now the ergodic theorem (Theorem 21) to the dynamical system $(S^{\mathbb{N}_0}, \mathcal{B}_\Pi(S), m_\iota, \Sigma)$ to conclude that, for any $\varepsilon_1 > 0$, there exists a uniform partition ι_0 of I such that

$$\left| m_\iota(P_\pi) - \frac{\left| \{\Sigma^n(s_0^\infty) \in P_\pi : 0 \le n \le M - L\} \right|}{M - L + 1} \right| < \varepsilon_1$$

for all $\pi \in S_L$ and almost all $s_0^\infty \in S^{\mathbb{N}_0}$, if $\|\iota\| \le \|\iota_0\|$ (and, consequently, $M \ge M_0$). The greater the window size L, the greater the sample size $M - L + 1$ (hence, the greater M) we need to estimate $m_\iota(P_\pi)$ with the same precision. Furthermore,

$$\left| q_\pi(L) - \frac{\left| \{\Sigma^n(s_0^\infty) \in P_\pi : 0 \le n \le M - L\} \right|}{M - L + 1} \right| < \varepsilon_2, \tag{8.8}$$

where, similar to (8.2),

$$q_\pi(L) = \frac{|Q_\pi(L)|}{\sum_{\pi \in S_L} |Q_\pi(L)|} = \frac{|Q_\pi(L)|}{M} \tag{8.9}$$

(since F_M is a cycle of length M), and the error $\varepsilon_2 = O(1/(M - L))$ stems from the different denominators in (8.8) and (8.9), and also from the last $L - 1$ points $s_{M-L+1}, \dots, s_{M-1}$, whose size-$L$ windows stretch outside the orbit segment s_0^{M-1}. All in all,

$$\left| h^*((X^\iota)_0^{L-1}) - h_\delta^{(L)}(F_M) \right|$$

$$= \frac{1}{L-1} \left| \sum_{\pi \in \mathcal{S}_L} m_\iota(P_\pi) \log m_\iota(P_\pi) - \sum_{\pi \in \mathcal{S}_L} q_\pi \log q_\pi \right|$$

$$\leq \frac{1}{L-1} \left(\varepsilon_1 |\mathcal{P}_L| \log |\mathcal{P}_L| + O\left(\frac{1}{M-L}\right) |\mathcal{P}_L| \log |\mathcal{P}_L| \right)$$

+ terms of higher order in M and L,

i.e., the permutation entropy of order $L \ll M$ of the process X^ι coincides approximately with the discrete entropy of order L of F_M, the permutation of $\{0, 1, \ldots, M-1\}$ obtained from f in the way explained before.

The first term of the error,

$$e_1 = \frac{\varepsilon_1}{L} |\mathcal{P}_L| \log |\mathcal{P}_L|,$$

can be made arbitrarily small by taking M sufficiently large. In fact, since $|\mathcal{P}_L| = O(L^{L+1/2}e^{-L}) = O(e^{L(\ln L - 1)+(1/2)\ln L})$ (in general, a rough estimate), it suffices to take (a) $M \geq \max\{M_0, -\ln \varepsilon_1 / \ln L\}$ to derive $e_1 = o(L^{-(M-L)})$. As for the second term,

$$e_2 = \frac{1}{L-1} O\left(\frac{1}{M-L}\right) |\mathcal{P}_L| \log |\mathcal{P}_L|,$$

we need (b) $M - L > O(L^{L+1/2}e^{-L} \ln L)$, i.e., $M - L > O(e^{L(\ln L - 1)+(1/2)\ln L + \ln \ln L})$, to make e_2 vanish when $M, L \to \infty$. Therefore if we set, say, $M = Ce^{L \ln L} =: \vartheta^{-1}(L)$ with $C > 0$ large enough so that (a) is also fulfilled, then

$$\frac{1}{\vartheta(M)-1} \sum_{n=2}^{\vartheta(M)} \left| h^*((X^\iota)_0^{n-1}) - h_\delta^{(n)}(F_M) \right| \leq e(M, \vartheta(M)),$$

where $e(M, L) = e_1 + e_2 +$ terms of higher order in M and L, and $e(M, \vartheta(M)) \to 0$ when $M \to \infty$. Letting now $\|\iota\| \to 0$ (hence $M \to \infty$), we get

$$h_\mu^*(f) = \lim_{\|\iota\| \to 0} h^*(X^\iota) = \lim_{M \to \infty} h_\delta(F_M),$$

provided $h_\mu^*(f)$ exists. Since f is ergodic by assumption, Theorem 10 implies

$$\lim_{M \to \infty} h_\delta(F_M) = h_\mu(f). \tag{8.10}$$

A final caveat. We have supposed that $\mathcal{O}_f(x_0)$ was generic for μ. In order to avoid that different orbits lead to (8.10) with different (ergodic) measures, we suppose furthermore that f is uniquely ergodic (i.e., f is continuous and it has only one invariant measure, see Sect. A.1).

This proves the one-dimensional version of the following theorem.

Theorem 18 *Let* $I \subset \mathbb{R}$ *be a closed interval and* $f{:}I \rightarrow I$ *a uniquely ergodic map. Furthermore, let* F_M *be the permutation of* $\{0, 1, \ldots, N - 1\}$ *obtained from* f *in the way explained above. Then* $\lim_{M\to\infty} h_\delta(F_M) = h_\mu(f)$, *where* μ *is the only* f*-invariant Borel measure on* I *and* $h_\mu(f)$ *is the metric entropy of* f *with respect to* μ.

The proof of the general case is analogous to the one-dimensional case. Theorem 18 justifies calling h_δ discrete entropy.

Example 20 In the following numerical simulations, we have used $M = 500,000$ and $2 \leq L \leq 9$. Figure 8.1 compares the discrete entropy with the Lyapunov exponent for the one-dimensional quadratic maps

$$f_a(x) = ax(1 - x), \quad 0 \leq x \leq 1, \quad 3.5 \leq a \leq 4.0. \tag{8.11}$$

Figure 8.2 compares the discrete entropy with the largest Lyapunov exponent for the two-dimensional quadratic maps

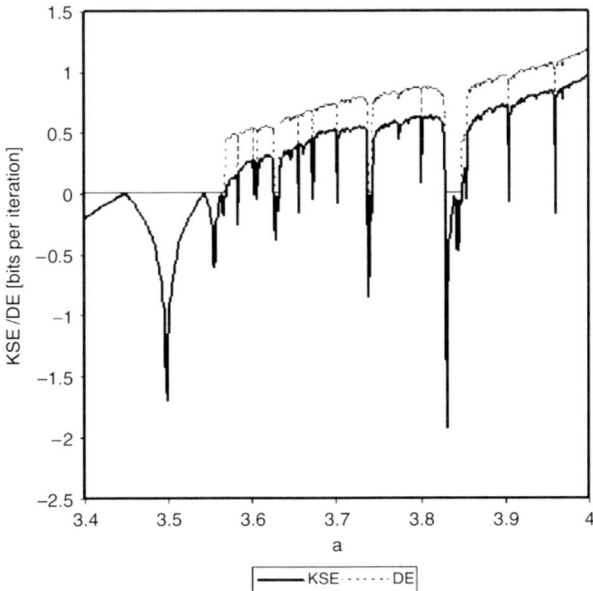

Fig. 8.1 Discrete entropy (*dashed line*) for the one-dimensional quadratic maps (8.11). The discrete entropy tracks the positive part of the Lyapunov exponent (*bold line*) with a uniform bias over the parameter values

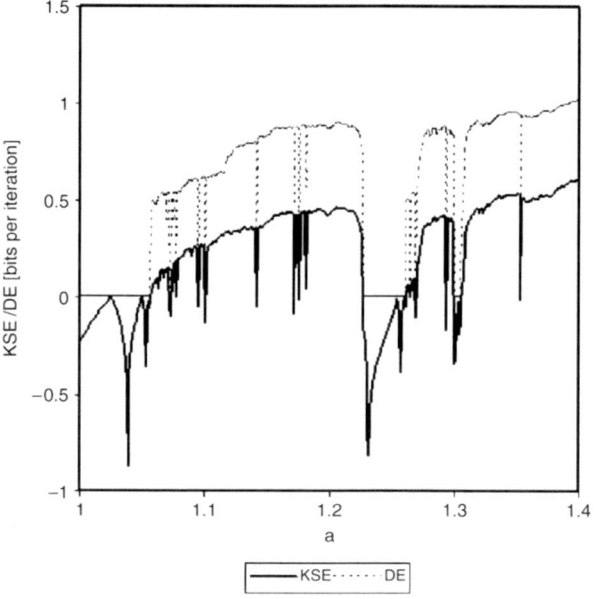

Fig. 8.2 Discrete entropy (*dashed line*) for the two-dimensional quadratic maps (8.12). The discrete entropy traces the positive part of the largest Lyapunov exponent (*bold line*) with a uniform bias over the parameter values

$$f_a(x, y) = (1 - ax^2 + 0.3y, x), \quad 0 \le x, y \le 1, \quad 1 \le a \le 1.4. \tag{8.12}$$

We observe that in both cases, the discrete entropy follows the profile of the positive part of the corresponding Lyapunov exponent with an approximately constant bias, due to the slow (and seemingly uniform) convergence of discrete entropy to its continuous counterpart.

8.3 Discrete Topological Entropy

As a matter of fact, the previous approach to discrete entropy admits some variations, in both concept and implementation. We shall elaborate here only on one of them, based on the topological permutation entropy of a piecewise monotone one-dimensional interval map $f: I \to I$ (Sect. 7.4).

Given a permutation F on $S_M = \{0, 1, \ldots, M - 1\}$, define the partition of S_M

$$\mathcal{Q}_n = \{Q_\pi \neq \emptyset : \pi \in \mathcal{S}_n\},$$

with $Q_\pi = \{s \in S_M : F^{\pi_0}(s) < \cdots < F^{\pi_{n-1}}(s)\}$ as in (8.1). Similar to (7.20) and (7.19), we propose the following definition.

Definition 6 We call

$$h_{\delta, \text{top}}^{(n)}(F) = \frac{1}{n - 1} \log |\mathcal{Q}_n|$$

the discrete topological entropy of the permutation F of order n, and

$$h_{\delta,top}(F) = \frac{1}{n_{max} - 1} \sum_{n=2}^{n_{max}} h_{\delta,top}^{(n)}(F)$$

($n_{max} = \max\{n:\mathcal{Q}_n \neq \emptyset\}$) the discrete topological entropy of F.

In order to treat $h_\delta(F)$, (8.6), and $h_{\delta,top}(F)$ on the same footing, one could refer to the first as discrete permutation entropy and write $h_{\delta,per}(F)$ instead.

Let $F_M:S_M \to S_M$ be the permutation obtained from f as explained in Sect. 8.2. The analogue of Theorem 18 for discrete topological entropy holds as well.

Theorem 19 *Let $f:I \to I$ be a uniquely ergodic and piecewise monotone map. Then $\lim_{M \to \infty} h_{\delta,top}(F_M) = h_{top}(f)$.*

Proof From the proof of Theorem 18, it follows that $q_\pi \to m_t(P_\pi)$ for every $\pi \in S_L$ as $\|t\| \to 0$ (or $M \to \infty$). Therefore, $|\mathcal{Q}_n| = |\mathcal{P}_n|$ in that limit, and we get $\lim_{M \to \infty} h_{\delta,top} = h_{top}^*(f) = h_{top}(f)$, the last equality following from Theorem 16. $\square$

From

$$h_{\delta,per}^{(n)}(F) = -\frac{1}{n-1} \sum_{\pi \in S_n} q_\pi \log q_\pi \leq \frac{1}{n-1} \log |\mathcal{Q}_n| = h_{\delta,top}^{(n)}(F)$$

for $n = 2, 3, \ldots, n_{max}$, we deduce that the discrete topological entropy is an upper bound of the discrete (permutation) entropy—the same as for their "continuous" counterparts.

Example 21 For the permutation

$$F = [3, 5, 1, 7, 0, 6, 2, 4] = (0, 3, 7, 4)(1, 5, 6, 2),$$

we get

$$h_{\delta,top}^{(2)}(F) = \log 2 = 1 \text{ bit/symbol},$$

$$h_{\delta,top}^{(3)}(F) = \frac{1}{2} \log 6 = 1.2925 \text{ bit/symbol},$$

$$h_{\delta,top}^{(4)}(F) = \frac{1}{3} \log 8 = 1 \text{ bit/symbol},$$

so that $h_{\delta,top}(F) = 1.0975$ bit/symbol. As for

$$\Gamma_8 = [4, 0, 5, 1, 6, 2, 7, 3] = (0, 4, 6, 7, 3, 1)(2, 5),$$

permutation (8.5) on $S_8 = \{0, 1, \ldots, 7\}$ with maximal discrete Lyapunov exponent, we find

$$h_{\delta,top}^{(2)}(\Gamma_8) = \log 2 = 1 \text{ bit/symbol},$$

$$h_{\delta,top}^{(3)}(\Gamma_8) = \frac{1}{2} \log 4 = 1 \text{ bit/symbol},$$

$$h_{\delta,top}^{(4)}(\Gamma_8) = \frac{1}{3} \log 6 = 0.8617 \text{ bit/symbol},$$

$$h_{\delta,top}^{(5)}(\Gamma_8) = \frac{1}{4} \log 6 = 0.6462 \text{ bit/symbol},$$

$$h_{\delta,top}^{(6)}(\Gamma_8) = \frac{1}{5} \log 6 = 0.5170 \text{ bit/symbol},$$

and $h_{\delta,top}(\Gamma_8) \approx 0.8050$ bit/symbol. Thus $h_{\delta,top}(\Gamma_8) < h_{\delta,top}(F)$. The same is true for the discrete permutation entropy: $h_{\delta,per}(\Gamma_8) = 0.7968$ bit/symbol $< h_{\delta,per}(F) = 1.0833$ bit/symbol.

Chapter 9
Detection of Determinism

In Chap. 2 we have illustrated the applications of ordinal patterns with four examples. In this chapter we present a further application, this time to the detection of determinism in noisy time series. Following the common usage of the term in applied science, "determinism" is meant here as the opposite to statistical independence, hence it includes colored noise as well. This application hinges on two basic properties of ordinal patterns: existence of forbidden patterns in the orbits of maps (Sects. 1.2, 3.3, and 7.7) and robustness to observational noise (Sects. 3.4.3, and 9.1). We shall actually present two detection methods.

Method I is based on the number of missing ordinal patterns. It proceeds by (i) counting the number of missing ordinal patterns in sliding, overlapping windows of size L along the data sequence, (ii) randomizing the sequence, and (iii) repeating (i) with the randomized sequence. Is the result of step (iii) clearly greater than the result of step (i), so may we conclude that the original noisy sequence has a deterministic component.

Method II is based on the distribution of the visible ordinal patterns. This method proceeds by (i) counting the number of ordinal patterns in sliding, non-overlapping windows of size L along the data sequence and (ii) performing a χ^2 test based on the results of (i), the null hypothesis being that the data are white noise. Hold the null hypothesis, so should all possible ordinal L-patterns be visible and evenly distributed over sufficiently many windows, at variance with what happens in the case of noisy deterministic data. In the latter case, the number of missing ordinal patterns is higher, its decay rate with L is slower, and the distribution of patterns is not necessarily uniform.

Both methods, as other applications of permutation entropy, are conceptually simple and computationally fast for moderate values of L. But not only this: Method II compares favorably to the popular Brock–Dechert–Scheinkman (BDS) independence test when applied to time series projected from the attractors of the Lorenz map and the time-delayed Hénon map. The bottom line is that determinism in noisy multivariate time series can be detected by observing a single component, a possibility that can come in handy in experimental situations.

Noisy univariate and multivariate time series have been intensively studied in the last few decades [1, 112]. Depending on the noise level of the data, one can expect to recover the full deterministic dynamics, to reconstruct the geometry of the noise-free

J.M. Amigó, *Permutation Complexity in Dynamical Systems,*
Springer Series in Synergetics, DOI 10.1007/978-3-642-04084-9_9,
© Springer-Verlag Berlin Heidelberg 2010

signal in some appropriate space, or just to ascertain the existence of an underlying determinism. The ordinal pattern-based methods described in this chapter falls in the third category. As a compensation for such a seemingly modest accomplishment, it has a remarkable success even with very high levels of noise. Besides the BDS method, which is based on the correlation dimension, other detection methods for determinism use the smoothness of the measure along reconstructed trajectories [164], functionals of probabilistic distributions [176], or the Higuchi fractal dimension on Poincaré sections [85].

9.1 Dynamical Robustness Against Observational Noise

Ordinal patterns are robust against small additive perturbations on account of being defined by inequalities. This property was called conditional robustness in Sect. 3.4.3. Yet, this property alone would not explain the persistence of forbidden patterns in the very noisy deterministic sequences that we are going to study in the next section. It turns out that, in deterministic sequences, there is a second mechanism for robustness, also in case of multi-dimensional maps—the dynamics itself. The result is an enhancement of the robustness of ordinal patterns against additive noise, which we call *dynamical robustness*. A simple explanation follows.

In the sequel we deal with a time series of the form

$$\xi_n = f^n(x_0) + w_n = x_n + w_n \tag{9.1}$$

($n \in \mathbb{N}_0$, or in practice $0 \le n \le N - 1$), where f is a self-map of the interval $[a, b] \subset \mathbb{R}$ and w_n are independent and uniformly distributed random variables (i.e., uniform white noise) in the interval $[-\eta, \eta]$. In order that the noise destroys a given allowed or forbidden pattern $\pi = \langle \pi_0, \ldots, \pi_{L-1} \rangle$ of the noise-free sequence $(x_n)_{n \in \mathbb{N}_0}$, it must happen that

$$x_{\pi_i} < x_{\pi_{i+1}}$$

but

$$x_{\pi_i} + w_{\pi_i} > x_{\pi_{i+1}} + w_{\pi_{i+1}}$$

for some $0 \le i \le L - 2$ and $w_{\pi_i}, w_{\pi_{i+1}} \in [-\eta, \eta]$. If η is small, this will be only possible if $x_{\pi_i} \approx x_{\pi_{i+1}}$, i.e., if $f^{\min\{\pi_i, \pi_{i+1}\}}(x_0)$ is an "approximately" periodic point with period $|\pi_i - \pi_{i+1}|$. We conclude that, indeed, the dynamics imposes an extra condition on $x_{\pi_i}, x_{\pi_{i+1}}$ so that a small amplitude perturbation can reverse their order.

To put some numbers on this argument, take $f(x) = 4x(1 - x)$, $0 \le x \le 1$, the logistic map. We know that for $\eta = 0$ this map has one forbidden 3-pattern, namely, $\langle 2, 1, 0 \rangle$ (Fig. 1.6). In other words, there exists no $x \in [0, 1]$ such that $f^2(x) < f(x) < x$. The pattern $\langle 2, 1, 0 \rangle$ can appear in the noisy sequence (9.1) by a single order reversal if the noise changes the order of x_n, x_{n+1} or the order of x_{n+1}, x_{n+2} in the allowed patterns

$$x_{n+2} < x_n < x_{n+1} \quad \text{or} \quad x_{n+1} < x_{n+2} < x_n,$$

respectively. In the first case, this requires $x_n \approx x_{n+1} = f(x_n)$, i.e., x_n must be close to any of the two fixed points of the map: $x = 0$ or $x = \frac{3}{4}$ (see Fig. 1.5). In the second case, the same applies to x_{n+1} and $x_{n+2} = f(x_{n+1})$. Therefore, it suffices to discuss the first case.

Consider the fixed point $x = 0$ and take $x_n = \delta > 0$. Then $x_{n+1} = f'(0)\delta + R\delta^2$, where R can be estimated with the remainder of the Taylor series. Since $\xi_n \in [x_n - \eta, x_n + \eta] =: I_n$, the inequality $\xi_{n+1} < \xi_n$ can be fulfilled only if the intervals I_n and I_{n+1} overlap, i.e., if

$$\delta \le \delta_0(\eta) = \frac{1 - f'(0) + \sqrt{(1 - f'(0))^2 + 8R\eta}}{2R}. \tag{9.2}$$

One can analogously estimate $\delta_+(\eta) > 0$ and $\delta_-(\eta) > 0$ such that if $x_n \in [\frac{3}{4} - \delta_-(\eta), \frac{3}{4} + \delta_+(\eta)]$, then x_n is sufficiently close to $x = \frac{3}{4}$ again in the sense that the inequality $\xi_{n+1} < \xi_n$ can hold for η small.

Thus, the probability $\Pr(\eta)$ for two consecutive orbit points $(x_n, x_{n+1}$ or $x_{n+1}, x_{n+2})$ to lie sufficiently close to either fixed point so as the pattern $\langle 2, 1, 0 \rangle$ becomes observable in a noisy orbit of the logistic map by means of a single order reversal is

$$\Pr(\eta) = \mu([0, \delta_0(\eta)]) + \mu([\tfrac{3}{4} - \delta_-(\eta), \tfrac{3}{4} + \delta_+(\eta)]),$$

where μ is the natural invariant measure for the logistic map,

$$\mu([c, d]) = \int_c^d \frac{dx}{\pi \sqrt{x(1 - x)}}$$

(see (1.20)). To make the argument even simpler, observe that once two consecutive orbit points in x_n, x_{n+1}, x_{n+2} are close to a fixed point, we may assume that the third one is around as well. In this case, the type of $\xi_n, \xi_{n+1}, \xi_{n+2}$ is going to depend basically on the type of w_n, w_{n+1}, w_{n+2}.

Consider now a string of length N, $\xi_0^{N-1} = \xi_0, \xi_1, \ldots, \xi_{N-1}$, along with the $\lfloor \frac{N}{3} \rfloor$ independent random vectors $\xi_n^{n+2} = \xi_n, \xi_{n+1}, \xi_{n+2}, n = 0, 3, 6, \ldots$. If we pick one of those vectors, the probability $\Pr(\langle 2, 1, 0 \rangle)$ that $\xi_{n+2} < \xi_{n+1} < \xi_n$ holds is then

$$\Pr(\langle 2, 1, 0 \rangle) \approx \Pr(\eta) \Pr\{w_{n+2} < w_{n+1} < w_n\})$$
$$= \Pr(\eta) \cdot \frac{1}{6}.$$

In order to verify these results, the probability P of finding at least once the pattern $\langle 2, 1, 0 \rangle$ in any of the $\lfloor \frac{N}{3} \rfloor$ windows $\xi_{3n}, \xi_{3n+1}, \xi_{3n+2}$ of the noisy time series $(\xi_n)_{n=0}^{N-1}$, (9.1), was calculated numerically. From the reasoning above, this probability should be close to $1 - (1 - \Pr(\eta)/6)^{\lfloor \frac{N}{3} \rfloor}$ for the logistic map contaminated with additive, uniform white noise of small amplitude η, whereas it should be $1 - (1 - 1/6)^{\lfloor \frac{N}{3} \rfloor}$ for uniform white noise only (i.e., $\xi_n = w_n$ in (9.1)). Clearly, the former probability is greater than the latter because $\Pr(\eta)$ is going to be very small. This is confirmed by Fig. 9.1.

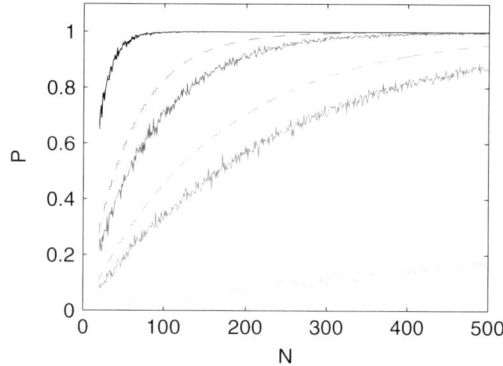

Fig. 9.1 Numerical computation (*continuous line*) and analytical estimation (*dashed*) of the probability P of finding the pattern $\langle 2, 1, 0 \rangle$ in any of the $\lfloor \frac{N}{3} \rfloor$ windows $\xi_{3n}, \xi_{3n+1}, \xi_{3n+2}$ of a time series of length N generated with the logistic map. The noise amplitude is $\eta = 0.0001$ (*light gray*), $\eta = 0.01$ (*gray*), $\eta = 0.1$ (*dark gray*). The *top curve* corresponds to uniform white noise. Clearly the probability P is smaller for a noisy, deterministic time series than for uniform white noise

9.2 Detection of Determinism I: Number of Missing Ordinal Patterns

We already know (Sect. 1.2) that if $(x_n)_{n \in \mathbb{N}_0}$ is a univariate time series generated by a piecewise monotone interval map f, then there exist ordinal patterns which are forbidden for f. The theoretical situation in higher dimensions is less satisfactory in that the existence of forbidden patterns has been proved so far only under the somewhat restrictive condition of expansiveness (Sect. 7.6). There is nevertheless numerical evidence that forbidden ordinal patterns are also a general feature of higher dimensional dynamics. Since, on the other hand, univariate and multivariate random sequences have no forbidden patterns with probability 1, we conclude that the existence of forbidden patterns can be used as a fingerprint of deterministic orbit generation. Here "random sequence" means generated by an unconstrained, stochastic process taking on values in an interval. In summary, the difference between deterministic and random time series is clear-cut from an ordinal-theoretical point of view: the former have forbidden patterns while the latter have not.

However, when it comes to exploit this forbidden pattern-based strategy to detect determinism, two important practical issues arise: finiteness and noise contamination. Finiteness produces *false forbidden patterns*, that is, ordinal patterns which are missing in a finite (segment of a) random sequence without constraints. Noise destroys forbidden patterns; for instance, a forbidden pattern of the "clean" sequence can turn visible because of additive random fluctuations. Let us mention in passing that were not for the observational noise, determinism could be easily ascertained, for example, with graphical methods. It is therefore interesting that ordinal patterns themselves provide the remedy to the two said issues. First of all, the number of false forbidden patterns of a fixed length always decreases with the length of the time series. Second, "true" forbidden patterns (i.e., forbidden patterns for an underlying deterministic dynamics) possess an additional dynamical robustness against additive noise (Sect. 9.1). This translates into a greater number of missing ordinal patterns in a noisy deterministic sequence than in a random one, and also to a slower decay rate with the length of the sequence. We shall shortly present numerical evidence that forbidden patterns persist in very noisy deterministic data—so noisy that the traditional methods [1, 112, 152] fail to uncover the underlying deterministic dynamics. But before coming to this point, let us dwell on some practical issues.

In practice one uses sliding windows of size L to comb a finite sequence $(x_n)_{n=0}^N$ for visible ordinal L-patterns. Note that a sequence of length N allows $N - L + 1$ windows of size L, for $2 \leq L \leq N$. Thus, in order to allow every possible ordinal pattern of length L to occur in a time series of length N, the condition $L! \leq N-L+1$ must hold. Moreover, in cases where undersampling might occur, $N \gg L! + L - 1$ should also hold. As a rule of thumb we chose $(L + 1)! \leq N$ in the numerical simulations below, although $L! \leq N$ would do also in our case (very noisy data). Furthermore, $(x_n)_{n=0}^N$ will be initial segments of variable length $N \leq N_{max} = 8000$, taken from a sequence $(x_n)_{n=0}^{N_{max}}$. All these constraints leave $L = 4, 5, 6$ as interesting choices for L. In general one takes also moderate values for L, not least because of the sharp increase of the function $L!$.

Under these provisos, suppose now that the ordinal pattern $\pi \in \mathcal{S}_L$ is missing in a finite noise-free time series. Of course, the odds that a *false* forbidden pattern persists in a random or deterministic sequence (or sample of sequences) will decrease exponentially with the number of data (see, e.g., Sect. 9.1). As a result, the number of false forbidden patterns in $(x_n)_{n=0}^N$ will decay as N increases up to N_{max}, the number of data at our disposal. Otherwise, if $(x_n)_{n=0}^{N_{max}}$ is a deterministic *noise-free* time series and π is a forbidden pattern, then π will be missing in $(x_n)_{n=0}^N$ for all $N \leq N_{max}$. In other words, the number of *true* forbidden patterns in $(x_n)_{n=0}^N$ does not depend on N.

Consider a fixed initial condition x and suppose that $\pi_{forb} = \langle \pi_0, \ldots, \pi_{L-1} \rangle$ is a forbidden pattern for f. Suppose furthermore that we switch on a discrete-time random perturbation w_k, $|w_k| \leq w_{max}$, such that π_{forb} is still missing in the finite sequence $(f^k(x) + w_k)_{k=0}^{N-1}$ (due to robustness). Observe that the *noisy* time series $\xi_k = f^k(x) + w_k$ can be viewed both as a perturbation of an underlying deterministic dynamics and as a random process correlated with the deterministic dynamics[1] f.

---

[1] Sometimes *colored noise* (i.e., a random process whose variables are statistically dependent) is numerically simulated in this way. For other methods, see, e.g., [113, 83].

If the orbit of x would be infinitely long, then the noisy time series had no missing patterns and π_{forb} would be visible with probability 1. In the finite-length case we are considering, this is in general not the case; rather, there is a threshold $\theta = \theta(N)$ (the greater N, the smaller θ) such that π_{forb} will do appear in $(\xi_k)_{k=0}^{N-1}$ only if $w_{\text{max}} > \theta$. We conclude that amplifying a random perturbation destroys progressively the forbidden patterns of the underlying deterministic dynamic.

In the following we are going to test numerically one of the properties discussed above, namely, the robustness of *true* forbidden patterns against additive random perturbations. In order to estimate the average number $\langle n(L, N) \rangle$ of missing ordinal L-patterns in a finite, noisy sequence of length N,

$$\xi_k = x_k + w_k, \quad 0 \leq k \leq N - 1,$$

with $x_{k+1} = f(x_k)$ and w_k a random process, we generate 100 samples of length $N_{\text{max}} = 8000$ and normalize the corresponding count of missing patterns of lengths $4 \leq L \leq 6$. To check the decay of $\langle n(L, N) \rangle$ with N, this parameter is allowed to vary in the range $(L + 1)! \leq N \leq N_{\text{max}}$. We highlight next a few results obtained with f being the logistic map and w_k being white noise uniformly distributed in the interval $[-w_{\text{max}}, w_{\text{max}}]$, $0 \leq w_{\text{max}} \leq 1$.

Figure 9.2 shows $\langle n(L, N) \rangle$ when (a) $w_{\text{max}} = 0.25$, (b) $w_{\text{max}} = 0.50$, and (c) $w_{\text{max}} = 1$ and $f^k(x) = 0$ (noise only), respectively. Note the different orders of magnitude of the vertical scales. Needless to say, $\langle n(L, N) \rangle$ decays with increasing N because the greater the N, the more unlikely that an L-pattern is missing in a noisy or random sequence of length N; this is a statistical effect. The important features for us are the magnitude of $\langle n(L, N) \rangle$ and its decay rate with N, since these two properties are tightly related to the forbidden patterns of the underlying deterministic dynamic via robustness: the smaller the w_{max}, the closer we are to the deterministic case, therefore, the more missing ordinal patterns and the slower their decrease with N.

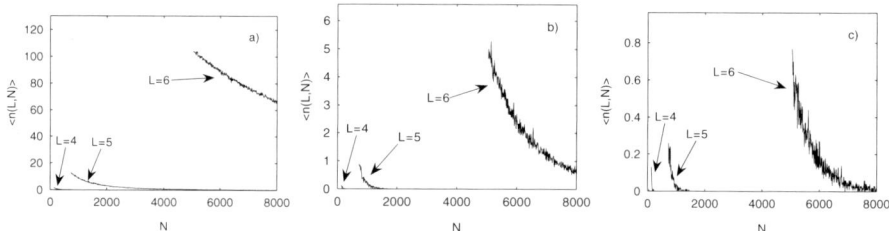

Fig. 9.2 Average number of missing ordinal patterns of length L found in a time series of length N, $\langle n(L, N) \rangle$, for noisy series of the logistic map with $w_{\text{max}} = 0.25$ (**a**), $w_{\text{max}} = 0.5$ (**b**), and for a series of uniformly distributed noise (**c**)

Figure 9.3 depicts ξ_{k+1} vs ξ_k in the previous cases (a) and (b). The higher order of magnitude of, e.g., $\langle n(6, N) \rangle$ in Fig. 9.2(b) as compared to Fig. 9.2(c) signalizes an underlying deterministic law, in spite of the fact that Fig. 9.3(b) hardly gives any clue about this.

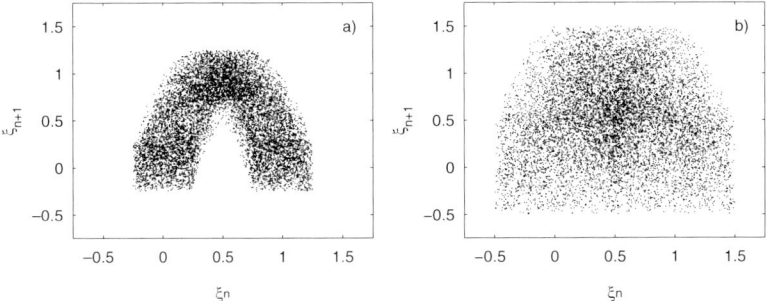

Fig. 9.3 Return map for noisy time series from the logistic map with $w_{max} = 0.25$ (**a**) and $w_{max} = 0.5$ (**b**). In the latter case, the high noise level does not allow to recognize the underlying deterministic dynamics. However, the number of missing ordinal patterns is sensibly higher than in the purely random case

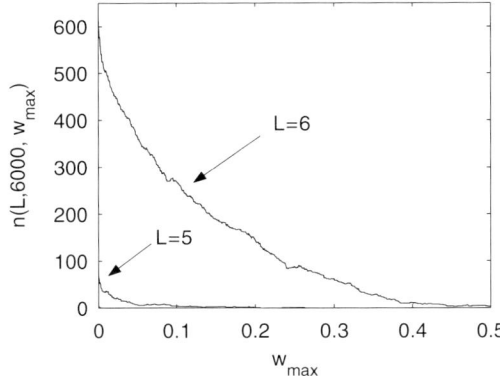

Fig. 9.4 Number of missing ordinal patterns of length L found in a noisy time series of the logistic map with length 6000 vs the uniform noise amplitude w_{max}

Finally, Fig. 9.4 nicely illustrates the resistance of the true forbidden patterns to disappear with increasing noise levels. In this figure, $N = 6000$, $L = 5, 6$, and $0 \leq w_{max} \leq 0.5$.

These numerical simulations suggest the following simple-minded, three-step method to discriminate noisy, deterministic, finite time series from random ones, at least when the noise is white.

(a) Compute the number of missing ordinal L-patterns of adequate length (say $(L + 1)! \leq N$) in sliding windows along the sequence. It is convenient to use segments of variable length N and to draw the corresponding curves, as in Fig. 9.2.

(b) Randomize the sequence, i.e., change the temporal structure of the data in a random way.

(c) Proceed as in step (a) with the randomized sequence.

If the results of (a) and (c) are about the same, the sequence is very likely not deterministic (or the observational noise is so strong as compared to the deterministic signal that the latter has been completely masked). Otherwise, the sequence stems from a deterministic one. Needless to say, the method is more reliable if a statistically significant sample of sequences can be obtained, for instance, by cutting a long sequence into shorter pieces. In the next section we discuss a more quantitative method.

9.3 Detection of Determinism II: Distribution of Visible Ordinal Patterns

Consider once more a univariate or multivariate time series of the form

$$\xi_n = f^n(x_0) + w_n, \tag{9.3}$$

$(0 \le n \le N - 1)$ where w_n is white noise, i.e., outcomes of an independent and identically distributed (i.i.d.) random process. In order to differentiate white noise from a noisy deterministic time series of form (9.3), the perhaps simplest tool consists in counting visible ordinal patterns before and after randomizing the time series under scrutiny; depending on whether the number of visible patterns remains about the same or decreases significantly, we may conclude that the series is random or deterministic, respectively. This is the method discussed in Sect. 9.2.

A more quantitative method calls for performing a chi-square test based on the count of visible ordinal patterns. The *null hypothesis* reads

$$H_0: \quad \text{the } \xi_n \text{ are i.i.d.} \tag{9.4}$$

From a statistical point of view, this method is going to be a test of independence since the alternative to H_0 includes also colored noise.

The method goes as follows. Take sliding windows of size $L \ge 2$, overlapping at a single point (i.e., the last point of a window is the first point of the next one) down the sequence $\xi_0^{N-1} = \xi_0, \ldots, \xi_{N-1}$. For brevity, we call them "non-overlapping" windows. The number of such windows is

$$K = \left\lfloor \frac{N-1}{L-1} \right\rfloor, \tag{9.5}$$

each comprising the entries

$$\mathbf{e}_k = \xi_{kL-k}, \ldots, \xi_{(k+1)L-(k+1)}, \quad 0 \le k \le K - 1.$$

Notice that if the variables $\xi_0, \xi_1, \ldots, \xi_{N-1}$ are independently drawn from the same probability distribution, then the ordinal L-patterns defined by the components of $\mathbf{e}_k \in \mathbb{R}^L$, which we denote by $\pi(\mathbf{e}_k) \in \mathcal{S}_L$, will also be independent and, moreover, uniformly distributed random variables. Therefore, if one or several ordinal patterns are missing in a sample obtained using non-overlapping windows, this might be a statistically significant signal that independence and/or the equality of the distribution are/is not fulfilled.

Given the non-overlapping windows $\{\mathbf{e}_k \in \mathbb{R}^L : k \geq 0\}$ corresponding to an arbitrarily long time series $\{\xi_n : n \geq 0\}$, suppose that some ordinal patterns of length L are missing in the initial segment $\xi_0, \xi_1, \ldots, \xi_{N-1}$. Let ν_π be the number of $\mathbf{e}_k$'s such that $\mathbf{e}_k$ is of type $\pi \in \mathcal{S}_L$ (i.e., $\pi(\mathbf{e}_k) = \pi$). Thus, $\nu_\pi = 0$ means that the L-pattern π has not been observed.

In order to accept or reject the null hypothesis H_0, (9.4), based on our observations, we apply a chi-square goodness-of-fit hypothesis test with statistic [135]

$$
\begin{aligned}
\chi^2(L) &= \sum_{\pi \in \mathcal{S}_L} \frac{(\nu_\pi - K/L!)^2}{K/L!} \\
&= \frac{L!}{K} \left(\sum_{\pi \in \mathcal{S}_L} \nu_\pi^2 - 2\frac{K}{L!} \sum_{\pi \in \mathcal{S}_L} \nu_\pi + \left(\frac{K}{L!}\right)^2 \sum_{\pi \in \mathcal{S}_L} 1 \right) \\
&= \frac{L!}{K} \sum_{\pi \in \mathcal{S}_L} \nu_\pi^2 - 2K + K \\
&= \frac{L!}{K} \sum_{\pi \in \mathcal{S}_L : \text{visible}} \nu_\pi^2 - K,
\end{aligned}
\tag{9.6}
$$

since (i) $\sum_{\pi \in \mathcal{S}_L} \nu_\pi = K$ and (ii) $\nu_\pi = 0$ if π is missing. Here $K/L!$ is the expected relative frequency of an ordinal L-pattern, if H_0 holds true. In the affirmative case, $\chi^2 = \chi^2(L)$ converges in distribution (as $K \to \infty$) to a chi-square distribution with $L! - 1$ degrees of freedom. Thus, for large K, a test with approximate level α is obtained by rejecting H_0 if $\chi^2 > \chi^2_{L!-1, 1-\alpha}$, where $\chi^2_{L!-1, 1-\alpha}$ is the upper $1 - \alpha$ critical point for the chi-square distribution with $L! - 1$ degrees of freedom [135]. In our case, the hypothetical convergence of χ^2 to the corresponding chi-square distribution may be considered sufficiently good if $\nu_\pi > 10$ for all visible L-patterns π, and

$$
\frac{K}{L!} > 5.
\tag{9.7}
$$

Notice that since this test is based on distributions, it could happen that a deterministic map has no forbidden L-patterns, thus $\nu_\pi \neq 0$ for all $\pi \in \mathcal{S}_L$; however, the null hypothesis be rejected because those ν_π's are not evenly distributed.

9.4 A Benchmark

A well-known benchmark for independence in time series is the Brock–Dechert–Scheinkman (BDS) test [38, 193], which is based on the correlation dimension. Since the numerical simulations below use the algorithm provided in [136], we follow this reference for the basics of the BDS test.

Let X_t, $t \geq 1$, be i.i.d. random variables, and

$$
I_\epsilon(x, y) = \begin{cases} 1 & \text{if } |x - y| < \epsilon, \\ 0 & \text{otherwise.} \end{cases}
$$

The probability that two length-m vectors are within ϵ can be estimated by the correlation sum

$$
C_{m,n}(\epsilon) = \frac{2}{n(n-1)} \sum_{s=1}^{n} \sum_{t=s+1}^{n} \prod_{j=0}^{m-1} I_\epsilon(X_{s-j}, X_{t-j}).
$$

It is shown in [38] that

$$
W_{m,n}(\epsilon) = \sqrt{n} \, \frac{C_{m,n}(\epsilon) - C_{1,n}^m(\epsilon)}{\sigma_{m,n}(\epsilon)}
$$

converges in distribution to a standard normal distribution. The normalization $\sigma_{m,n}(\epsilon)$ is given by

$$
\sigma_{m,n}^2(\epsilon) = 4 \left[B^m + 2 \sum_{j=1}^{m-1} B^{m-j} C^{2j} + (m-1)^2 C^{2m} - m^2 BC^{2m-2} \right],
$$

where C is consistently estimated by $C_{1,n}(\epsilon)$ and B can be estimated by

$$
B_n(\epsilon) = \frac{6}{n(n-1)(n-2)} \sum_{t=1}^{n} \sum_{s=t+1}^{n} \sum_{r=s+1}^{n} h_\epsilon(X_t, X_s, X_r),
$$

$$
h_\epsilon(i, j, k) = \frac{1}{3} \left[I_\epsilon(i, j)I_\epsilon(j, k) + I_\epsilon(i, k)I_\epsilon(k, j) + I_\epsilon(j, i)I_\epsilon(i, k) \right].
$$

A statistically significant non-zero value of $W_{m,n}(\epsilon)$ is evidence for determinism in the univariate time series $\{X_t : t \geq 1\}$.

This method relies on the selection of the parameters m and ϵ. Following the usual procedure [140], we take $\epsilon = 0.9^j$ with $j = 0, 1, 2, \ldots$. The criterion to say whether a combination of m and ϵ is "adequate" call for evaluating if a random time series is accepted as deterministic using this test the number of cases prescribed by the significance level of the test α.

9.5 Numerical Simulations

As underlying deterministic time series we use projections on the first coordinate of orbits generated by the Lorenz and time-delayed Hénon maps (this amounts in practice to using the standard lexicographical order). The additive noise w_n is modeled as Gaussian white noise,

$$\mathbb{E}(w_m \cdot w_n) = \sigma^2 \delta_{mn}$$

($\mathbb{E}$ stands for expectation value), with different standard deviations σ. Simulations with uniformly distributed noise yield similar results.

Two kinds of results are going to be presented in the two next sections: (i) Plots of the number of missing ordinal patterns as in Sect. 9.2 and (ii) plots of the distribution of the χ^2 statistic. Although the first ones provide only qualitative information, they can eventually complement the information provided by the second ones, as we shall see in the case of the Lorenz map. The specifics of plots (i) and (ii) are as follows.

(i) Let N_{max} denote the length of the data sequence under scrutiny and let $n(L, N)$ be the number of missing L-patterns in the initial segment $\xi_0, \xi_1, \ldots, \xi_{N-1}$ of variable length $N \leq N_{max}$. The numbers $n(L, N)$ are determined with *overlapping* sliding windows of sizes $4 \leq L \leq 7$. In order to make the most of sequences of length $N_{max} = 8000$, we take this time

$$L! \lesssim N \leq N_{max}.$$

An average number $\langle n(L, N) \rangle$ is then estimated from 100 sequences.

(ii) *Non-overlapping* windows are used for the chi-square test of independence based on the distributions of ordinal L-patterns, with statistic (9.6)

$$\chi^2 = \chi^2(L) = \frac{L!}{K} \sum_{\pi \in S_L : \text{visible}} v_\pi^2 - K. \tag{9.8}$$

Here, $K = \left\lfloor \frac{N-1}{L-1} \right\rfloor$ is the number of non-overlapping windows of size L in a data sequence of length N, (9.5). The window sizes in the simulations are $L = 4, 5$. For $L = 4$, the acceptance/rejection thresholds of the null hypothesis (9.5) at levels $\alpha = 0.10, 0.05$ are

$$\chi^2_{23, 0.90} = 32.01, \quad \chi^2_{23, 0.95} = 35.17, \tag{9.9}$$

respectively. For $L \geq 5$, corresponding to degrees of freedom over 100, the following approximation for the thresholds $\chi^2_{L!-1, 1-\alpha}$ is used [135]:

$$\chi^2_{L!-1, 1-\alpha} \approx (L! - 1) \left(1 - \frac{2}{9(L! - 1)} + z_{1-\alpha} \sqrt{\frac{2}{9(L! - 1)}} \right)^3,$$

where $z_{1-\alpha}$ is the upper $1 - \alpha$ critical point for the standard normal distribution, $\mathcal{N}(0, 1)$; in particular, $z_{0.90} = 1.282$ and $z_{0.95} = 1.645$. Thus,

$$\chi^2_{119, 0.90} = 139.15, \quad \chi^2_{119, 0.95} = 145.46. \tag{9.10}$$

Remember from (9.7) that $5L! \lesssim K$ should hold for the chi-square test to be statistically significant. Therefore

$$5L! \lesssim \frac{N}{L - 1},$$

i.e., $N \gtrsim 5(L - 1)L!$. In consequence we take sequences of length $N = 1000$ for $L = 4$ and $N = 8000$ for $L = 5$. To plot the χ^2-value distribution, a sample of $10,000$ sequences was used.

The numerical results are summarized in the following two sections.

9.5.1 The Lorenz Map

The *Lorenz map* [193] is defined as

$$x_{n+1} = x_n y_n - z_n, \quad y_{n+1} = x_n, \quad z_{n+1} = y_n. \tag{9.11}$$

It has an attractor with *Kaplan–Yorke dimension* $D_{KY} = 2$ [193]. Assuming the well-tested Kaplan–Yorke conjecture $D_{KY} = D_1$, where D_1 is the *information dimension*, then the *fractal dimension* D_0 satisfies

$$D_0 \geq D_1 = 2.$$

Figure 9.5 shows the return map $\xi_{n+1} = x_{n+1} + w_{n+1}$ vs $\xi_n = x_n + w_n$ for a typical orbit of the Lorenz map on its attractor and additive Gaussian white noise w_n with $\sigma = 0.25$ ($SNR^2 \simeq 10$ dB). The geometry of the attractor has been completely washed out by the noise, but the underlying determinism can still be detected because of the different count of missing ordinal patterns before (Fig. 9.6) and after (Fig. 9.7) switching off the deterministic signal. Not only the count of missing ordinal patterns is different in these two cases, but also their decay rate with N. The different behavior in Fig. 9.6 of the curve $L = 4$, on the one hand, and the curves $L \geq 5$, on the other hand, strongly indicates that the Lorenz map has no forbidden 4-patterns.

Figure 9.8 shows the distribution of the statistic χ^2, (9.8), obtained from 10,000 projections x_0^{N-1} of orbits of the Lorenz map, contaminated with additive Gaussian noise with $\sigma = 0.25, 0.50$ (SNR $\simeq 10, 4.0$ dB, respectively). Since the rejection

---

[2] SNR is short for "signal-to-noise ratio" and dB is short for "decibel."

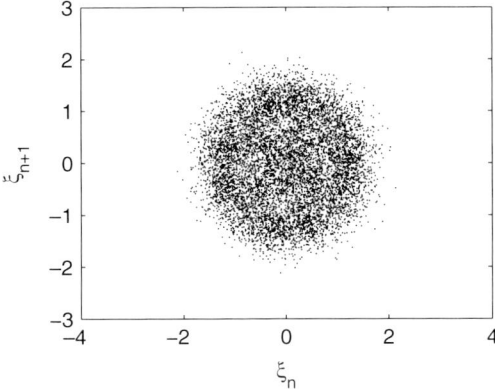

Fig. 9.5 Return map for a time series of the Lorenz map contaminated with Gaussian white noise with $\sigma = 0.25$ (SNR $\simeq 10$ dB). The structure of the underlying chaotic attractor has been totally blurred. However, the count of missing ordinal patterns is sensibly higher than in the purely random case

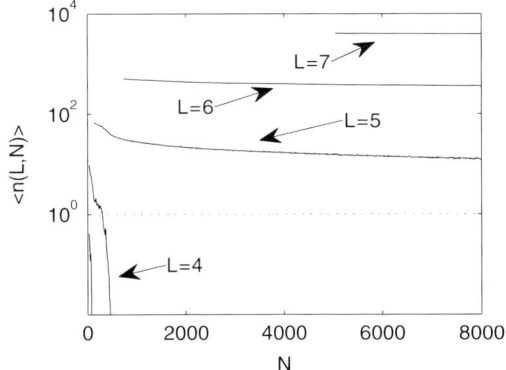

Fig. 9.6 Average number of missing ordinal patterns of length L found in a time series of length N, $\langle n(L, N) \rangle$ (in logarithmic scale), for a noisy series of the Lorenz map with $\sigma = 0.25$ (SNR $\simeq 10$ dB)

threshold of the null hypothesis H_0 (9.4) at level $\alpha = 0.05$ is $\chi^2_{23,0.95} = 35.17$ in (a) and $\chi^2_{119,0.95} = 145.46$ in (b), see (9.9), the χ^2 test clearly detects determinism. It is worth noticing that the rejection of H_0 in case (a) is due to the non-uniform distribution of ν_π since, according to Fig. 9.6, all 4-patterns are visible in noisy time series generated by the Lorenz map with $N \gtrsim 500$ and $\sigma = 0.25$.

Finally, the comparison with the BDS test is shown in Fig. 9.9. There we show the probability P of rejecting the null hypothesis (9.4) for the 27 possible adequate BDS tests on a time series $\xi_0^{N-1} = (x_n + w_n)_{n=0}^{N-1}$ of length $N = 1000$, where now w_n is Gaussian white noise with $0 \leq \sigma \leq 2$. In the same figure it is also

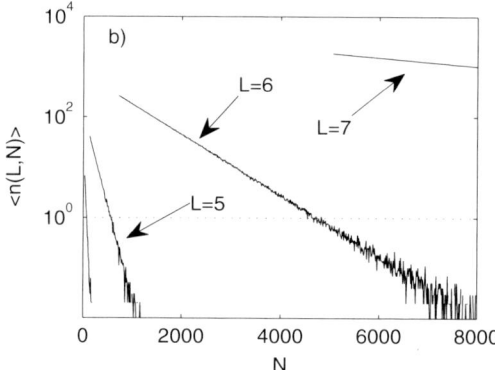

Fig. 9.7 Average number of missing ordinal patterns of length L found in a time series of length N, $\langle n(L,N) \rangle$ (in logarithmic scale), for time series of Gaussian white noise with $\sigma = 0.25$

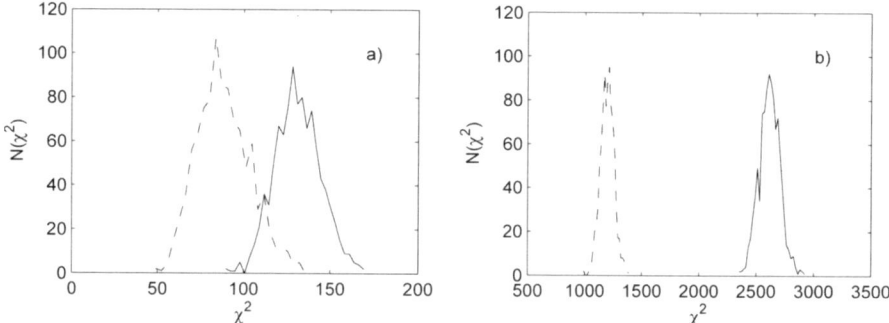

Fig. 9.8 Distribution $N(\chi^2)$ of χ^2 for $10,000$ noisy sequences generated with the Lorenz map, for $L = 4$, $N = 1000$, $\sigma = 0.25$ (*continuous line*) and $\sigma = 0.50$ (*dashed line*) (SNR $\simeq 10$, 4.0 dB, respectively) (**a**) and for $L = 5$, $N = 8000$, $\sigma = 0.25$ (*continuous line*) and $\sigma = 0.50$ (*dashed line*) (SNR $\simeq 10$, 4.0 dB, respectively) (**b**)

plotted the probability P of rejecting the null hypothesis using the chi-square test with the same level $\alpha = 0.05$. Notice that the chi-square test correctly rejects the null hypothesis with higher probability than the BDS test in the high-noise regime ($\sigma \geq 1$), and its performance is comparable to the best one of the BDS test in the low-noise regime ($\sigma \leq 1$). Put in a different way, the probability of a false positive is higher with the BDS test. We conclude also from Fig. 9.9 that the BDS test performance strongly depends on the combinations of ϵ and m; for some combinations, this method wrongly accepts the null hypothesis even for small values of σ.

9.5.2 The Delayed Hénon Map

The *time-delayed Hénon map* [194] is defined as

$$x_n = 1 - ax_{n-1}^2 + bx_{n-d}, \qquad (9.12)$$

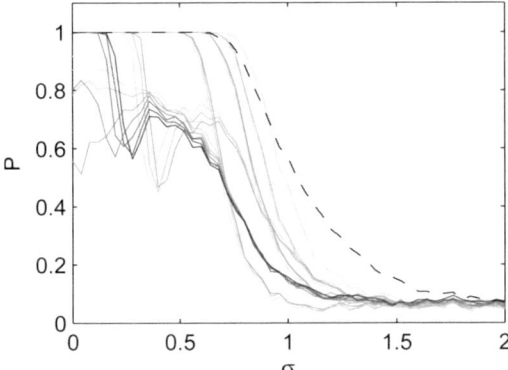

Fig. 9.9 The *continuous lines* indicate the probability of rejecting the null hypothesis H_0 ("the time series is i.i.d.") for a time series projected from the Lorenz map's attractor, contaminated with Gaussian white noise with σ up to $\sigma = 2$, when applying the BDS test with level $\alpha = 0.05$. In total, 27 tests for different combinations of ϵ and m were performed. The lighter the *gray color* is, the bigger is the value of ϵ used (see text for details). The *dashed line* indicates the probability of rejecting H_0 when using the chi-square test based on missing ordinal patterns, with the same level $\alpha = 0.05$. The chi-square test correctly rejects the null hypothesis more often than the BDS test

where a, b are real constants and $d \geq 1$. For $d = 1$, the time-delayed Hénon map is equivalent to the logistic map $x_{n+1} = A x_{n-1}(1 - x_{n-1})$, with [194]

$$A = \frac{b - 1}{2a} \pm \frac{1}{2a}\sqrt{(b - 1)^2 + 4a}.$$

For $d = 2$ and $a = 1.4$, $b = 0.3$, we recover the familiar two-dimensional dissipative Hénon map.

For $a = 1.6$ and $b = 0.1$, Sprott [194] finds the following linear relation between D_{KY} and d over the range $1 \leq d \leq 100$:

$$D_{KY} \cong 0.192d + 0.699.$$

The Kaplan–Yorke conjecture implies now

$$D_0 \geq D_1 = D_{KY} \cong 0.192d + 0.699$$

for the fractal dimension D_0 of the attractor, $1 \leq d \leq 100$. In particular, $D_0 \geq 1.083$ for $d = 2$, $D_0 \geq 10.299$ for $d = 50$, and $D_0 \geq 19.899$ for $d = 100$. Thus, this family of maps provides attractors with a wide range of fractal dimensions.

Figure 9.10 shows the return map ξ_{n+1} vs ξ_n for a typical orbit on the attractor of the time-delayed Hénon map with $d = 50$, both in the absence of noise, $\xi_n = x_n$ (a) and corrupted with Gaussian white noise, $\xi_n = x_n + w_n$, with $\sigma = 0.5$ (SNR $\simeq$ 1.3 dB) (b). Again, the geometry of the attractor has been completely blurred by the

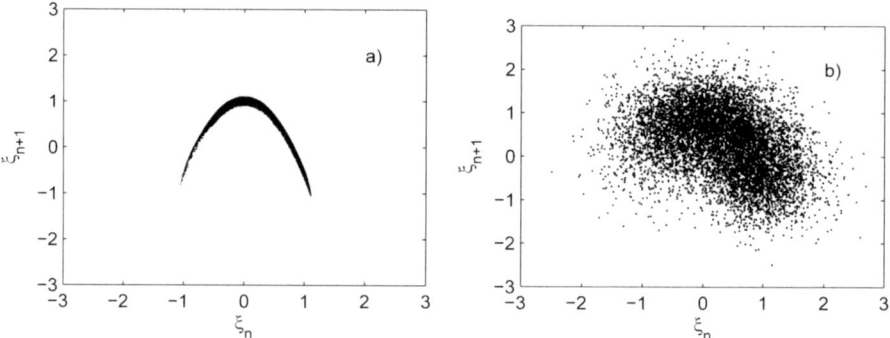

Fig. 9.10 Return map for a time series of the time-delayed Hénon map with $d = 50$ in the absence of noise (**a**) and contaminated with Gaussian white noise with $\sigma = 0.5$ (SNR $\simeq 1.3$ dB) (**b**). The structure of the underlying chaotic attractor has been totally blurred. Here again the count of missing ordinal patterns is sensibly higher than in the purely random case

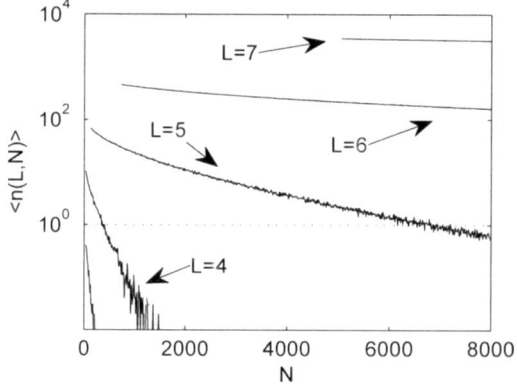

Fig. 9.11 Average number of missing ordinal patterns of length L found in a time series of length N, $\langle n(L, N) \rangle$ (in logarithmic scale), for a noisy series of the time-delayed Hénon map with $\sigma = 0.5$ (SNR $\simeq 1.3$ dB)

presence of the noise. However, it can be seen in Fig. 9.11 that also in this case, the number of missing ordinal L-patterns found in a time series of length N, $\langle n(L, N) \rangle$, is sensibly larger than in the white noise-only case, Fig. 9.7.

Figure 9.12(a)–(c) depicts the comparison of the chi-square test with the BDS test for $d = 2$, $d = 50$, and $d = 100$, respectively. Again, the probability of a false positive is higher with the BDS test. Since we are interested in the detection of determinism, we may conclude that the chi-square test, based on the distribution of visible ordinal patterns, is more reliable.

In conclusion, the (conditional + dynamical) robustness against additive noise of the forbidden patterns makes them a practical tool to distinguish deterministic, noisy time series from white noise. It is in this sense that we claim that forbidden patterns can be used to detect determinism in noisy time series—determinism as opposite to

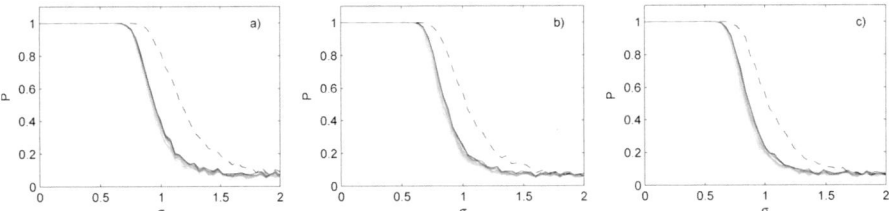

Fig. 9.12 Comparison of the chi-square test and the BDS test applied to projections of the time-delayed Hénon map with $d = 2$ (**a**), $d = 50$ (**b**), $d = 100$ (**c**), and Gaussian white noise with $0 \leq \sigma \leq 2$. The *continuous lines* indicate the probability of rejecting the null hypothesis H_0 ("the time series is i.i.d.") when applying the BDS test with level $\alpha = 0.05$. In total, 27 tests with different combinations of ϵ and m were performed. The lighter the *gray color* is, the bigger is the value of ϵ. The *dashed line* indicates the probability of rejecting H_0 when using the chi-square test with the same level $\alpha = 0.05$. Clearly, the chi-square test rejects the null hypothesis more often than the BDS for all noise values and for the three values of d

statistical independence. In fact, determinism is usually equated to statistical dependence among the observations in applications. On the other hand, the discrimination of deterministic, noisy time series from colored noise seems problematic, although some interesting methods have been proposed; see, e.g., [119] for a method based on nonlinear predictability.

Chapter 10
Space–Time Dynamics

All applications of ordinal analysis hitherto had to do with time series analysis or abstract dynamical systems. A remaining challenge is to expand the applications to physical systems.

In order to tackle the viability of this program, we are going to study the permutation complexity of two simple models of spatially extended physical systems: cellular automata (CA) and coupled map lattices (CMLs). CA were presented in Sect. 1.5. CMLs can be considered as a generalization of the CA; they retain the space coarse graining of the CA, but the state variable take on real values. Despite their apparent simplicity, these are the preferred models when studying the emergence of collective phenomena (such as turbulence, space–time chaos, symmetry breaking, ordering) in systems of many particles interacting nonlinearly. Indeed, their ability to reproduce complex phenomena in, say, fluid dynamics and solid state physics, is impressive. For this reason, they are the ideal choice for our purpose.

10.1 Spatially Extended Systems

Dynamical systems discrete in time as well as in space have been studied to understand physical phenomena while keeping the technical burden at a minimum. The discrete space can be an infinite lattice of dimension 1 (which can be identified with $\mathbb{Z}$) or a finite lattice with periodic or fixed boundary conditions. At each site i of the lattice there is a local variable $x_t(i)$ taking on values in a set S called the *state space*, at every time $t \in \{0, 1, \ldots\} = \mathbb{N}_0$. The change of the state variable $x_t(i)$ from time t to time $t + 1$ depends only on the variables in some fixed vicinity of i at time t.

Unless otherwise explicitly stated, we assume in this chapter the following restrictions for simplicity and computational convenience.

(i) Periodic boundary conditions:

$$x_t(0) = x_t(N) \text{ and } x_t(N + 1) = x_t(1) \tag{10.1}$$

for all $t \geq 0$. These conditions amount to the N sites lying on a ring.

J.M. Amigó, *Permutation Complexity in Dynamical Systems*,
Springer Series in Synergetics, DOI 10.1007/978-3-642-04084-9_10,
© Springer-Verlag Berlin Heidelberg 2010

(ii) Nearest neighbors interaction, i.e.,

$$x_{t+1}(i) = f(x_t(i-1),\ x_t(i),\ x_t(i+1)), \tag{10.2}$$

where $1 \le i \le N$.

Depending on the state space S, there are two well-known instances of such space–time systems: one-dimensional *cellular automata* (CA) if S is finite and one-dimensional *coupled map lattices* (CMLs) if S is an interval of $\mathbb{R}$. Given the formal similarity between both systems (see below for details), it comes as no surprise that they exhibit similar dynamical phenomena, like coherent traveling structures and space–time chaos [58, 110, 46, 47]. Perhaps more surprisingly is the fact that one-dimensional CMLs can be completely described in terms of symbolic dynamical concepts [170, 171]. Along similar lines, we are going to show that CA and CMLs can be handled in a satisfactory way with techniques based on ordinal patterns. In particular, (i) two so-called regularity parameters to be defined below seem to be useful for discriminating Wolfram's complexity classes in the case of CA and (ii) the number of admissible ordinal patterns in the configurations of CMLs separates space–time chaos from regular pattern dynamics.

CA and CMLs are not only related with each other but, in turn, are also related to networks—a subject of much interest in current research [162]. The main difference is the connectivity: while CA and CMLs feature near-neighbor interactions, networks allow also for long-range interactions. For a multidisciplinary introduction to dynamics on complex networks, see [132]. Networks of coupled maps have been studied, e.g., in [145, 104] with reference to synchronization. Whether ordinal analysis is also useful in this more general spatially extended systems is an open question as yet. Nevertheless, in view of the results reported in Sect. 2.4 on synchronization, we conjecture that ordinal analysis will be helpful to characterize the different synchronization regimes.

10.1.1 Cellular Automata

We refer to Sect. 1.1.5 for the generalities on cellular automata (CA). According to restriction (10.2), we consider local maps f with a neighborhood of size 1; furthermore, the state space will be $S = \{0, 1\}$, thus $f:\{0, 1\}^3 \rightarrow \{0, 1\}$. Technically, CA correspond to continuous, shift-commuting maps F from a full shift to itself; F is the global transition map induced by f on the configuration space Ω. Thus (Ω, F) is a continuous dynamical system. More generally one can also consider continuous, shift-commuting maps between subshifts of finite type (i.e., shift-invariant subsets of a full shift obtained after excluding a finite set of fixed blocks of symbols) [92, 123].

For brevity, one-dimensional binary CA with a neighborhood of size 1 will be called *elementary*. Elementary CA can be labeled as follows. Given the local rule

$$f(p,\ q,\ r) = \beta,$$

where p, q, r, $\beta \in \{0, 1\}$, order lexicographically the eight different configurations in the neighborhood $\mathcal{U}_1(i) = \{i - 1, i, i + 1\}$, to wit:

$$(0, 0, 0), (0, 0, 1), (0, 1, 0), (0, 1, 1), \ldots, (1, 1, 1).$$

If $\beta_0, \beta_1, \ldots, \beta_7 \in \{0, 1\}$ are the corresponding values of β, then the cellular automaton with the local rule f can be unambiguously identified by the number

$$\text{ID} = \sum_{i=0}^{7} \beta_i 2^i \in \{0, 1, \ldots, 255\}.$$

In other words, there are 256 different elementary CA.

Alternatively, one can argue as follows. To define a local rule, one must specify the update state of the central cell given all possible configurations of its local neighborhood. Since there are eight such configurations and two update states, the number of possible assignments is $2^8 = 256$.

For example, the cellular automaton with local rule

$$
\begin{aligned}
f(0, 0, 0) &= 0, & f(1, 0, 0) &= 0, \\
f(0, 0, 1) &= 1, & f(1, 0, 1) &= 1, \\
f(0, 1, 0) &= 1, & f(1, 1, 0) &= 1, \\
f(0, 1, 1) &= 1, & f(1, 1, 1) &= 0
\end{aligned}
$$

is coded as the decimal number

$$\text{ID} = 0 \times 2^0 + 1 \times 2^1 + 1 \times 2^2 + 1 \times 2^3 + 0 \times 2^4 + 1 \times 2^5 + 1 \times 2^6 + 0 \times 2^7$$
$$= 110.$$

Conversely, the local rule $f(p, q, r) = \beta$ of an elementary cellular automaton can be obtained from its identification number ID in a recursive form:

$$\beta_0 = \text{ID} \bmod 2,$$
$$\beta_i = \frac{\text{ID} - \beta_0 - \cdots - \beta_{i-1} 2^{i-1}}{2^i} \bmod 2,$$

$1 \leq i \leq 7$. Let us emphasize that in order to determine the evolution of these CA, all we need are the eight bits β_i—no closed formula for f is necessary. An explicit eight-parameter rule to construct a map $f : \{0, 1\}^3 \rightarrow \{0, 1\}$ delivering the right update states β_i for each local configuration can be found in [54, Table 4].

Stephen Wolfram studied exhaustively the asymptotic behavior of all 256 elementary CA. For each local rule and each initial configuration, he calculated the time evolution of the cellular automaton till it exhibited a stable pattern of behavior. Out of all these simulations, Wolfram proposed to classify the elementary cellular automata in four classes [206, 207]. In order of increasing complexity, these classes are the following:

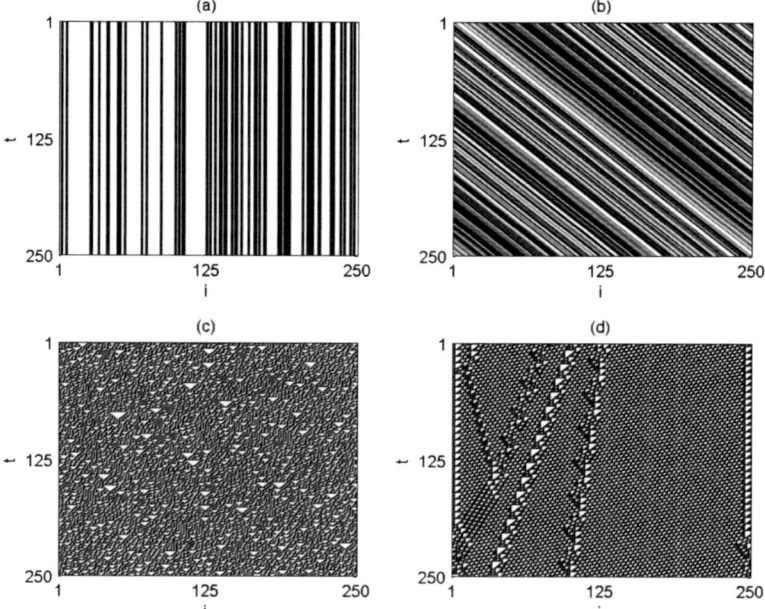

Fig. 10.1 Typical trajectories of elementary CA belonging to the complexity classes W1 (**a**), W2 (**b**), W3 (**c**), and W4 (**d**). The number of cells represented is $N = 250$. Time elapses top to bottom ($T = 250$ iterations represented)

(W1) The configurations converge to a fixed point; Fig. 10.1(a).
(W2) Time evolution yields a sequence of simple stable or periodic structures; Fig. 10.1(b).
(W3) The behavior is "chaotic"; Fig. 10.1(c).
(W4) Time evolution yields localized structures that move around and interact with each other in very complicated ways; Fig. 10.1(d).

A word of caution for the CA practitioners. Real cellular automata are finite deterministic machines, so their configuration space is finite. This means that their evolution is periodic, albeit the period can be very large—so large that this fact may be ignored in simulations.

10.1.2 Coupled Map Lattices

A CML is a discrete-time dynamical system with discrete space and *continuous* states. So one can think of CMLs as generalizations of CA [50], or rather as an intermediate between CA and partial differential equations. CMLs were introduced by Kaneko [106, 107] as a simple test bed for spatiotemporal complexity (turbulence, convection, etc.). For the theoretical aspects of CMLs the reader is referred to the papers of Bunimovich and Sinai, e.g., [42–44].

In dimension 1 the most common choices for the evolution rule (10.2) are

$$x_{t+1}(i) = (1 - \varepsilon)f(x_t(i)) + \varepsilon f(x_t(i - 1))$$

and

$$x_{t+1}(i) = (1 - \varepsilon)f(x_t(i)) + \frac{\varepsilon}{2}\left[f(x_t(i - 1)) + f(x_t(i - 1))\right], \qquad (10.3)$$

which correspond to the so-called *one-way* and *diffusive* CMLs, respectively. Here $0 \le \varepsilon \le 1$ so as all coupling coefficients are positive, $i = 1, \ldots, N$ label the sites, and f is a self-map of the state space $I \subset \mathbb{R}$. When the *coupling constant* ε is small, the oscillators will be practically independent of each other, hence the CML will behave similar to an ensemble of uncoupled oscillators. At the other end, strongly coupled oscillators will evolve more or less in a synchronized fashion. Between both cases, we expect to see a variety of behaviors as, so to speak, locally organized dynamics percolates along the lattice. It is the interplay between simple local properties (in our case, the coupling between neighboring oscillators) and the emergence of a complex dynamics on a global scale, what makes the study of CMLs, cellular automata, and the like, so rewarding (Fig. 10.2). For more general evolution rules, see e.g., [192].

The diffusive CML—the only one we consider henceforth— is the discrete analogue of the reaction–diffusion equation with a symmetrical interaction. Additional complexity can be added by allowing the map f to depend on a parameter. Following [108], we shall take the nonlinear ansatz

$$f(x) = 1 - ax^2, \quad x \in [-1, 1] \qquad (10.4)$$

and call $a \in (0, 2]$ the *nonlinearity* of f. Observe that if $x_0(i) \in [-1, 1]$ for $1 \le i \le N$, then $x_t(i) \in [-1, 1]$ for all $t \ge 1$ and $1 \le i \le N$.

Researchers on this field use to borrow terms from continuum physics like ordered or unordered phase, phase transition, local of global defects. According to [108], the *logistic coupled lattice* (10.3) (10.4) exhibits six major "phases":

(K1) Frozen random patterns; Fig. 10.3(a)
(K2) Pattern selection and suppression of chaos; Fig. 10.3(b)
(K3) Brownian motion of defects; Fig. 10.3(c)
(K4) Defect turbulence; Fig. 10.3(d)
(K5) Pattern competition intermittency; Fig. 10.3(e)
(K6) Fully developed turbulence; Fig. 10.3(f)

These six phases are shown on an a–ε diagram in Fig. 10.2; see [108] for details. Two-dimensional CMLs have been investigated, e.g., in [210, 23, 71].

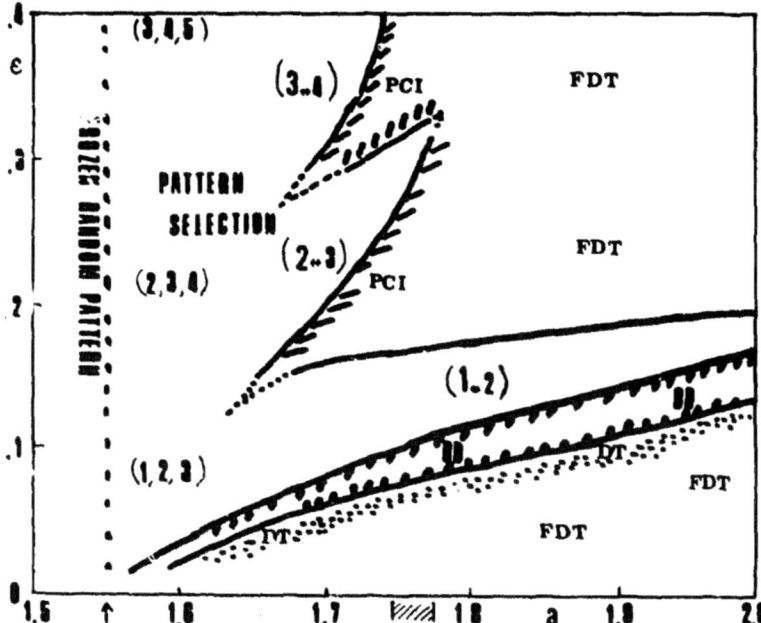

Fig. 10.2 [Reproduced with permission from [108].] Phase diagram of the coupled logistic map (10.4) (*a* varies along the *horizontal axis*, ε along the *vertical*). Here BD, DT, PCI, and FDT are the abbreviations of Brownian motion of defect, defect turbulence, pattern competition intermittency, and fully developed turbulence, respectively. The numbers such as 1,2,3 represent the selected domain sizes

10.2 Applications of Permutation Complexity to Spatiotemporal Dynamics

In this section we are going to show that the ordinal pattern-based approach to time series analysis and abstract dynamical systems works out also with one-dimensional binary cellular automata and one-dimensional coupled logistic lattices. This is a first step to extend ordinal analysis to space–time dynamics.

10.2.1 Topological Entropy of CA

The spatiotemporal complexity of a cellular automaton can be measured by the topological entropy. In Sect. 1.1.5 we mentioned that

$$h_{\text{top}}(F) = \lim_{w \to \infty} \lim_{t \to \infty} \frac{1}{t} \log R(w, t), \tag{10.5}$$

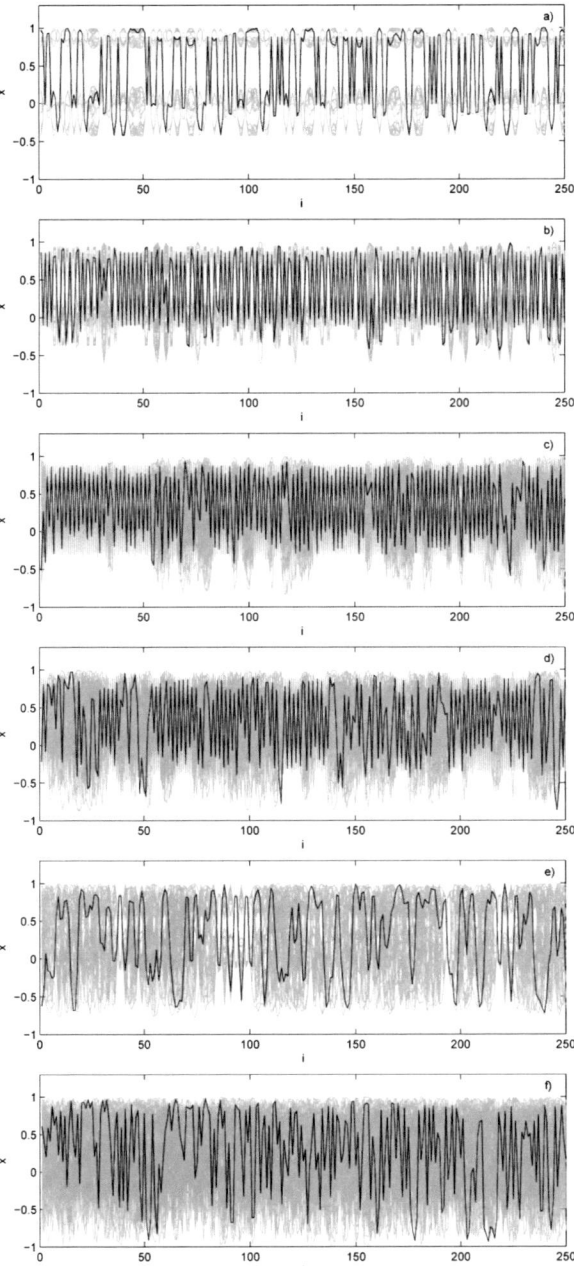

Fig. 10.3 CML space–time plots for (**a**) frozen random patterns ($a = 1.44$, $\varepsilon = 0.1$), (**b**) pattern selection and suppression of chaos ($a = 1.65$, $\varepsilon = 0.1$), (**c**) Brownian motion of defects ($a = 1.86$, $\varepsilon = 0.1$, (**d**) defect turbulence ($a = 1.89$, $\varepsilon = 0.1$), (**e**) pattern competition intermittency ($a = 1.8$, $\varepsilon = 0.3$), and (**f**) fully developed turbulence ($a = 2$, $\varepsilon = 0.3$). Each black line shows the CML state at time $n = 500$, the grey background is the superposition of states at $1 \leq n \leq 499$

where $F: S^{\mathbb{Z}} \to S^{\mathbb{Z}}$ and $R(w, t)$ is the number of distinct rectangles of width w and height (temporal extent) t occurring in a space–time evolution diagram of $(S^{\mathbb{Z}}, F)$; see (1.21) and Fig. 1.4.

Another possibility consists in using the topological permutation entropy $h^{*}_{\mathrm{top}}(F)$ instead. We shall shortly claim that, under some provisos, the result is going to be the same. But even in a general situation we might wish to link the spatiotemporal complexity of a cellular automaton to the permutation complexity of its time evolution as measured by the topological permutation entropy (in practice, by one or several entropy rates of finite order), or by other quantities based on ordinal patterns. Examples of the latter eventuality are provided by the parameters $\chi^{2}_{\mathrm{time}}(L)$ and $\chi^{2}_{\mathrm{space}}(L)$ presented below, absolute and relative frequency distributions of ordinal patterns, and any probability functional whose value is estimated by means of ordinal patterns.

Theorem 15 states that $h^{*}_{\mathrm{top}}(f) = h_{\mathrm{top}}(f)$ for any *positively expansive* self-map f of an n-dimensional simple domain. We could argue at this point that the proof of Theorem 15 does not rely on any particular property of compact sets in $\mathbb{R}^{n}$, in order to infer

$$h_{\mathrm{top}}(F) = h^{*}_{\mathrm{top}}(F) \tag{10.6}$$

for any positively expansive map F on a compact metric space, in particular when F is the global transition map of a one-dimensional cellular automaton. But for our purposes it will suffice to equate $h_{\mathrm{top}}(F)$ with the topological entropy of a topologically conjugate *interval* map. A cellular automaton is said to be *expansive* (correspondingly, *positively expansive*) when its global transition map F is expansive (correspondingly, positively expansive). It is interesting to point out that (i) positively expansive CA only exist in dimension 1 [188] (while expansive interval maps only exist in dimensions greater than 1 [19, Thm. 2.2.31]) and (ii) positively expansive CA are topologically conjugate to one-sided full shifts [62].

So, let us show how to calculate $h_{\mathrm{top}}(F)$ by means of the topological entropy of a two-dimensional interval map. Set $\Omega = S^{\mathbb{Z}}$, where $S = \{0, 1, \ldots, |S| - 1\}$ in the case of a one-dimensional cellular automaton with $|S|$ states, and define similar to (4.20) the map $\phi_{|S|} = (\phi^{-}_{|S|}, \phi^{+}_{|S|}) : S^{\mathbb{Z}} \to [0, 1]^{2}$,

$$\phi_{|S|} : \mathbf{x}_{t} \mapsto (\phi^{-}_{|S|}(\mathbf{x}^{-}_{t}), \phi^{+}_{|S|}(\mathbf{x}^{+}_{t})), \tag{10.7}$$

where $\mathbf{x}_{t} = (x_{t}(i))_{n \in \mathbb{Z}}$, $\mathbf{x}^{-}_{t} = (x_{t}(-i))_{i \in \mathbb{N}}$ is the left sequence of $\mathbf{x}_{t}$, $\mathbf{x}^{+}_{t} = (x_{t}(i))_{i \in \mathbb{N}_{0}}$ is the corresponding right sequence, the component maps $\phi^{-}_{|S|} : S^{\mathbb{N}} \to [0, 1]$, $\phi^{+}_{|S|} : S^{\mathbb{N}_{0}} \to [0, 1]$ are given by

$$\phi^{-}_{|S|}(\mathbf{x}^{-}_{t}) = \sum_{i=1}^{\infty} \frac{x_{t}(-i)}{|S|^{i}}, \quad \phi^{+}_{|S|}(\mathbf{x}^{+}_{t}) = \sum_{i=0}^{\infty} \frac{x_{t}(i)}{|S|^{i+1}}, \tag{10.8}$$

and the bisequences $\mathbf{x}_t = (\mathbf{x}_t^-, \mathbf{x}_t^+)$ are lexicographically ordered as in (4.19). We already know (Sect. 4.3) that the map $\phi_{|S|}$ is an order isomorphism ($[0, 1]^2$ being lexicographically ordered), up to a measure zero set $\mathcal{N}$ which comprises those bisequences whose left and/or right sequences terminate in $1, 0^\infty$ or $0, (|S| - 1)^\infty$. Furthermore, it is easy to check that $\phi_{|S|}$ is a homeomorphism from $S^{\mathbb{Z}} \backslash \mathcal{N}$ to its range. In other words, the continuous dynamical systems (Ω, F) and $([0, 1]^2, \phi_{|S|} \circ F \circ \phi_{|S|}^{-1})$ are topologically conjugate (modulo 0), hence

$$h_{\text{top}}(F) = h_{\text{top}}(\tilde{F}). \tag{10.9}$$

where $\tilde{F} := \phi_{|S|} \circ F \circ \phi_{|S|}^{-1} : [0, 1]^2 \to [0, 1]^2$ is an interval map.

Suppose, moreover, that F is positively expansive. In this case the same holds for $\tilde{F}$ since positive expansiveness is a topological conjugacy invariant (Sect. B.3.1). Then

$$h_{\text{top}}(\tilde{F}) = h_{\text{top}}^*(\tilde{F}) \tag{10.10}$$

according to Theorem 15. The bottom line from (10.9) and (10.10) is

$$h_{\text{top}}(F) = h_{\text{top}}^*(\tilde{F}) \tag{10.11}$$

for positively expansive (one-dimensional) CA (Ω, F). Finally, to go from (10.11) to (10.6), we only need to invoke that topological permutation entropy is an invariant of order isomorphy (here embodied by the homeomorphism $\phi_{|S|}$); see Theorem 14.

A convenient shortcut in actual calculations is the following. The lexicographical order of bisequences $\mathbf{x} \in S^{\mathbb{Z}}$ and points $(x, y) \in [0, 1]^2$ is determined by the right sequences $\mathbf{x}^+$ and ordinates y, respectively. This means that if the right sequences of a finite orbit $F^t(\mathbf{x}_0)$, $0 \le t \le T$, are all different (as usual in numerical simulations), then we may restrict attention to the ordinates of the order-isomorphic orbit $\phi_{|S|} \circ F^t(\mathbf{x}_0) = \phi_{|S|}(\mathbf{x}_t)$. From (10.7) we learn that the ordinate of $\phi_{|S|} \circ F^t(\mathbf{x}_0)$ is

$$\phi_{|S|}^+(F^t(\mathbf{x}_0)^+) = \phi_{|S|}^+(\mathbf{x}_t^+) = \sum_{i=0}^{\infty} \frac{\mathbf{x}_t(i)}{|S|^{i+1}}. \tag{10.12}$$

To check numerically the coincidence of topological permutation entropy and topological entropy for positively expansive CA, we resort to *linear* automata. A one-dimensional CA is said to be linear if its local rule is of the form

$$f(s_t(i - l), s_t(i - l + 1), \dots, s_t(i + l)) = \sum_{j=-l}^{j=l} \lambda_j s_t(i + j) \bmod |S|. \tag{10.13}$$

For a one-dimensional linear CA, (10.5) yields a closed formula for the topological entropy [62]: if $p_1^{m_1} \cdots p_h^{m_h}$ is the prime factor decomposition of $|S|$, and

$$P_i = \{0\} \cup \{j: \gcd(\lambda_j, p_i) = 1\}, \; L_i = \min P_i, \; R_i = \max P_i,$$

then

$$h_{\text{top}}(F) = \sum_{i=1}^{h} m_i(R_i - L_i) \log p_i. \tag{10.14}$$

Furthermore, it can be proved [141, Theorem 3.2] that a one-dimensional *linear* CA (10.13) is positively expansive if and only if

$$\gcd(|S|, \lambda_{-l}, \ldots, \lambda_{-1}) = \gcd(|S|, \lambda_1, \ldots, \lambda_l) = 1. \tag{10.15}$$

From (10.15) it follows that the local rule $f:\{0, 1\}^3 \to \{0, 1\}$ with

$$
\begin{aligned}
f(s_t(i-1), s_t(i), s_t(i+1)) &= s_t(i-1) + s_t(i+1) \bmod 2 \\
&= s_t(i-1) \oplus s_t(i+1) \tag{10.16}
\end{aligned}
$$

$(\lambda_{-1} = \lambda_1 = 1, |S| = 2)$ defines a *positively expansive* CA. According to (10.14),

$$h_{\text{top}}(F) = 2 \log 2 = 2 \text{ bit/symbol}.$$

The topological permutation entropy of the automaton defined by the local rule (10.16) can be now estimated via the ordinal patterns of its global map $F:\{0, 1\}^{\mathbb{Z}} \to \{0, 1\}^{\mathbb{Z}}$ or alternatively via the ordinal patterns of the interval map $\tilde{F} = \phi_2 \circ F \circ \phi_2^{-1}:[0, 1]^2 \to [0, 1]^2$. As explained above, it suffices to keep account of the ordinal patterns defined by the ordinates of $\phi_2 \circ F^t(\mathbf{x})$, namely, $\phi_2^+(\mathbf{x}_t^+)$, (10.12).

Figure 10.4 shows different aspects of the cellular automaton (10.16): (a) the time evolution of cells $1 \le i \le 250$; (b) the ordinates $\phi_2^+(\mathbf{x}_t^+)$ of the finite orbit $\phi_2(F^t(\mathbf{x}_0)) = \tilde{F}^t(\phi_2(\mathbf{x}_0))$, $0 \le t \le 250$, where $x_0(1), \ldots, x_0(250)$ were chosen randomly and extended periodically in both directions; (c) the return map $\phi_2^+(\mathbf{x}_t^+)$ vs $\phi_2^+(\mathbf{x}_{t+1}^+)$ (this graph has seemingly a fractal structure); and (d) the convergence of the topological permutation entropy rates of order L,

$$h_{\text{top}}^*(L, \tilde{F}) = -\frac{1}{L} \log |\{\pi \in \mathcal{S}_L : P_{\pi} \ne \emptyset\}|,$$

to $h_{\text{top}}(F) = 2$ bit/symbol, with the length of the ordinal patterns. This convergence is fast, also in computation.

10.2.2 Complexity Classes of Elementary CA

Elementary CA with periodic boundary conditions were also extensively studied in a series of papers by Chua and collaborators. According to [55],

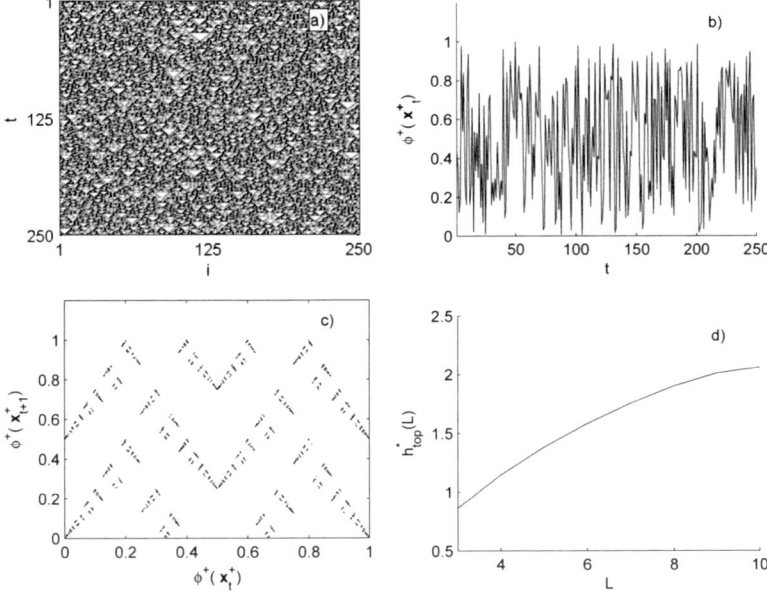

Fig. 10.4 Different aspects of a positively expansive CA (see text). Plot (**d**) shows the convergence of the topological permutation entropy of the automaton to its topological entropy

(1) the cellular automaton with local rule

$$f(p, q, r) = \frac{1}{2}[1 + \text{sign}(2p + 4q + 2r - 5)],\qquad(10.17)$$

ID $= 200$, is an instance of *class W1*;

(2) the cellular automaton with local rule

$$f(p, q, r) = p\qquad(10.18)$$

(corresponding to the right shift on $\{0, 1\}^{\mathbb{Z}}$), ID $= 240$, belongs to *class W2*;

(3) the cellular automaton with local rule

$$f(p, q, r) = p + q + r + qr \bmod 2,\qquad(10.19)$$

ID $= 30$, is *class W3*; and

(4) the cellular automaton with local rule

$$f(p, q, r) = (1 + p)qr + q + r \bmod 2,\qquad(10.20)$$

ID $= 110$, belongs to *class W4*. Moreover, this automaton is surely universal in the sense that it can emulate a universal Turing machine [207, p. 1115].

In order to discriminate these four classes, we propose two parameters inspired in the statistic χ^2, used in Sects. 9.3 and 9.5 for detecting determinism in noisy time series. The rationale is as follows. Since the statistic χ^2 is based on ordinal pattern distribution, being small for i.i.d. random processes and large for deterministic processes, we expect that it can also discriminate irregular from regular configurations as time evolves. For this reason, we call them *regularity parameters*.

(a) *Temporal regularity parameter* $\chi^2_{\text{time}}(L)$. In numerical simulations, let

$$\mathbf{x}_t = (x_t(i))_{i=1}^N = x_t(1), x_t(2), \ldots, x_t(N)$$

be the configuration of cells $1 \leq i \leq N$ at time t, $0 \leq t \leq T$. Calculate now $\chi^2(L)$, (9.8), for the multivariate time series

$$\mathbf{x}_0, \mathbf{x}_1, \ldots, \mathbf{x}_T \tag{10.21}$$

using, say, lexicographical order. Alternatively, transform each $\mathbf{x}_t$ into a dyadic rational,

$$\phi(\mathbf{x}_t) = \sum_{i=1}^N \frac{x_t(i)}{2^i} \in [0, 1), \tag{10.22}$$

and calculate $\chi^2(L)$ for the univariate time series

$$\phi(\mathbf{x}_0), \phi(\mathbf{x}_1), \ldots, \phi(\mathbf{x}_T), \tag{10.23}$$

since sequences (10.21) and (10.23) are order isomorphic. Call $\chi^2_{\text{time}}(L)$ the result.

(b) *Spatial regularity parameter* $\chi^2_{\text{space}}(L)$. We want now to calculate the regularity of the univariate time series consisting of the state variables at time t,

$$x_t(1), x_t(2), \ldots, x_t(N),$$

and average the results over all times, $0 \leq t \leq T$. There is a catch though. Statistic (9.8) correspond to i.i.d. random variables taking on real values. In the finite-state case we are considering now, some symbols will necessarily repeat as soon as the length of the sequence exceeds the number of states. For binary variables this implies that not all 2^L *ordinal* patterns of an i.i.d. binary sequence are equiprobable. Indeed, all the $L + 1$ words of length L

$$(0, 0, \ldots, 0, 0), (0, 0, \ldots, 0, 1), (0, 0, \ldots, 1, 1), \ldots, (0, 1, \ldots, 1, 1), (1, 1, \ldots, 1, 1)$$

are of type $\pi_0 = \langle 0, 1, 2, \ldots, L - 1 \rangle$, while each of the remaining $2^L - L - 1$ words defines a distinct ordinal pattern. Therefore, the chi-square statistic χ^2 for windows of size L takes the following form for binary sequences:

$$\chi^2(L) = \frac{\left(\nu_0 - \frac{L+1}{2^L}\right)^2}{(L+1)/2^L} + (2^L - L - 1)\frac{\left(\nu_1 - \frac{1}{2^L}\right)^2}{1/2^L} \tag{10.24}$$

$$= \frac{\left(2^L \nu_0 - L - 1\right)^2}{2^L(L+1)} + \left(1 - \frac{L+1}{2^L}\right)(2^L \nu_1 - 1)^2,$$

where ν_0 is the number of times the pattern $\pi_0 = \langle 0, 1, 2, \ldots, L - 1 \rangle$ has been observed in the sequence and ν_1 is the number of patterns $\pi \in S_L, \pi \neq \pi_0$, observed in the same sequence, when using non-overlapping sliding windows. In sum, in order to obtain the spatial regularity $\chi^2_{\text{space}}(L)$, calculate the parameter $\chi^2(L)$, (10.24), of the univariate time series $(x_t(i))_{i=1}^N$ for each time $0 \leq t \leq T$ and average over them:

$$\chi^2_{\text{space}}(L) = \left\langle \chi^2(L) \right\rangle.$$

In our numerical simulations we chose $N = 250$. To avoid too small samples, we take $L \leq 4$ for $\chi^2_{\text{space}}(L)$. For $\chi^2_{\text{time}}(L)$ we may choose L larger, provided that T is sufficiently long. Furthermore, in order to let transients die out, we forgo the first 5000 iterations.

For the four representatives of the complexity classes W1–W4 given above (ID = 200, 240, 30, and 110), we have simulated their time evolution, starting from 100 randomly chosen initial configurations. When the resulting values of $\chi^2_{\text{time}}(5)$ are plotted against $\chi^2_{\text{space}}(4)$ we see, Fig. 10.5, that they cluster in different, non-overlapping regions.

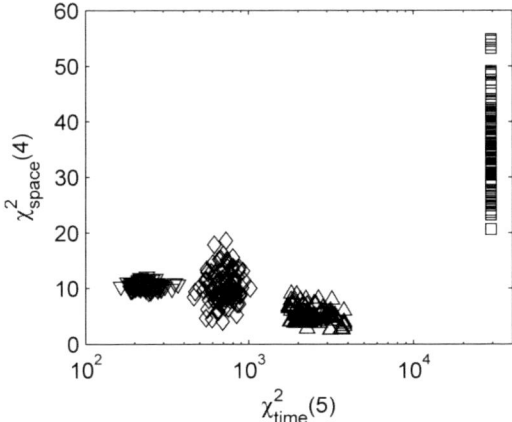

Fig. 10.5 Values of $\chi^2_{\text{time}}(5)$ and $\chi^2_{\text{space}}(4)$ for four CA of different complexity classes and 100 random initial configurations. Symbol assignment: Classes W1 ($\square$), W2 ($\diamond$), W3 (∇), and W4 ($\triangle$)

We have repeated the same exercise with a few more CA and the results are similar, although the clusters of different CA belonging to the same complexity class may lie in different parts of the χ^2_{time}–χ^2_{space} diagram. All this hints that regularity parameters capture the basic features of the different complexity classes of elementary CA.

For the study of the complexity of CA rules by other methods, see, e.g., [103].

10.2.3 Phases of CMLs

The basic difference between CA and CMLs concerns the state space and eventually the appearance of free parameters in the second case (e.g., the nonlinearity a in (10.4)). Therefore, we expect that the same tools used in the last section to study the spatiotemporal complexity of CA will also be useful for CMLs. We shall use the logistic coupled lattice as study case.

So, consider a one-dimensional logistic coupled lattice with N sites (extended periodically in both directions), pick an initial configuration $(x_0(i))_{i=1}^N$, $x_0(i) \in [0, 1]$, and let it evolve during $T_0 = 5000$ time steps according to the diffusive rule (10.3)–(10.4). From T_0 on we assume that the lattice exhibits its asymptotic dynamics.

A first proposal to quantify the complexity of a CML, inspired in the calculation of the topological entropy of positively expansive CA, is the following. At each iteration of the CML, define the symbolic sequence

$$(s_t(i))_{i=1}^N = s_t(1), s_t(2), \ldots, s_t(N) \equiv \mathbf{s}_t, \tag{10.25}$$

where

$$s_t(i) = \begin{cases} 0 & \text{if } x_t(i) \leq 0, \\ 1 & \text{if } x_t(i) > 0. \end{cases} \tag{10.26}$$

In this way we get a finite multivariate binary sequence $\mathbf{s}_0, \mathbf{s}_1, \ldots, \mathbf{s}_T$; alternatively, we might prefer to work with the order-isomorphic sequence $\phi(\mathbf{s}_0), \phi(\mathbf{s}_1), \ldots, \phi(\mathbf{s}_T)$, of the dyadic rationals

$$\phi(\mathbf{s}_t) = \sum_{i=1}^N \frac{s_t(i)}{2^i} \in [0, 1). \tag{10.27}$$

At this point we could count the number of visible ordinal patterns of length L, $N(L)$, of the sequence $\mathbf{s}_0, \mathbf{s}_1, \ldots, \mathbf{s}_T$ or, equivalently, $\phi(\mathbf{s}_0), \phi(\mathbf{s}_1), \ldots, \phi(\mathbf{s}_T)$, and estimate their metric or topological permutation entropy. Other even simpler possibility consists in representing $N(L)$ on the $(a\text{-}\varepsilon)$-plane. This has been done in Fig. 10.6 for $L = 5$ and $N = 250$. As for the nonlinearity a and the coupling constant ε, they are allowed throughout to take 75 values uniformly distributed in similar ranges as in Fig. 10.2, namely, $[1.4, 2]$ and $[0, 0.5]$, respectively. Remarkably, there are two zones of dark/light gray colors in Fig. 10.6 that roughly correspond with the zones

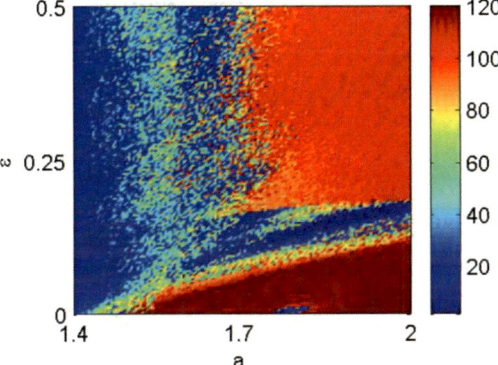

Fig. 10.6 Number of visible ordinal 5-patterns for the logistic coupled lattice as a function of a and ε, obtained from the symbolic sequence $\{\phi(\mathbf{s}_t)\}_{t=1}^T$

of space–time chaos and regularity sketched in Fig. 10.2. Note that higher values of $N(L)$ correspond to more complex dynamics.

One further possibility out of many others is to calculate the number of visible ordinal patterns, $N(L)$, in each univariate sequence $\mathbf{x}_t = (x_t(i))_{i=1}^N$ and to average the $T+1$ results. In our case, the value of L has to be small because of the condition $L! \ll N = 250$ (so as every ordinal L-pattern has a chance to appear in sliding windows along $\mathbf{x}_t$). The result is shown in Fig. 10.7; note that this figure gives information complementary to that provided by Fig. 10.6. A global increase of regularity (thus a decrease of $N(L)$) is observed as the strength of the coupling ε grows, as expected, but drastic transitions are also observed, corresponding to changes in the dynamics observed previously.

As a benchmark we consider next plots of Lyapunov exponents; these have been used to study various features of CMLs, like synchronization [18]. Figure 10.8

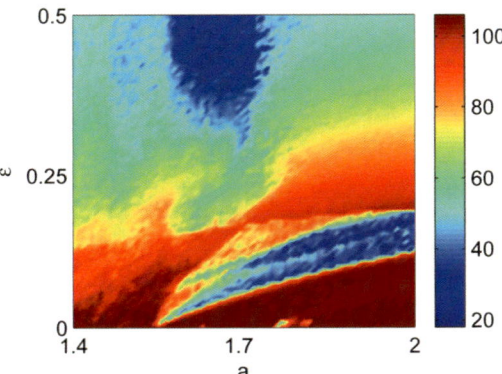

Fig. 10.7 Number of visible ordinal 5-patterns for the logistic coupled lattice as a function of a and ε, obtained from $\mathbf{x}_t$ and averaged over t

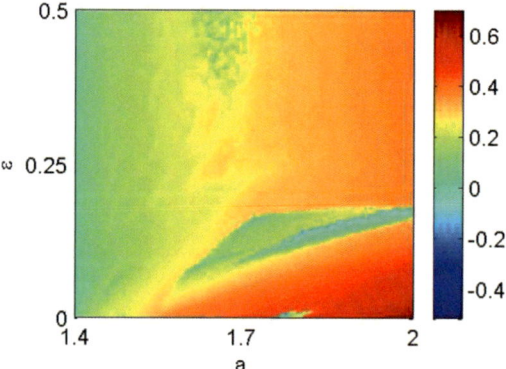

Fig. 10.8 Calculation of the largest Lyapunov exponent of a CML as a function of a and ε

shows a plot of the largest Lyapunov exponent λ calculated for the logistic coupled lattice (10.3)–(10.4) using Wolf's algorithm [204]. It can be observed there that the boundaries between the different phases of the CML sketched by Kaneko coincide roughly with abrupt changes in the value of λ. These results are coherent with the results observed in our calculations of $N(L)$ in Figs. 10.6 and 10.7. Let us point out that the separation between the domains of fully developed turbulence and the rest of phases can be distinguished more clearly in the $N(L)$ plots.

For the sake of completeness we consider also a chain of 60 coupled oscillators,

$$\dot{u}_i = 0.5 - 4v_i + \kappa(u_{i+1} + u_{i-1} - 2u_i),$$
$$\dot{v}_i = -v_i + 2\max\{u_i - 8\cos t - 16, 0\}, \tag{10.28}$$

with periodic boundary conditions. If we make a stroboscopic map of the variable u_i and plot $u_i(2\pi n)$ against $u_i(2\pi(n+1))$, points lie approximately on a one-dimensional curve with a critical point at $u_c \approx 6.6$ (Fig. 10.9). Thus, each period 2π we assign to the stroboscopic map of system (10.28), $\{u_i(2\pi n)\}_{i=1}^{60}$, a string of symbols following the usual procedure ($s_i(n) = 0$ if $u_i(2\pi n) < u_c$, and $s_i(n) = 1$ otherwise), and count the number of visible ordinal patterns of the ensuing binary multivariate time series $\mathbf{s}_n = (s_1(n), \ldots, s_{60}(n))$. Figure 10.10 represents the number of ordinal 4-patterns, $N(4)$, of such series as a function of the coupling constant κ. The inlets in this figure are space–time plots of $\{u_i(2\pi n)\}_{i=1}^{60}$ for $n = 1, .., 200$ and three values of κ: $\kappa = 0.008$, $\kappa = 0.1$, and $\kappa = 0.18$ (left to right). Observe that the decrease of $N(L)$ with κ parallels the diminution of dynamical complexity, in particular the regularization of the dynamics and/or the reduction of chaotic domains (i.e., the number of consecutive sites with chaotic dynamics). We conclude that ordinal analysis might also be suitable to characterize the complexity of oscillator chains.

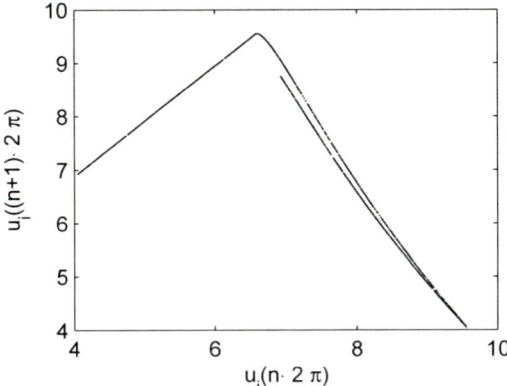

Fig. 10.9 Return map observed by plotting $u_i(n2\pi)$ against $u_i((n+1)2\pi)$ for any of the oscillators of the chain with $\kappa = 0$. Its unimodal appearance allows using symbolic sequences

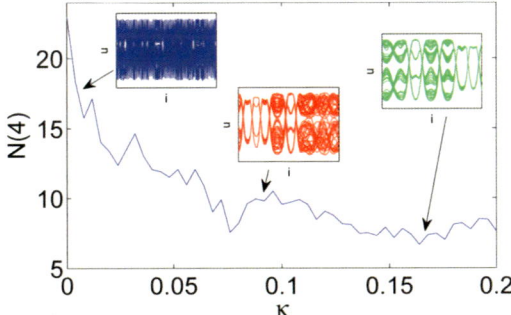

Fig. 10.10 Number of ordinal patterns of length $L = 4$, $N(4)$, found in a time series of length 200. The numbers are actually averages over the results for 20 initial conditions. The decrease of $N(4)$ is consistent with the decrease in complexity of the space–time dynamics shown in the *inlets*, which are space–time plots $\{u_i(\pi 2n)\}_{i=1}^{60}$ for $n = 1, .., 200$ and three values of κ: $\kappa = 0.008$, $\kappa = 0.1$, and $\kappa = 0.18$ (*left* to *right*)

Needless to say, the tools that can be chosen to measure the complexity of a CML are manifold. In the next section we study the use of regularity parameters.

10.2.4 Spatiotemporal Regularity of CMLs

Lastly, we consider the same temporal and spatial regularity parameters proposed for CA. But since the entries of the time series are now real numbers, the parameter $\chi^2(L)$ is given by (9.8) also when calculating $\chi^2_{space}(L)$.

Similarly as in Sect. 10.2.2, we have simulated the evolution of six logistic coupled lattices with $N = 250$ sites, each starting from 100 different random initial configurations. The corresponding parameters a and ε were chosen as in Fig. 10.3, so each lattice was in one of the six phases listed in Sect. 10.1.2. Figure 10.11

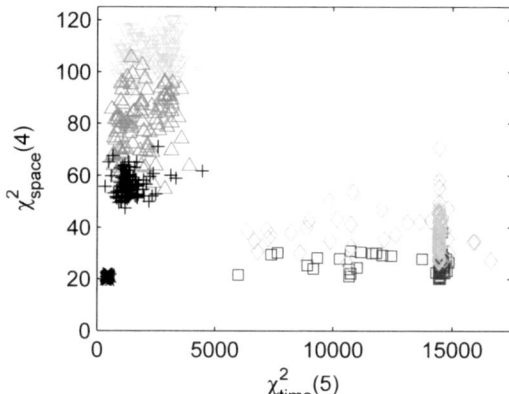

Fig. 10.11 Values of $\chi^2_{\text{time}}(5)$ and $\chi^2_{\text{space}}(4)$ for six logistic coupled lattices in different phases and 100 random initial configurations. Symbol assignment: frozen random patterns ($\square$), pattern selection and suppression of chaos ($\diamond$), Brownian motion of defects (∇), defect turbulence $\triangle$, pattern competition intermittency (+), and fully developed turbulence ($\times$). Different colors in the symbols are used when convenient

summarizes the results again on the plane $\chi^2_{\text{time}}(5)$ vs $\chi^2_{\text{space}}(4)$. The values cluster in different zones, but this time they overlap. The results are coherent with the types of dynamics described by Kaneko [108]. In some cases, overlapping might be due to multistability, i.e., depending on the initial conditions the type of dynamics may greatly vary.

Chapter 11
Conclusion and Outlook

Ordinal (or permutation-based) analysis of dynamical systems originates from the properties of the order relations and order isomorphisms. Thereby it is assumed that the state space of the systems is equipped with a total ordering. The order relations among consecutive elements in the orbits of deterministic or random dynamical systems are then codified in the form of ordinal patterns. The ordinal patterns themselves—whether admissible or forbidden—together with other "higher level" tools based on them, like permutation entropy rates, discrete entropy, frequency or probability distributions, regularity parameters, build the main repertoire of ordinal analysis. Since the sort of properties addressed by ordinal analysis and captured by its tools are not the same as in the usual measure-theoretical and topological approaches, we proposed the term "permutation complexity" to distinguish them.

In the foregoing chapters we have reviewed the theoretical and practical aspects of ordinal analysis. Among the first ones, let us highlight the study of metric (Chap. 6) and topological (Chap. 7) permutation entropies, together with the relation to their standard counterparts. Among the applications, some of them are well established, like the estimation of entropy (Sect. 2.1), complexity analysis of time series (Sect. 2.2), or detection of determinism (Chap. 9). Others like the complexity analysis of spatially extended systems (Chap. 10) are still in an initial stage. An important message to keep regarding all ordinal pattern-based applications is their robustness against observational noise—an asset when analyzing real systems. In particular, deterministic generation is responsible for the persistence of forbidden patterns in very noise data, as shown in Sect. 9.1. Robustness makes ordinal analysis a practical tool.

The reader might be tempted to dismiss ordinal analysis of dynamics as an uninteresting equivalent to well-known symbolic dynamics. In fact, ordinal patterns of dynamical systems do maintain equivalent results with symbolic dynamics, such as the metric and topological entropies we discussed in Chaps. 6 and 7, respectively, but in other ways, there are major distinctions, which are just starting to be explored for permutations. For instance, the canonical tent map and the Bernoulli shift ($f(x) = 2x \mod 1$) are isomorphic under a conventional analysis and in symbolic dynamics are equivalent to an i.i.d. source of white bits. However, under permutation-based analysis, once the state is imbued with total ordering, the class of order isomorphisms is different. Both conventional symbolic dynamics, assuming a

J.M. Amigó, *Permutation Complexity in Dynamical Systems,*
Springer Series in Synergetics, DOI 10.1007/978-3-642-04084-9_11,
© Springer-Verlag Berlin Heidelberg 2010

generating partition of a map, and ordinal analysis are useful discrete representations of what would otherwise be a dynamical system in continuous space. However, the symbolic dynamics which results from a conventional partitioning is not fundamentally distinguishable from a noisy system; both result in conventional information sources on a discrete alphabet with a positive Shannon entropy. By contrast, the ordinal analysis does show a fundamental distinction between deterministic chaos and noisy systems. With chaos there is a rich structure of forbidden patterns among the ordinal patterns of different length and a hierarchy of consequent derived forbidden patterns (Chap. 3), the nature of which is not shared with conventional symbolic dynamics. More closely impacting the present work, the number of allowed permutations can scale superexponentially, which is fundamentally faster than the exponential scaling which must eventually happen with a noise-free deterministic chaotic system.

As in any research field, work on theory and applications of ordinal analysis is in progress, meaning that the picture is far from complete. In the course of the exposition, we have pointed out different questions which are waiting for answers. I summarized next the most important ones.

One of the basic open problems refers to the relation between a map and the structure of its forbidden patterns. Some natural questions that arise in this context are the following:

- Understand how the allowed or forbidden ordinal patterns (especially the root patterns) depend on the map.
- Given a map, determine the length of its shortest forbidden pattern.
- Describe and/or enumerate (exactly or asymptotically) any of the above classes of ordinal patterns.
- Given a finite or infinite set of, say, root forbidden patterns, find a map with the corresponding ordinal pattern structure.
- More generally, characterize those hierarchies of ordinal patterns for which there exist maps realizing them.

Of course, some of these questions can be answered graphically for simple maps and short pattern lengths. What we seek though are general results, possibly emanating from the structure of periodic points. We reported partial successes along this line for the shifts (Chap. 4) and signed shifts (Chap. 5), but the general case seems exceedingly hard. Even the ordinal structure of a general subshift of finite type (order isomorphic to some piecewise linear maps) seems to be beyond the techniques used in those chapters. A list of more advanced research topics would include the relation of forbidden patterns with the kneading invariants of one-dimensional interval maps or, say, with the directional entropy of cellular automata.

Other interesting (albeit theoretical) problem is the exact relation between the original definition of permutation entropy by Bandt et al. [29], and the definition given in Chaps. 6 and 7. Technically, the difference boils down to the order of two limiting processes (ever longer ordinal patterns and ever finer partitions) in a double limit. In particular, the results of Sects. 6.2 and 6.3 show that both definitions of metric permutation entropy overlap for one-dimensional, piecewise ergodic maps, and numerical simulations advocate a more general coincidence. In any case, the usual computations, with an arithmetic precision fixed by default or by the numerical format chosen, implement our "Kolmogorov-like" approach to permutation entropy.

For practical applications, the numerical tools of the type we discussed in Chap. 9 serve as a way of distinguishing chaos-like dynamics from noise, at least in simulations. This may be useful in the detection of emergent "coherent structures" similar to low-dimensional chaos in what otherwise might be a high degree of freedom system which could be rather noise-like. We comment on the unique property of permutations having a discrete "algebraic" nature permitting some rapid computational methods, without the requirement of estimating a generating partition for each dynamics. We feel that the appropriate tools for analysis of the typically short *observed* time series will require more sophisticated statistical thinking and methods still, just as high-quality estimation of entropies from low-alphabet information sources can be a difficult problem despite the apparent simplicity of the definitions themselves.

In Chap. 9 we also showed that the forbidden pattern-based technique outperforms one of the standard methods for detecting statistical dependence. Similar conclusions were reached in the ordinal analysis of synchronization in [159], see Sect. 2.4. This exercise—comparing a pattern-based technique with the traditional methods—is missing in other applications of ordinal analysis to time series like entropy estimation or complexity study. If the applications refer to natural systems, then the possibilities are virtually unlimited. Real time series appeared only in Sect. 2.2 ("Permutation complexity"), where we considered biomedical data, a recurrent topic in the literature. But, of course, other kinds of real data have also been studied (see Sect. 2.2).

Apart from the future lines of research related to the above-mentioned open problems, other lines of research refer to more recent topics and other follow-up investigations. In Chap. 10 we showed that ordinal analysis provides quantitative tools for and insights into the dynamics of space–time dynamics. This brief account was meant as a corroboration of performances shown in other contexts, as well as a stimulus to further research. Clearly, a survey of permutation complexity in cellular automata and coupled map lattices is a broad field that will require time and ingenuity, especially in the unexplored dimensions 2 and higher. Add to this general networks of coupled map lattices, and you get a long-term research program! But the great challenge is the complexity analysis of physical systems. Simple models, like cellular automata and coupled map lattice, provide a bridge to this more ambitious objective, in that they model non-trivial physical phenomena while being amenable to discrete methods. The situation resembles the study of complex dynamical systems via symbolic dynamics—a quite remarkable technique. The author believes that the interplay between complex dynamical systems and discrete methods is a promising approach also in the case of physical systems. Chapter 10 reported on progress in this direction from the ordinal front. New chapters will follow.

Annex A
Mathematical Framework

This annex is a summary of the mathematical background needed for this book.

A.1 Dynamical Systems

In this book we only consider two kinds of "discrete-time" dynamical systems: continuous and measure-preserving systems. Roughly speaking, the first are the basic objects of topological dynamics and the second ones play a major role in the study of statistical properties.

Definition 7 A continuous (or topological) dynamical system is a pair (M, f), where M is a topological space and $f : M \to M$ a continuous map.

Let Ω be a non-empty set, $\mathcal{B}$ a sigma-algebra of subsets of Ω, and $\mu : \mathcal{B} \to \mathbb{R} \cup \{+\infty\}$ a positive measure on the measurable space $(\Omega, \mathcal{B})$. A typical example of measurable space is a topological space endowed with the *Borel sigma-algebra*, i.e., the sigma-algebra generated by the open sets. The *measure space* $(\Omega, \mathcal{B}, \mu)$ is called a finite-measure space if $\mu(\Omega) < \infty$. A measurable map (function, transformation) $f : \Omega \to \Omega$ is said to preserve the measure μ, or to be μ-preserving, if $\mu(f^{-1}(B)) = \mu(B)$ for all $B \in \mathcal{B}$. Equivalently, the measure μ is said to be f-invariant. Sometimes $(\Omega, \mathcal{B}, \mu)$ is called the *state space* of the dynamic f.

Definition 8 Let $(\Omega, \mathcal{B}, \mu)$ be a finite-measure space and $f : \Omega \to \Omega$ a μ-preserving map. Then $(\Omega, \mathcal{B}, \mu, f)$ is called a measure-preserving dynamical system.

If $(\Omega, \mathcal{B}, \mu, f)$ is a measure-preserving dynamical system, we can assume without loss of generality that $\mu(\Omega) = 1$, i.e., that $(\Omega, \mathcal{B}, \mu)$ is a *probability space*. In this light, Ω is the space of elementary events, $\mathcal{B}$ comprises all outcomes we might be interested in, and $\mu(B)$ is the probability of the outcome $B \in \mathcal{B}$.

Given a measurable map $f : \Omega \to \Omega$, it is very difficult in practice to prove that f preserves the measure μ since, in general, not all elements $B \in \mathcal{B}$ are explicitly known. In general, all we know is a semi-algebra $\mathcal{S}$ generating $\mathcal{B}$. For example, if $\mathcal{B}$ is the Borel sigma-algebra of the interval $[0, 1] \subset \mathbb{R}$ with the standard topology, then $\mathcal{S}$ can be taken to be the collection of all subintervals of $[0, 1]$, or just the collection

J.M. Amigó, *Permutation Complexity in Dynamical Systems*,
Springer Series in Synergetics, DOI 10.1007/978-3-642-04084-9,
© Springer-Verlag Berlin Heidelberg 2010

of subintervals of the forms $[0, b]$ and $(a, b]$, $0 \leq a < b \leq 1$. It can be proved [202] that if (i) $\mathcal{S}$ is a semi-algebra which generates $\mathcal{B}$ and (ii) for every $A \in \mathcal{S}$, $f^{-1}(A) \in \mathcal{B}$ and $\mu(f^{-1}(A)) = \mu(A)$, then f preserves the measure μ.

Exercise 13 Prove that $\mathcal{S} = \{[a, b): 0 \leq a < b < 1\}$ is a semi-algebra of subsets of the interval $[0, 1)$ that generates the Borel sigma-algebra of $[0, 1)$.

Example 22 Suppose $\Omega = [0, 1)$, $\mathcal{B}$ is the Borel sigma-algebra of $[0, 1)$, and λ is the Lebesgue measure on $[0, 1)$. Furthermore, let $f: \Omega \to \Omega$ be the map given by $f(x) = Nx \bmod 1$, where $N \in \mathbb{Z}$, $|N| \geq 2$. Then f preserves λ. Indeed, for every half-open interval $[a, b) \subset [0, 1)$,

$$f^{-1}([a, b)) = \bigcup_{i=0}^{N-1} \left[\frac{a + i}{N}, \frac{b + i}{N} \right)$$

if $N \geq 2$ and

$$f^{-1}([a, b)) = \bigcup_{i=1}^{|N|} \left(\frac{i - b}{|N|}, \frac{i - a}{|N|} \right]$$

if $N \leq -2$. Hence,

$$\lambda \left(f^{-1}[a, b) \right) = \sum_{i=0}^{N-1} \frac{b - a}{N} = \sum_{i=1}^{|N|} \frac{b - a}{|N|} = b - a = \lambda([a, b)).$$

Example 23 Let the measure space $(\Omega, \mathcal{B}, \mu)$ be as in the previous example and $f: \Omega \to \Omega$ be given now by $f(x) = x + r \bmod 1$, with $r > 0$. This transformation preserves also the Lebesgue measure λ since, for every $[a, b) \subset [0, 1)$,

$$
\begin{aligned}
f^{-1}([a, b)) &= [a - r, b - r) & &\text{if } a \geq r, \\
f^{-1}([a, b)) &= [a + 1 - r, b + 1 - r) & &\text{if } b \leq r, \\
f^{-1}([a, b)) &= [0, b - r) \cup [a + 1 - r, 1) & &\text{if } a < r < b.
\end{aligned}
$$

In any case,

$$\lambda \left(f^{-1}([a, b)) \right) = b - a = \lambda([a, b)).$$

A perhaps more natural way of dealing with this example views f as a rotation on the circle. The f-invariance of λ is then straightforward.

More generally, the Lebesgue measure on $\mathbb{R}^n$ is invariant under translations and rotations in $\mathbb{R}^n$. More sophisticated examples of invariant measures include the Haar measure on a locally compact topological group, the map being the action of the group. In the next section we will meet invariant measures on product spaces.

Exercise 14 Let $f:[0, 1) \to [0, 1)$ be the Gauss transformation,

$$f(x) = \begin{cases} 0 & \text{if } x = 0, \\ \frac{1}{x} \ (\text{mod } 1) & \text{if } x \neq 0. \end{cases}$$

Show that f preserves the measure

$$\mu(B) = \frac{1}{\ln 2} \int_B \frac{dx}{1+x}, \qquad (A.1)$$

where B is a Borel set of $[0, 1)$. Hint:

$$f^{-1}([a, b)) = \bigcup_{n=1}^{\infty} \left(\frac{1}{b+n}, \frac{1}{a+n} \right].$$

Krylov and Bogolioubov showed that invariant measures exist under quite general conditions.

Theorem 20 [202] *Let Ω be a compact metric space and $f:\Omega \to \Omega$ a continuous map. Then there exists an f-invariant probability measure μ on $(\Omega, \mathcal{B})$, where $\mathcal{B}$ is the Borel sigma-algebra of Ω.*

In general, there can exist more than one f-invariant measure and, besides, some of them can be rather "pathological." For instance, if δ_p is the Dirac measure at p, i.e.,

$$\delta_p(B) = \begin{cases} 1 & \text{if } p \in B \\ 0 & \text{if } p \notin B \end{cases},$$

$B \in \mathcal{B}$, and x is a period-n point for f, then

$$\mu(B) = \frac{1}{n} \sum_{k=0}^{n-1} \delta_{f^k(x)}(B)$$

($f^0(x) := x$ and $f^i(x) = f(f^{i-1}(x))$ for $i \geq 1$) is an atomic measure supported on the points $\{x, f(x), \ldots, f^{n-1}(x)\}$. A set $E \subset \Omega$ is said to be the (unique) *support* of μ if (i) E is closed in Ω, (ii) $\mu(E \cap U) > 0$ if $E \cap U \neq \emptyset$ and U is open in Ω, and (iii) $\mu(E') = 0$, where $E' = \Omega \backslash E$ is the complement of E.

In general, the ordered set $\{f^i(x):i \geq 0\}$ is called the orbit or trajectory of the point (state, initial condition, etc.) $x \in \Omega$ under the "discrete-time" dynamic f and denoted by $\mathcal{O}_f(x)$. In the case of invertible maps, one writes $\mathcal{O}_f^+(x) = \{f^i(x):i \geq 0\}$ for the "forward" orbit, while orbit means $\mathcal{O}_f(x) = \{f^i(x):i \in \mathbb{Z}\}$.

It can happen that for almost all x in a set $U \subset \Omega$ with positive Lebesgue measure, its orbit is bounded and, moreover, the sequences of probability measures

$$\frac{1}{n} \sum_{k=0}^{n-1} \delta_{f^k(x)}$$

converge weakly to a measure μ, i.e., for almost all $x \in U$ and any continuous map $\varphi : \Omega \to \Omega$,

$$\lim_{k \to \infty} \frac{1}{n} \sum_{k=0}^{n-1} \varphi(f^k(x)) = \int_\Omega \varphi d\mu$$

holds. Then μ is an f-invariant measure that is usually called the natural or physical measure for its relevance in physics and computer simulations [72].

An important issue in measure-preserving dynamical systems is the existence of absolutely continuous invariant measures. A measure μ on a topological space Ω is said to be *absolutely continuous* (with respect to the Lebesgue measure dx), if $\mu(dx) = \rho(x)dx =: d\mu$, where the *density function* $\rho : \Omega \to \Omega$ (also called the Radon–Nikodym derivative of μ with respect to the Lebesgue measure, $d\mu/dx$) is continuous. For example, if μ is measure (A.1) on the interval $[0, 1)$ endowed with the Borel sigma-algebra, then

$$\mu(dx) = \frac{1}{\ln 2} \frac{dx}{1+x} \quad \text{or} \quad \frac{d\mu}{dx} = \frac{1}{\ln 2} \frac{1}{1+x}.$$

In general there are few results on the existence of absolutely continuous invariant measures. In the case of self-maps of one-dimensional intervals, there are some general conditions that appear in the usual theorems on existence of such measures.

Recall that a *partition* of a measure space $(\Omega, \mathcal{B}, \mu)$ is a disjoint collection of elements of $\mathcal{B}$ whose union is Ω.

Definition 9 Let $\alpha = \{I_i\}_{i=1}^d$ be a partition of the interval $I = [a, b] \subset \mathbb{R}$ into subintervals I_i. Given the map $f : I \to I$, assume that $f|_{I_i}$ is C^k ($k \geq 1$) for each i.

(a) f is said to be C^k *piecewise expanding* if there exists $\lambda > 1$ such that $|f'(x)| > \lambda$ for all $x \in I_i$ and each i.
(b) f is said to be C^k *Markov* if $f(\mathring{I}_i) \supset \mathring{I}_j$ whenever $f(\mathring{I}_i) \cap \mathring{I}_j \neq \emptyset$ ("Markov property"), where $\mathring{I}_i$ stands for the interior of I_i, $1 \leq i \leq d$. In this case, α is called a *Markov partition* for f. The matrix $A = (A_{ij})_{1 \leq i, j \leq d}$ with

$$A_{i,j} = \begin{cases} 1 & \text{if } f(\mathring{I}_i) \supset \mathring{I}_j, \\ 0 & \text{if } f(\mathring{I}_i) \cap \mathring{I}_j = \emptyset, \end{cases} \tag{A.2}$$

is called the transition matrix for f.

See, for instance, [37, Chap. 5] and [105] for results concerning the existence of absolutely continuous invariant measures for piecewise expanding and/or Markov transformations (complying with additional conditions).

Exercise 15 Prove that the logistic map $g(x) = 4x(1-x), 0 \leq x \leq 1$, has an invariant measure with density function

$$\rho(x) = \frac{1}{\pi \sqrt{x(1-x)}}, \tag{A.3}$$

i.e., $\int_0^1 \rho(x)dx = 1$, and

$$\int_{[a,b]} \rho(x)dx = \int_{g^{-1}[a,b]} \rho(x)dx,$$

for all $0 \leq a < b \leq 1$. Figure A.1 shows the plot of the function $\rho(x)$. Is $g(x)$ piecewise expanding? Is $g(x)$ Markovian?

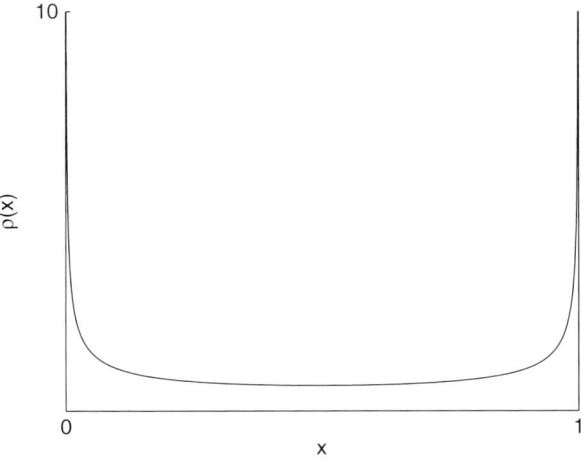

Fig. A.1 The density $\rho(x)$, (A.3)

Once we know that invariant measures are rather abundant objects, suppose that $f:\Omega \to \Omega$ is such that $f^{-1}(B) = B$ for some $B \in \mathcal{B}$. Then $f^{-1}(\Omega \backslash B) = \Omega \backslash B$ and the action of f on Ω can be decomposed into two disjoint pieces: $f|_B$ and $f|_{\Omega \backslash B}$. If f is indecomposable in the previous sense, one says that f is ergodic.

Definition 10 Let $(\Omega, \mathcal{B}, \mu, f)$ be a measure-preserving dynamical system. The map f is said to be ergodic if

$$f^{-1}(B) = B, \quad B \in \mathcal{B} \Rightarrow \mu(B) = 0 \quad \text{or} \quad \mu(B) = 1$$

Alternatively, μ is said to be an ergodic measure for f. Also, the dynamical system $(\Omega, \mathcal{B}, \mu, f)$ is said to be ergodic.

Thus an ergodic measure cannot be decomposed as a (properly weighted or "convex") sum of invariant measures. It might seem that this definition is a far cry from the original Boltzmann's *Ergodenhypothese*, which states that the trajectory of a closed thermodynamic system in the phase space (spanned by the coordinates and conjugate canonical momenta of its constituent particles) covers densely and uniformly the "energy shell," that is, the hypersurface in phase space defined by the restriction that the energy of the system is constant. But it was on the way to lying Boltzmann's proposal on a mathematically sound basis that G. Birkhoff introduced the concept of ergodicity in its modern version. Birkhoff's seminal *ergodic theorem* states the following.

Theorem 21 [202] *If $(\Omega, \mathcal{B}, \mu, f)$ is an ergodic dynamical system, then*

$$\lim_{n \to \infty} \frac{1}{n} \sum_{i=0}^{n-1} \varphi(f^i(x)) = \int_\Omega \varphi d\mu \quad a.e. \tag{A.4}$$

for all $\varphi \in L^1(\mu)$.

As usual, "a.e." is shorthand for "almost everywhere" with respect to the relevant measure (μ here) and $L^1(\mu)$ is the space of μ-integrable functions. The property assumed by the *Ergodenhypothese* goes by the name of *topological transitivity* in the theory of discrete dynamical systems. A continuous self-map f of a compact metric space Ω is called topologically transitive if there exists some $x \in \Omega$ such that $\mathcal{O}_f(x)$ is dense in Ω (if f is invertible, then $\mathcal{O}_f(x)$ also includes the "backward" iterates $f^{-n}(x)$, $n \in \mathbb{N}$).

Let χ_B denote the *characteristic function* of the set $B \in \mathcal{B}$,

$$\chi_B(x) = \begin{cases} 1 & \text{if } x \in B \\ 0 & \text{if } x \notin B \end{cases}.$$

The substitution $\varphi = \chi_B$ in (A.4) yields then

$$\frac{1}{n} \sum_{i=0}^{n-1} \chi_B(f^i(x)) \to \mu(B) \quad a.e.,$$

when $n \to \infty$. This means that if $(\Omega, \mathcal{B}, \mu, f)$ is ergodic, then the orbit of almost every initial condition $x \in \Omega$ visits the region B of the state space with asymptotic frequency $\mu(B)$. This resembles the law of large numbers in statistics and, in fact, there are plenty of deep relations between ergodic theory and statistics [31, 67].

Let Ω be a compact metrizable space Ω, and $\mathcal{B}$ the Borel sigma-algebra on Ω. A continuous map $f : \Omega \to \Omega$ is called *uniquely ergodic* if there is only one f-invariant Borel probability measure on Ω. A map f is uniquely ergodic if and only if it has exactly one invariant measure. If f is uniquely ergodic and μ is its invariant measure, then (A.4) holds for all continuous transformations φ and all $x \in \Omega$ [202].

Ergodicity is just but the first step in a series of notions measuring the statistical properties of the orbits generated by the dynamic: ergodicity, mixing, completely positive entropy, etc. Here we will recall only the definition of strong mixing.

Definition 11 The measure-preserving dynamical system $(\Omega, \mathcal{B}, \mu, f)$ is called (strong) mixing if

$$\lim_{n \to \infty} \mu(f^{-n}(A) \cap B) = \mu(A)\mu(B) \tag{A.5}$$

for all $A, B \in \mathcal{B}$.

In contrast to (A.5), f is ergodic if and only if

$$\lim_{n \to \infty} \frac{1}{n} \sum_{i=0}^{n-1} \mu(f^{-i}(A) \cap B) = \mu(A)\mu(B) \tag{A.6}$$

for all $A, B \in \mathcal{B}$. Hence mixing is a stronger condition than ergodicity. In practice it suffices to check (A.6) and (A.5) for $A, B \in \mathcal{S}$, a semi-algebra that generates $\mathcal{B}$.

Sufficient conditions for the existence of *ergodic* absolutely continuous invariant measures can be found, e.g., in [52, Chap. 5] . Mixing piecewise C^2 expanding Markov maps have unique ergodic invariant measures [105].

As in any other area of mathematics, the notion of isomorphism is central. It specifies when two dynamical systems are to be considered equivalent from the point of view of the properties that matter in this theory.

Definition 12 Given the measure-preserving dynamical systems $(\Omega_1, \mathcal{B}_1, \mu_1, f_1)$ and $(\Omega_2, \mathcal{B}_2, \mu_2, f_2)$, we say that f_1 is (metrically) isomorphic to f_2 if there exist $B_1 \in \mathcal{B}_1$, $B_2 \in \mathcal{B}_2$ with $\mu_1(B_1) = \mu_2(B_2) = 1$ such that (i) $f_1(B_1) \subset B_1, f_2(B_2) \subset B_2$ and (ii) there is an invertible, measure-preserving map $\phi: B_1 \to B_2$ with $\phi \circ f_1(x) = f_2 \circ \phi(x)$ for all $x \in B_1$.

The dynamical systems $(\Omega_1, \mathcal{B}_1, \mu_1, f_1)$ and $(\Omega_2, \mathcal{B}_2, \mu_2, f_2)$ are said to be isomorphic. Sometimes ϕ is called an isomorphism "modulo 0" or just "mod 0" (shorthand for modulo measure zero sets), but usually we dispense with measure zero sets without stating it explicitly. In the more general case that ϕ is measure preserving but only surjective, $(\Omega_2, \mathcal{B}_2, \mu_2, f_2)$ is called a *factor* of $(\Omega_1, \mathcal{B}_1, \mu_1, f_1)$ (or $(\Omega_1, \mathcal{B}_1, \mu_1, f_1)$ a *cover* of $(\Omega_2, \mathcal{B}_2, \mu_2, f_2)$) via the factor map ϕ. Two isomorphic maps are obtained from each other by a change of coordinates, so that properties that are independent of such changes of coordinates are invariant. Isomorphism invariants include ergodicity and mixing.

There is a broader (and more technical) concept called *conjugacy* that embraces isomorphism. Both concepts are though equivalent in virtually all probability spaces that one encounters in applications (e.g., compact metric spaces). Indeed, as it turns out, there is essentially only one type of probability space, called a *Lebesgue space*, which is characterized as being measure-theoretically isomorphic to the union of an interval of $\mathbb{R}$ endowed with Lebesgue measure, with at most countably many points of positive measure (called atoms) [177, 202]. In a Lebesgue space, set maps are always induced by point maps. Conjugacy and isomorphy coincide for a Lebesgue space, so both terms can be used interchangeably in that case.

Example 24 The symmetric tent map $\Lambda:[0, 1] \rightarrow [0, 1]$,

$$\Lambda(x) = \begin{cases} 2x & 0 \leq x \leq \frac{1}{2} \\ 2 - 2x & \frac{1}{2} \leq x \leq 1 \end{cases}, \tag{A.7}$$

preserves the Lebesgue measure $\lambda(dx) = dx$. If, furthermore, $\mu(dx) = \frac{1}{\pi \sqrt{x(1-x)}} dx$ is the natural invariant measure of the logistic map $g:[0, 1] \rightarrow [0, 1]$, $g(x) = 4x(1 - x)$ (see (A.3)), then $\phi:([0, 1], \lambda) \rightarrow ([0, 1], \mu)$ given by

$$\phi(x) = \sin^2 \left(\frac{\pi}{2} x \right) \tag{A.8}$$

is invertible, measure preserving, and it satisfies $g \circ \phi = \phi \circ \Lambda$. Hence, Λ and g are conjugate.

Exercise 16 Show that

$$x_k = \sin^2 (2^k \xi),$$

$\xi \in \mathbb{R}$, is a solution of the logistic recursion (or finite difference equation)

$$x_{k+1} = 4x_k(1 - x_k), \quad k \geq 0,$$

$x_k \in [0, 1]$, with initial condition $x_0 = \sin^2 \xi$.

A.2 Shift Systems

Shift systems are dynamical systems which due to their importance as models and prototypes are considered separately in this section. In the simplest and most usual version, the elements of the shift spaces are one-sided or two-sided sequences of N symbols or "letters". Sometimes one has to consider also sequences with elements from an arbitrary (countable or uncountable) "alphabet," and this requires some degree of sophistication. We set out from this more general situation.

First of all, let us recall the definition of a *product measurable space*. For our purposes it is sufficient to consider products of countably many copies of a measurable space $(\Omega, \mathcal{B})$. As index set $\mathbb{K}$ we take without restriction $\mathbb{K} = \mathbb{N}_0 := \{0\} \cup \mathbb{N}$ or $\mathbb{K} = \mathbb{Z}$. Then, $\Pi_{k \in \mathbb{K}}(\Omega, \mathcal{B}) = (\Omega^{\mathbb{K}}, \mathcal{B}_{\Pi}(\Omega))$, where

$$\Omega^{\mathbb{K}} = \{(\omega_k)_{k \in \mathbb{K}} : \omega_k \in \Omega\}$$

is the set of all one-sided sequences

$$(\omega_k)_{k\in\mathbb{N}_0} = \omega_0, \ldots, \omega_k, \ldots$$

if $\mathbb{K} = \mathbb{N}_0$, or the set of all two-sided sequences (also called bisequences or doubly infinite sequences)

$$(\omega_k)_{k\in\mathbb{Z}} = \ldots, \omega_{-k}, \ldots, \omega_0, \ldots, \omega_n, \ldots$$

if $\mathbb{K} = \mathbb{Z}$, and $\mathcal{B}_\Pi(\Omega)$ is the sigma-algebra generated by the semi-algebra $\mathcal{S}$ of *cylinder sets*

$$\prod_{j\in\mathbb{F}} A_j \times \prod_{k\notin\mathbb{F}} \Omega = \{(\omega_k)_{k\in\mathbb{K}}:\omega_j \in A_j \text{ for } j \in \mathbb{F}\}, \qquad (A.9)$$

where $\mathbb{F} \subset \mathbb{K}$ is finite and $A_j \in \mathcal{B}$ for $j \in \mathbb{F}$. If $\mathbb{K} = \mathbb{N}_0$ (correspondingly, $\mathbb{K} = \mathbb{Z}$), then we can take $\mathbb{F} = \{0, 1, \ldots, n\}$ (correspondingly, $\mathbb{F} = \{-n, \ldots, 0, \ldots, n\}$), $n \in \mathbb{N}_0$, in (A.9) without restriction.

In most applications we have in mind (for instance, to information theory), $(\Omega, \mathcal{B}) = (S, 2^S)$ with $S = \{0, \ldots, N{-}1\}$, $N \geq 2$, and 2^S denoting as usual the family of all subsets of S. In this case, the set of all one-sided sequences of the symbols $0, 1, \ldots, N{-}1$,

$$S^{\mathbb{N}_0} = \{(s_n)_{n\in\mathbb{N}_0}:s_n \in S\}, \qquad (A.10)$$

is called the (one-sided) *sequence space on N symbols*. Depending on the context, the set of symbols S may receive different names. In the setting of information theory, S is called an *alphabet*, its elements are called *letters*, and sequences $\mathbf{s} = (s_n)_{n\in\mathbb{N}_0}$ are called *messages*. In dynamics, S is sometimes called the state space and its elements, states. Segments (or *words*) of symbols of length L, like $s_k, s_{k+1}, \ldots, s_{k+L-1}$, will be shortened as s_k^{k+L-1}.

If S is thought to be a topological space (eventually endowed with the discrete topology), then $S^{\mathbb{N}_0}$ can be promoted to a topological space by means of the product topology, which is generated by the corresponding cylinder sets

$$C_{a_0,\ldots,a_n} = \{\mathbf{s} \in S^{\mathbb{N}_0}:s_k = a_k, 0 \leq k \leq n\}, \qquad (A.11)$$

where $a_0, \ldots, a_n \in S$. (The general definition (A.9) with $A_j = \{a_j\}$ leads to the same topology.) The product topology makes $S^{\mathbb{N}_0}$ compact, perfect (i.e., it is closed and all its points are accumulation points), and totally disconnected. Such topological spaces are sometimes called *Cantor sets* because they are homeomorphic to Cantor's ternary set in the unit interval. By definition, the product sigma-algebra, $\mathcal{B}_\Pi(S)$, is generated by the cylinder sets (A.11) and comprises all Borel sets of $S^{\mathbb{N}_0}$.

Moreover, $S^{\mathbb{N}_0}$ is a metrizable space. In fact, there are several (non-equivalent) metrics compatible with the topology of $S^{\mathbb{N}_0}$, the perhaps most popular being

$$d_K(\mathbf{s}, \mathbf{s}') = \sum_{n=0}^{\infty} \frac{\delta(s_n, s_n')}{K^n}, \tag{A.12}$$

where $\delta(s_n, s_n') = 1$ if $s_n \neq s_n'$, $\delta(s_n, s_n) = 0$ and $K > 2$. Observe that given $\mathbf{s} \in C_{a_0,\dots,a_n}$, then $d_K(\mathbf{s}, \mathbf{s}') < \frac{1}{K^n}$ if $\mathbf{s}' \in C_{a_0,\dots,a_n}$, and $d_K(\mathbf{s}, \mathbf{s}') \geq \frac{1}{K^n}$ if $\mathbf{s}' \notin C_{a_0,\dots,a_n}$, thus $C_{a_0,\dots,a_n} = B_{d_K}(\mathbf{s}; \frac{1}{K^n})$, the open ball of radius K^{-n} and center $\mathbf{s}$ in the metric space $(S^{\mathbb{N}_0}, d_K)$. Moreover, every point in $B_{d_K}(\mathbf{s}; \frac{1}{K^n})$ is a center, a property known from non-Archimedean normed spaces (e.g., the rational numbers with p-adic norms [115]).

Exercise 17 1. Prove that the cylinder sets (thus the open balls) are also closed in the product topology. Open and closed sets are sometimes called clopen sets.
2. Prove that the cylinder sets are not connected (i.e., they can be written as a disjoint union of open sets).

Shifting all the symbols of a one-sided sequence to the left one place and dropping the first symbol define a self-map of one-sided sequence spaces which plays an important role in both theory and applications. Formally, the (one-sided) *shift* $\Sigma : S^{\mathbb{N}_0} \to S^{\mathbb{N}_0}$ is defined as

$$\Sigma(s_0, s_1, s_2, \dots) = (s_1, s_2, s_3, \dots), \tag{A.13}$$

that is, $\Sigma(\mathbf{s}) = \mathbf{s}'$ with $s_n' = s_{n+1}$. Since $\Sigma^{-1} C_{a_0,\dots,a_n} = \bigcup_{a \in S} C_{a,a_0,\dots,a_n}$, Σ is continuous on $(S^{\mathbb{N}_0}, d_K)$, each point $\mathbf{s} \in S^{\mathbb{N}_0}$ having exactly N preimages under Σ. Furthermore, Σ has N fixed points: $\mathbf{s} = a_0^{\infty}, 0 \leq a \leq N - 1$.

In order to make a measure-preserving dynamical system out of $S^{\mathbb{N}_0}$, $\mathcal{B}_\Pi(S)$, and Σ, only a Σ-invariant measure is missing. All probability measures on $(S^{\mathbb{N}_0}, \mathcal{B}_\Pi(S))$ that make Σ a measure-preserving transformation are obtained in the following way [202]. For any $n \geq 0$ and $a_i \in S, 0 \leq i \leq n$, let a real number $p_n(a_0, \dots, a_n)$ be given such that (i) $p_n(a_0, \dots, a_n) \geq 0$, (ii) $\sum_{a_0 \in S} p_0(a_0) = 1$, and (iii) $p_n(a_0, \dots, a_n) = \sum_{a_{n+1} \in S} p_{n+1}(a_0, \dots, a_n, a_{n+1})$. If we define now

$$m(C_{a_0,\dots,a_n}) = p_n(a_0, \dots, a_n),$$

then m can be extended to a probability measure on $(S^{\mathbb{N}_0}, \mathcal{B}_\Pi(S))$. The resulting dynamical system $(S^{\mathbb{N}_0}, \mathcal{B}_\Pi(S), m, \Sigma)$ is called the *one-sided shift system*.

If instead of considering (one-sided) sequences $\mathbf{s} = (s_n)_{n \in \mathbb{N}_0}, s_n \in S = \{0, \dots, N-1\}$, we consider two-sided sequences $\mathbf{s} = (s_n)_{n \in \mathbb{Z}}$, we are in the realm of the *two-sided sequence spaces on N symbols*,

$$S^{\mathbb{Z}} = \{(s_n)_{n \in \mathbb{Z}} : s_n \in S\}.$$

The corresponding (invertible) *two-sided shift* on $S^{\mathbb{Z}}$ is defined as $\Sigma : \mathbf{s} \mapsto \mathbf{s}'$ with $s_n' = s_{n+1}$, $n \in \mathbb{Z}$. (Although not strictly correct, we use the same letter Σ for one-sided and two-sided shifts.) The cylinder sets are given now as

$$C_{a_{-n},\dots,a_0,\dots,a_n} = \{\mathbf{s} \in S^{\mathbb{Z}} : s_k = a_k, |k| \leq n\}$$

and

$$d_K(\mathbf{s}, \mathbf{s}') = \sum_{n \in \mathbb{Z}} \frac{\delta(s_n, s'_n)}{K^{|n|}},$$

$K > 3$, is a metric for $S^{\mathbb{Z}}$. The dynamical system $(S^{\mathbb{Z}}, \mathcal{B}_\Pi(S), m, \Sigma)$ is called the *two-sided shift system*.

Exercise 18 Prove that the cylinder set $C_{a_{-n},\dots,a_0,\dots,a_n}$ of $S^{\mathbb{Z}}$ coincides with the open ball $B_{d_K}(\mathbf{s}; K^{1-n})$, where $\mathbf{s}$ is any point of $C_{a_{-n},\dots,a_0,\dots,a_n}$.

Example 25 (a) Let $\mathbf{p} = (p_0, p_1, \dots, p_{N-1})$, $N \geq 2$, be a probability vector with non-zero entries (i.e., $p_i > 0$ and $\sum_{i=0}^{N-1} p_i = 1$). Set

$$p_n(a_0, a_1, \dots, a_n) = p_{a_0} p_{a_1} \cdots p_{a_n}.$$

The resulting measure on $(S^{\mathbb{K}}, \mathcal{B}_\Pi(S))$ is called the Bernoulli measure defined by $\mathbf{p}$. The dynamical system $(S^{\mathbb{K}}, \mathcal{B}_\Pi(S), m, \Sigma)$, where m is the Bernoulli measure defined by the probability vector $\mathbf{p}$, is called a one-sided (if $\mathbb{K} = \mathbb{N}_0$) or two-sided (if $\mathbb{K} = \mathbb{Z}$) $\mathbf{p}$-*Bernoulli shift*.

(b) Let $\mathbf{p} = (p_0, p_1, \dots, p_{N-1})$ be a probability vector as in (a) and $P = (p_{ij})_{0 \leq i,j \leq N-1}$ an $N \times N$ stochastic matrix (i.e., $p_{ij} \geq 0$ and $\sum_{j=0}^{N-1} p_{ij} = 1$) such that $\sum_{i=0}^{N-1} p_i p_{ij} = p_j$. Set then

$$p_n(a_0, a_1, \dots, a_n) = p_{a_0} p_{a_0 a_1} p_{a_1 a_2} \cdots p_{a_{n-1} a_n}.$$

The resulting measure on $(S^{\mathbb{K}}, \mathcal{B}_\Pi(S))$ is called the Markov measure defined by $(\mathbf{p}, P)$. The dynamical system $(S^{\mathbb{K}}, \mathcal{B}_\Pi(S), m, \Sigma)$, where m is the Markov measure defined by the probability vector $\mathbf{p}$ and the stochastic matrix P, is called a one-sided (if $\mathbb{K} = \mathbb{N}_0$) or two-sided (if $\mathbb{K} = \mathbb{Z}$) $(\mathbf{p}, P)$-*Markov shift*. A $\mathbf{p}$-*Bernoulli shift* can be considered as a $(\mathbf{p}, P)$-Markov shift by taking $p_{ij} = p_j$.

Simple as they might seem, one-sided and two-sided shifts exhibit most of the basic properties of ergodic theory, like ergodicity and strong mixing. In particular, they are easily shown to be *chaotic* in the sense of Devaney [69], i.e., they are sensitive to initial conditions, are strong mixing, and their periodic points are dense. Let us recall at this point the notion of sensitivity to initial conditions.

Definition 13 Given a metric space (M, d), a map $f:M \to M$ is said to be sensitive to initial conditions if there exists $\delta > 0$, called a sensitivity constant, such that for every $x \in \Omega$ and $\varepsilon > 0$ there exists $y \in \Omega$ with $d(x, y) < \varepsilon$ and $d(f^n(x), f^n(y)) \geq \delta$ for some $n \in \mathbb{N}$.

Equivalently, a continuous self-map of a compact metric space is said to be chaotic if it is topologically transitive (that is, it has a dense orbit) and its periodic points are dense [91].

Exercise 19 Prove that the one- and two-sided shifts on N symbols are sensitive to initial conditions, are topological transitive, and their periodic points are dense.

Example 26 Let $\Omega = [0, 1]$, $\mathcal{B}$ the Borel sigma-algebra of $[0, 1]$, λ the corresponding Lebesgue measure, and $E_2 : x \mapsto 2x \pmod 1$ the so-called dyadic map. The dynamical system $([0, 1], \mathcal{B}, \lambda, E_2)$ is then isomorphic (up to a measure zero set) to the one-sided $(\frac{1}{2}, \frac{1}{2})$-Bernoulli shift on the symbols $\{0, 1\} = S$. An isomorphism $\phi : S^{\mathbb{N}_0} \to [0, 1]$ is given by

$$(x_0, x_1, \ldots, x_k, ..) \mapsto \sum_{k=0}^{\infty} x_k 2^{-(k+1)}. \tag{A.14}$$

Of course, the map ϕ is not injective in strict sense because the sequences $(x_0, \ldots, x_{n-1}, 0, 1^{\infty})$ and $(x_0, \ldots, x_{n-1}, 1, 0^{\infty})$ are sent to the same point (the upper label "∞" means indefinite repetition); indeed,

$$\sum_{k=0}^{n-1} x_k 2^{-(k+1)} + \sum_{k=n+1}^{\infty} 2^{-(k+1)} = \sum_{k=0}^{n-1} x_k 2^{-(k+1)} + 2^{-(n+1)}.$$

However, since the set of sequences eventually terminating in an infinite string of 0's or 1's is countable, we conclude that $(S^{\mathbb{N}_0}, \mathcal{B}_{\Pi}(S), m, \Sigma)$ and $([0, 1], \mathcal{B}, \lambda, E_2)$ are conjugate modulo 0, i.e., the diagram

$$\begin{array}{ccc} \Sigma : \{0, 1\}^{\mathbb{N}_0} & \to & \{0, 1\}^{\mathbb{N}_0} \\ \phi \downarrow & & \downarrow \phi \\ E_2 : [0, 1] & \to & [0, 1] \end{array}$$

is commutative almost everywhere: $E_2 = \phi \circ \Sigma \circ \phi^{-1}$. Observe that there is otherwise a topological obstruction that prevents $S^{\mathbb{N}_0}$ and $[0, 1]$ from being homeomorphic: the first is (homeomorphic to) a Cantor set while, certainly, the second is not.

Exercise 20 Prove that the map $\phi : S^{\mathbb{N}_0} \to [0, 1]$ defined in (A.14) is measure preserving, i.e., $m(\phi^{-1}(I)) = \lambda(I)$ for any interval $I \subset [0, 1]$. It suffices to consider "dyadic" intervals, i.e., intervals of the forms $[0, k_2/2^n]$ and $(k_1/2^n, k_2/2^n]$, $0 \le k_1 < k_2 \le 2^n$, $n \in \mathbb{N}$.

Let us mention in passing the dyadic map $x \mapsto 2x \pmod 1$ is just the first member of the family of *expanding maps* of the circle:

$$E_N : x \mapsto Nx \pmod 1,$$

where N is an integer of absolute value greater than 1. In a way similar to Example 26 one can show that $([0, 1], \mathcal{B}, \lambda, E_N)$ and the $(\frac{1}{N}, \ldots, \frac{1}{N})$-Bernoulli shift are conjugate for $N \ge 2$. In this case, map (A.14) is replaced by $(x_0, x_1, \ldots) \mapsto \sum_{k=0}^{\infty} x_k N^{-(k+1)}$.

Exercise 21 What transformation induces on the sequence space $\{0, 1\}^{\mathbb{N}_0}$ the expanding map E_{-2} via map (A.14)?

A.3 Stochastic Processes and Sequence Spaces

A stochastic (or random) process is a mathematical model for the occurrence of random phenomena as time goes on. This is the case, for example, when a random experiment is repeated over and over again. Put in a formal way, a *stochastic process* is a collection of random variables $\mathbf{X} = \{X_t\}_{t \in \mathcal{T}}$ on a common probability space $(\Omega, \mathcal{B}, \mu)$, called the *sample space*, taking on values in a measurable space $(S, \mathcal{A})$, called the *state space*. Technically this means that $X_t : \Omega \to S$ is a measurable map for all $t \in \mathcal{T}$, i.e., $X_t^{-1}(A) \in \mathcal{B}$ for all $A \in \mathcal{A}$. The index $t \in \mathcal{T}$ is conveniently interpreted as time, the usual choices for $\mathcal{T}$ being (i) $\mathcal{T} = \mathbb{R}$ or $\mathbb{R}_+ = [0, \infty]$, in which case $\mathbf{X}$ is called a continuous-time stochastic process or (ii) $\mathcal{T} = \mathbb{N}_0$ or $\mathbb{Z}$, in which case $\mathbf{X}$ is called a discrete-time stochastic process. The map $t \mapsto X_t(\omega)$ is the realization (sample path, trajectory, etc.) of the process $\mathbf{X}$ associated with the fixed sample point $\omega \in \Omega$. As usual in probability theory and statistics, a realization of a random variable X will be denoted by the same letter in small caps: $X(\omega) = x$.

The stochastic process $\mathbf{X}$ is characterized by its joint (finite-dimensional) probability distributions

$$\mu\{\omega \in \Omega : X_{t_1}(\omega) \in A_1, \ldots, X_{t_r}(\omega) \in A_r\} = \Pr\{X_{t_1} \in A_1, \ldots, X_{t_r} \in A_r\},$$

where $r \geq 1$, $t_1, \ldots, t_r \in \mathcal{T}$ and $A_1, \ldots, A_r \in \mathcal{A}$. If, furthermore, $\mathcal{T}$ is such that $\mathcal{T} + t \in \mathcal{T}$ for any $t \in \mathcal{T}$ (think of $\mathcal{T} = [0, \infty)$ or $\mathcal{T} = \mathbb{N}_0$) and the distribution of the random vector $(X_{t_1+t}, X_{t_2+t}, \ldots, X_{t_r+t})$ does not depend on t for any $r \geq 1$, $t_1, \ldots, t_r \in \mathcal{T}$, then the process $\mathbf{X}$ is called *stationary*. Stationary stochastic processes are also called *information sources* because they are used in information theory to model data sources.

In this book we consider mostly discrete-time, finite-state, one-sided stochastic processes modeling, say, finite-alphabet information sources or arising as symbolic dynamics after dividing the state space of a dynamical system. In this case we use the following notation for the joint probability distributions of the *discrete* random variables $X_0, \ldots, X_n$ with states in (without restriction) $S = \{0, 1, \ldots, N - 1\}$:

$$\mu\{\omega \in \Omega : X_0(\omega) = x_0, \ldots, X_n(\omega) = x_n\} = \Pr\{X_0 = x_0, \ldots, X_n = x_n\}$$
$$= p(x_0, \ldots, x_n), \tag{A.15}$$

and the corresponding notations for the conditional probabilities, etc. Occasionally, these finite-state processes will arise as discretizations or quantizations $\mathbf{X}^\Delta$ of processes $\mathbf{X}$ taking values in a finite interval $I \subset \mathbb{R}^q$ endowed with the Lebesgue measure. Formally this means that there exists a (usually uniform) partition $\delta = \{\Delta_1, \ldots, \Delta_{|\delta|}\}$ of I into a finite number of Lebesgue-measurable subsets (say, subintervals), such that $X_n^\Delta = a_j$ if $X_n^\Delta \in \Delta_j$, where $a_j \in \Delta_j$ is usually set by the precision with which the outputs of $\mathbf{X}$ are measured.

Example 27 A finite-state stochastic process $\mathbf{X} = \{X_n\}_{n \in \mathbb{N}_0}$ is called a Markov process or Markov chain if

$$\Pr\{X_n = x_n | X_{n-1} = x_{n-1}, \dots, X_0 = x_0\} = \Pr\{X_n = x_n | X_{n-1} = x_{n-1}\},$$

$n \geq 1$, where $x_0, \dots, x_n \in S = \{0, \dots, N-1\}$. If, moreover, the conditional probability $\Pr\{X_n = x_n | X_{n-1} = x_{n-1}\}$ does not depend on n, then the Markov process $\mathbf{X}$ is called time homogeneous or time invariant. In this case,

$$P_{i,j} := \Pr\{X_n = j | X_{n-1} = i\},$$

$0 \leq i, j \leq N - 1$, is called the transition matrix. We call a probability vector $\mathbf{p} = (p_0, \dots, p_{N-1})$ an invariant, stationary, or equilibrium probability for $\mathbf{X}$ if $\mathbf{p} = \mathbf{p}P$, that is, if $\mathbf{p}$ is a left eigenvector of P with eigenvalue 1.

Any *stationary* discrete-time stochastic process $\mathbf{X} = \{X_n\}_{n \in \mathbb{K}}$ on a probability space $(\Omega, \mathcal{B}, \mu)$ with state space $(S, \mathcal{A})$ corresponds in a standard way to a shift system $(S^{\mathbb{K}}, \mathcal{B}_\Pi(S), m, \Sigma)$, where $(S^{\mathbb{K}}, \mathcal{B}_\Pi(S))$ is the product measurable space $\Pi_{k \in \mathbb{K}}(S, \mathcal{A})$, via the map $\Phi : \Omega \to S^{\mathbb{K}}$ defined by $(\Phi(\omega))_n = X_n(\omega)$. Here the measure m is the induced or transported probability on the space of possible outputs, $\mathcal{B}_\Pi(S)$, of the random process $\mathbf{X}$:

$$m(B) = \mu(\Phi^{-1}B), \quad B \in \mathcal{B}_\Pi(S), \tag{A.16}$$

that is, $m = \mu \circ \Phi^{-1}$ (note that $\Phi^{-1}B \in \mathcal{B}$ because each X_n is measurable). Moreover, because of the stationarity of $\mathbf{X}$, the probability measure m is shift invariant on cylinder sets and hence on all of $\mathcal{B}_\Pi(S)$.

We will also refer to the shift systems $(S^{\mathbb{K}}, \mathcal{B}_\Pi(S), m, \Sigma)$ as the (sequence space) *model* of the stochastic process or information source $\mathbf{X}$; if S is finite, then we may speak of a *sequence space model*. Models allow to focus on the random process itself as given by the probability distribution of its outputs, dispensing with a perhaps complicated underlying probability space. Depending on the setting or the process being modeled, some particular choices for S and/or $\mathbb{K}$ may be more convenient. For instance, one-sided random processes (i.e., $\mathbb{K} = \mathbb{N}_0$) provide better models than the two-sided processes $\{X_n\}_{n \in \mathbb{Z}}$ for physical information sources that must be turned on at some time. Also, if the source is digital, a finite state space S is the right choice.

Finally, since each information source has associated a dynamical system— its sequence space model—we can eventually assign dynamical properties to the sources. Thus, we say that a source $\mathbf{X}$ is *ergodic*, *mixing*, etc., if its sequence space model $(S^{\mathbb{K}}, \mathcal{B}_\Pi(S), m, \Sigma)$ possesses those properties.

Annex B
Entropy

In this annex we review only the Shannon, Kolmogorov–Sinai, and topological entropies. Standard references include [91, 169, 202].

B.1 Shannon Entropy

One of the most important characterizations one can attach to a random variable and to a stochastic process is its entropy and entropy rate, respectively. We refer to Annex A, Sect. A.3, for the basics of random processes.

B.1.1 The Entropy of a Discrete Random Variable

Let X be a random variable with sample space $(\Omega, \mathcal{B}, \mu)$ and finite state space S. If φ is a real-valued map on S, $\varphi{:}S \to \mathbb{R}$, then $\varphi \circ X = \varphi(X)$ is a random variable with finitely many states $\varphi(S) \subset \mathbb{R}$. The expectation value or average of $\varphi(X)$ will be denoted by $\mathbb{E}\varphi(X)$,

$$\mathbb{E}\varphi(X) = \sum_{x \in S} p(x)\varphi(x),$$

where $p(x)$ is the probability function of X (see (B.21) with $n = 0$).

Definition 14 The (Shannon) entropy of a discrete random variable X on a probability space $(\Omega, \mathcal{B}, \mu)$ is defined by

$$H(X) = -\sum_{x \in S} p(x) \log p(x) = \mathbb{E} \log \frac{1}{p(X)}. \tag{B.1}$$

Whenever convenient, we will write $H_\mu(X)$ to make clear which measure enters into the definition of entropy. Alternatively, one may write $H(p)$ since the entropy depends actually on the probability function $p(x)$ and not on the values taken by X.

(The previous observations hold also for the definitions of different kinds of entropy we will encounter in the sequel.) The logarithm in (B.1) may be taken to any base greater than 1. If the base 2 is used, the entropy comes in units of *bits* (shorthand for "binary digits"). Another usual choice for the logarithm base is Euler's number $e \approx 2.7182818\ldots$, in which case the units of the entropy are called *nats*. Unless otherwise stated, we will henceforth assume the entropy to be in units of bits. Recall that one can change from one logarithmic base a to another base b by means of the formula $\log_b p = \log_b a \log_a p$. By convention, $0 \times \log 0 := \lim_{x \to 0+} x \log x = 0$. Note that $H(X) \geq 0$ because $0 < p(x) \leq 1$ implies $-\log p(x) = \log \frac{1}{p(x)} \geq 0$. On the other hand if $|S|$ denotes the cardinality of the state space S, then $H(X) \leq \log |S|$, as can be easily proved, e.g., using Lagrange multipliers, the highest entropy corresponding to random variables with equiprobable outcomes, that is, $p(x) = 1/|S|$ for all $x \in S$. Observe that Boltzmann's equation (6.1) is nothing else but the entropy for such a flat probability function, $H(X) = \log |S|$, except for the notation (S means entropy in (6.1), while we use S to denote the state space throughout the book) and the physical constant k_B.

Example 28 Suppose that a random variable X takes values $0, 1$ with probabilities $p(0) = p, p(1) = 1 - p(0) = 1 - p$. Then

$$H(X) = -p \log p - (1 - p) \log (1 - p) = H(p). \tag{B.2}$$

The function $H(p)$ is plotted in Fig. B.1. We see that $H(p)$ vanishes when $p = 0$ or $p = 1$, i.e., when the outcome is certain, and it is maximal when $p = 1/2$, i.e., when the uncertainty about the outcome is maximal: $H(1/2) = \log 2 = 1$ bit.

The entropy of a discrete random variable can be given different meanings; see [22] for three interesting interpretations. In information theory one defines

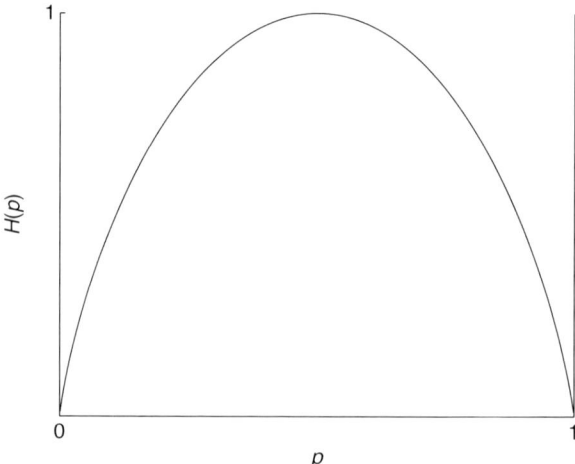

Fig. B.1 The function $H(p)$, (B.2)

$I(X) = -\log p(X)$ to be the *information* of a random variable X with probability function $p(x)$, $-\log p(x)$ being the information conveyed by the outcome $X = x$. Observe that the more rare the event x (that is, the more unlikely the observation of the event x), the more information is gained from its occurrence; one can argue that the most probable events are the less informative ones since their occurrence comes as no surprise. According to Definition 14, $H(X)$ is then the expected value of the information of X: $H(X) = \mathbb{E}I(X)$. Furthermore, if we agree that uncertainty means lack of information, then the entropy can be interpreted as the average uncertainty associated with a random variable or random experiment. In this light, equiprobable events correspond to maximal uncertainty about the outcome.

We turn now to the problem of characterizing the uncertainty associated with more than one random variable.

The *relative entropy* or *Kullback–Leibler distance* between two probability mass functions $p(x)$ and $q(x)$, $x \in S$, is defined as

$$D(p \,\|\, q) = \sum_{x \in S} p(x) \log \frac{p(x)}{q(x)}. \tag{B.3}$$

In this definition, the convention (based on continuity arguments) that $0 \log \frac{0}{q} = 0$ and $p \log \frac{p}{0} = \infty$ is used. From definition (B.3) it follows that $D(p \,\|\, q) \geq 0$ and $D(p \,\|\, q) = 0$ if and only if $p = q$ [59]. On the other hand (and despite of its name), $D(p \,\|\, q)$ is not symmetric in p, q and does not satisfy the triangle inequality. Nonetheless, it is often useful to think of $D(p \,\|\, q)$ as a "distance" between the distributions p and q. The relative entropy $D(p \,\|\, q)$ is a measure of the inefficiency of assuming that the distribution of the random variable X is q when the true distribution is p. For example, if we knew the true distribution p of X, then we could construct a code with average code-word length $H(p)$ (see Sect. 1.1.1, (1.2)). If, instead, we use the code for a distribution q, we would need $H(p) + D(p \,\|\, q)$ bits on the average to describe the random variable X.

Let X and Y be two random variables on a common sample space $(\Omega, \mathcal{B}, \mu)$ but, in general, with different finite state spaces S_1 and S_2, respectively. This corresponds to a situation where two different observations or measurements (with finite precision) are made at the same random experiment. If X and Y have the joint probability function

$$p(x, y) = \mu\{\omega \in \Omega : X(\omega) = x, Y(\omega) = y\} = \Pr(X = x, Y = y)$$

$(x \in S_1, y \in S_2)$, then the *joint entropy* of X and Y is defined as

$$H(X, Y) = -\sum_{x \in S_1} \sum_{y \in S_2} p(x, y) \log p(x, y) = \mathbb{E} \log \frac{1}{p(X, Y)}. \tag{B.4}$$

It is easy to prove that

$$H(X, Y) \leq H(X) + H(Y).$$

The generalization of (B.4) to $n \geq 2$ random variables is straightforward and needs no further elaboration.

The joint probability function $p(x, y)$ and the conditional probability function

$$p(y \,|x) = \frac{p(x, y)}{p(x)}$$

allow the definition of two instrumental concepts in information theory: the conditional entropy and the mutual information. The *conditional entropy* of Y given X is

$$H(Y \,|X) = - \sum_{x \in S_1} \sum_{y \in S_2} p(x, y) \log p(y \,|x) = \mathbb{E} \log \frac{1}{p(Y \,|X)}, \qquad (\text{B.5})$$

and the *mutual information* of X and Y is

$$\begin{aligned}
I(X;Y) &= H(X) - H(X \,|Y) = H(Y) - H(Y \,|X) \\
&= H(X) + H(Y) - H(X, Y) \\
&= I(Y;X), \qquad\qquad\qquad\qquad\qquad\qquad (\text{B.6})
\end{aligned}$$

where we have used the so-called *chain rule* [59]:

$$H(X, Y) = H(X) + H(Y \,|X) . \qquad (\text{B.7})$$

Note that $H(Y \,|X)$ is the average of the uncertainties

$$H(Y \,|X = x) = - \sum_{y \in S_2} p(y \,|x) \log p(y \,|x)$$

weighted with the probabilities $p(x)$, $x \in S_1$. As for the mutual information of two random variables, $I(X;Y)$ is the information about X conveyed by Y (i.e., the information about the realization of X knowing the realization of Y), which is the same as the information about Y conveyed by X, (B.6). Alternatively,

$$I(X;Y) = \mathbb{E} \log \frac{p(X, Y)}{p(X)p(Y)}.$$

Let us mention in passing that the *capacity* of a discrete memoryless channel with input X, output Y, and transition probability $p(Y \,|X)$ is defined as

$$C = \max_{p(x)} I(X;Y),$$

where the maximum is taken over all possible input distributions $p(x)$.

Again, the generalization of these concepts to $n_1 + n_2$ random variables $X_0, \ldots,$ X_{n_1-1} and $Y_0, \ldots, Y_{n_2-1}$ is straightforward. In particular, the (joint) entropy of the random vector $X_0^{n-1} = X_0, \ldots, X_{n-1}$, where, say, all components can take the same states $x_i \in S$, is given by

$$H(X_0, \ldots, X_{n-1}) = - \sum_{x_0, \ldots, x_{n-1} \in S} p(x_0, \ldots, x_{n-1}) \log p(x_0, \ldots, x_{n-1})$$

$$= \mathbb{E} \log \frac{1}{p(X_0, \ldots, X_{n-1})},$$

where $p(x_0, \ldots, x_{n-1})$ is the joint probability function of $X_0, \ldots, X_{n-1}$.

Exercise 22 By iteration of the two-variable rules $p(X, Y) = p(X)p(Y | X)$ and (B.7) prove the general chain rule for the joint entropy: given the random variables $X_0, \ldots, X_{n-1}$ with a joint probability function $p(x_0, \ldots, x_{n-1})$, then

$$p(X_0, \ldots, X_{n-1}) = \prod_{i=0}^{n-1} p(X_i | X_{i-1}, \ldots, X_0) \tag{B.8}$$

and

$$H(X_0, \ldots, X_{n-1}) = \sum_{i=0}^{n-1} H(X_i | X_{i-1}, \ldots, X_0), \tag{B.9}$$

with the conventions $p(X_0 | X_{-1}) := p(X_0)$ and $H(X_0 | X_{-1}) := H(X_0)$.

B.1.2 The Entropy Rate of a Discrete-Time Finite-State Stochastic Process

Definition 15 The entropy rate of a finite-state random process $\mathbf{X} = \{X_n\}_{n \in \mathbb{N}_0}$ on a probability space $(\Omega, \mathcal{B}, \mu)$ is defined by

$$h(\mathbf{X}) = \lim_{n \to \infty} \frac{1}{n} H(X_0, \ldots, X_{n-1}), \tag{B.10}$$

provided the limit exists.

Sometimes the terms

$$h(X_0, \ldots, X_{n-1}) = \frac{1}{n} H(X_0, \ldots, X_{n-1})$$

$(n \geq 2)$ are called the *entropy rates of order n* of $\mathbf{X}$. Hence, $h(X_0, \ldots, X_{n-1})$ or, more compactly written, $h(X_0^{n-1})$ is the average uncertainty per symbol (time unit, channel use, etc. depending on the interpretation of n) about n consecutive outcomes of the random experiment modeled by $\mathbf{X}$. If we repeat the experiment an arbitrarily long number of times, these average uncertainty rates eventually converge to a limit—Shannon's entropy rate $h(\mathbf{X})$.

Although $h(X_0^{n-1})$ and, consequently, $h(\mathbf{X})$ are actually entropy *rates*, the term "rate" is generally omitted—also in other types of entropy. We follow sometimes this common usage, since this does not lead to misunderstandings.

Lemma 12 *For a stationary stochastic process* $\mathbf{X} = \{X_n\}_{n \in \mathbb{N}_0}$, *the sequence of conditional entropies* $H(X_n | X_{n-1}, \ldots, X_0)$ *is decreasing.*

Proof Indeed,

$$
\begin{aligned}
H(X_{n+1} | X_n, \ldots, X_1, X_0) &\leq H(X_{n+1} | X_n, \ldots, X_1) \\
&= H(X_n | X_{n-1}, \ldots, X_0),
\end{aligned}
$$

where the inequality follows from the fact that conditioning reduces uncertainty, and the equality follows from the stationarity of $\mathbf{X}$.   □

Theorem 22 *For a stationary stochastic process* $\mathbf{X} = \{X_n\}_{n \in \mathbb{N}_0}$,

$$
h(\mathbf{X}) = \lim_{n \to \infty} H(X_n | X_{n-1}, \ldots, X_0). \tag{B.11}
$$

Proof First of all, limit (B.11) converges because, according to Lemma 12, the positive sequence $H(X_n | X_{n-1}, \ldots, X_0)$ is decreasing. Furthermore, by the chain rule (B.9),

$$
h(X_0, \ldots, X_n) = \frac{1}{n+1} \sum_{i=0}^{n} H(X_i | X_{i-1}, \ldots, X_0).
$$

By Cesáro's mean theorem ("If $a_n \to a$ and $b_n = \frac{1}{n+1} \sum_{i=0}^{n} a_i$, then $b_n \to a$"),

$$
h(\mathbf{X}) = \lim_{n \to \infty} h(X_0, \ldots, X_n) = \lim_{n \to \infty} H(X_n | X_{n-1}, \ldots, X_0).
$$

□

From Lemma 12 it follows that the convergence of the entropy rates of order n, $h(X_0, \ldots, X_{n-1})$, to $h(\mathbf{X})$ is monotonically decreasing:

$$
h(X_0) \geq h(X_0, X_1) \geq \cdots \geq h(X_0, \ldots, X_{n-1}) \geq \cdots . \tag{B.12}
$$

Thus, when estimating the entropy rate of a stationary random process by its entropy rate of order n, the estimation always exceeds the true value. Intuitively speaking, with increasing n we see more and more correlations among the variables $X_0, \ldots, X_{n-1}$ and this reduces our uncertainty about the next observation X_n. We turn back to this point in Example 31.

In an information-theoretical setting and in applications (Sect. A.3), one can think of a stationary stochastic process $\mathbf{X} = \{X_n\}_{n\in\mathbb{N}_0}$ as a data source. Its realizations are then the messages output by the source. This is illustrated in Fig. B.2. Here x_0 can be considered the current and last letter of the message, the other letters having been output in the past, the greater the index, the earlier in time.

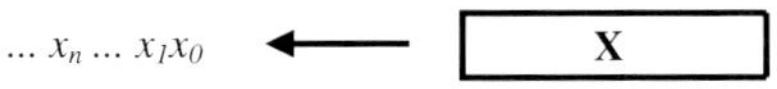

$$\ldots x_n \ldots x_1 x_0 \quad\longleftarrow\quad \boxed{\quad\mathbf{X}\quad}$$

Fig. B.2 A data source $\mathbf{X}$ outputs a message x_0^∞

B.2 Kolmogorov–Sinai Entropy

B.2.1 Deterministic Systems

A partition of a probability space $(\Omega, \mathcal{B}, \mu)$ is a collection $\alpha = (A_i)_{i\in J}$ of disjoint sets $A_i \in \mathcal{B}$, with a countable index set J, such that $\bigcup_{i\in J}\mu(A_i) = 1$. If J is finite, α is called a finite partition. If α is a finite partition of $(\Omega, \mathcal{B}, \mu)$, then the collection of all elements of $\mathcal{B}$ which are unions of elements of α is a finite sub-sigma-algebra of $\mathcal{B}$ which we denote by $\mathcal{B}(\alpha)$. We write $\alpha \leq \beta$, where α, β are two finite partitions of $(\Omega, \mathcal{B}, \mu)$, to mean that each element of α is a union of elements of β. In this case, β is called a *refinement* of α. We have $\alpha \leq \beta$ iff $\mathcal{B}(\alpha) \subset \mathcal{B}(\beta)$.

Definition 16 Let $\alpha = \{A_1, \ldots, A_{|\alpha|}\}$ be a finite partition of $(\Omega, \mathcal{B}, \mu)$. The entropy of the partition α is the number

$$H_\mu(\alpha) = -\sum_{i=1}^{|\alpha|} \mu(A_i) \log \mu(A_i).$$

The same considerations concerning the base of the logarithm we made after the definition of Shannon's entropy, Definition 14, apply here as well. By the same token, $H(\alpha)$ is a measure of the information gained (or the uncertainty removed) by performing a random experiment whose outcomes have probabilities $\mu(A_1), \ldots, \mu(A_{|\alpha|})$.

Sometimes it is convenient to quantify the "coarseness" of a partition. Roughly speaking, if we assign a "size" to each $A \in \alpha$, then we can take the maximum of those sizes as the coarseness of α. The resulting parameter is called the *norm of the partition* α and denoted by $\|\alpha\|$. In metric spaces (X, d), one can take $\|\alpha\| = \max_{A\in\alpha}\mathrm{diam}(A)$, where $\mathrm{diam}(A) = \sup\{d(x, y): x, y \in A\}$ is called the "diameter" of A.

If $f: \Omega \to \Omega$ is a measure-preserving function on the probability space $(\Omega, \mathcal{B}, \mu)$, we denote by $f^{-n}\alpha$ the partition $\{f^{-n}A_1, \ldots, f^{-n}A_{|\alpha|}\}$. Furthermore, given two finite partitions $\alpha = \{A_1, \ldots, A_{|\alpha|}\}$ and $\beta = \{B_1, \ldots, B_{|\beta|}\}$ of $(\Omega, \mathcal{B}, \mu)$, we denote by $\alpha \vee \beta$ their *least common refinement*,

$$\alpha \vee \beta = \{A \cap B : A \in \alpha, B \in \beta, \mu(A \cap B) > 0\}.$$

More general refinements, like

$$\alpha \vee f^{-1}\alpha \vee \cdots \vee f^{-(n-1)}\alpha = \bigvee_{i=0}^{n-1} f^{-i}\alpha,$$

are defined recursively.

Definition 17 Let $(\Omega, \mathcal{B}, \mu, f)$ be a measure-preserving dynamical system. If α is a finite partition of $(\Omega, \mathcal{B}, \mu)$, then

$$h_\mu(f, \alpha) = \lim_{n \to \infty} \frac{1}{n} H_\mu \left(\bigvee_{i=0}^{n-1} f^{-i}\alpha \right) \tag{B.13}$$

is called the metric entropy of f with respect to α.

In this setting, consider now a finite-state random process $\mathbf{X}^\alpha = \{X_n^\alpha\}_{n \in \mathbb{N}_0}$, with $X_n^\alpha : \Omega \to S = \{0, \dots, |\alpha| - 1\}$, defined as follows:

$$X_n^\alpha(\omega) = i \quad \text{iff} \quad f^n(\omega) \in A_i \in \alpha. \tag{B.14}$$

Note that $X_{n+1} = X_n \circ f$, thus $X_n = X_n \circ f^n$. Then

$$\Pr\left\{X_0^\alpha = i_0, \dots, X_n^\alpha = i_n\right\} = \mu \left\{\omega \in \Omega : \omega \in A_{i_0}, f(\omega) \in A_{i_1}, \dots, f^n(\omega) \in A_{i_n}\right\}$$
$$= \mu \left\{A_{i_0} \cap \cdots \cap f^{-n} A_{i_n}\right\}, \tag{B.15}$$

$n \geq 0$, and similarly,

$$\Pr\left\{X_k^\alpha = i_0, \dots, X_{n+k}^\alpha = i_n\right\} = \mu \left\{f^{-k}(A_{i_0} \cap \cdots \cap f^{-n} A_{i_n})\right\}$$
$$= \Pr\left\{X_0^\alpha = i_0, \dots, X_n^\alpha = i_n\right\}$$

because of the f-invariance of μ. We conclude that $\mathbf{X}^\alpha$ is a stationary process, which is called the *symbolic dynamics* of $(\Omega, \mathcal{B}, \mu, f)$ with respect to the partition ("coarse graining" or "quantization") α. Depending on the context, $\mathbf{X}^\alpha$ is also called a *coding map* (dynamical systems) or a collection of *simple observations* with respect to f with precision $\|\alpha\|$ (information theory). Moreover, it follows from (B.15) that

$$h_\mu(f, \alpha) = h_\mu(\mathbf{X}^\alpha). \tag{B.16}$$

This not only proves that limit (B.13) does exist but also that the *entropy rates of order n* of f with respect to α,

$$h_\mu^{(n)}(f,\alpha) = \frac{1}{n} H_\mu \left(\bigvee_{i=0}^{n-1} f^{-i}\alpha \right),$$

decrease to $h_\mu(f,\alpha)$ when $n \to \infty$ (remember (B.12)).

Definition 18 Let $(\Omega, \mathcal{B}, \mu, f)$ be a measure-preserving dynamical system and α a finite partition of $(\Omega, \mathcal{B}, \mu)$. Then,

$$h_\mu(f) = \sup_\alpha h_\mu(f,\alpha) \tag{B.17}$$

is called the metric entropy (or just, the entropy) of the map f with respect to μ.

Sometimes $h_\mu(f)$ is called the Kolmogorov–Sinai entropy or the measure-theoretic entropy too. To streamline the notation, the subscript μ may be dropped from $H_\mu(\alpha)$, $h_\mu(f,\alpha)$, and $h_\mu(f)$, as we generally do, if the probability measure is clear from the context.

The isomorphic invariance is one of the fundamental properties of entropy.

Theorem 23(a) If the dynamical systems $(\Omega_1, \mathcal{B}_1, \mu_1, f_1)$ and $(\Omega_2, \mathcal{B}_2, \mu_2, f_2)$ are isomorphic, then $h(f_1) = h(f_2)$.
(b) If $(\Omega_2, \mathcal{B}_2, \mu_2, f_2)$ is a factor of $(\Omega_1, \mathcal{B}_1, \mu_1, f_1)$, then $h(f_2) \le h(f_1)$.

It should be obvious from definitions (B.13) and (B.17) that the exact calculation of $h(f)$ from scratch is, in general, unfeasible. There are though a few results that, depending on the specifics of the dynamical system in question, can come to the rescue. We mention a few next.

A finite partition α of $(\Omega, \mathcal{B}, \mu)$ is called a *generating partition* or a *generator for a μ-preserving transformation* $f : \Omega \to \Omega$ if (i)

$$\bigvee_{n=-\infty}^{\infty} f^{-n} \mathcal{B}(\alpha) = \mathcal{B} \text{ (modulo μ-zero sets)} \tag{B.18}$$

when f is invertible (i.e., f is an automorphism) or (ii)

$$\bigvee_{n=0}^{\infty} f^{-n} \mathcal{B}(\alpha) = \mathcal{B} \text{ (modulo μ-zero sets)} \tag{B.19}$$

when f is non-invertible (i.e., f is an endomorphism). This means that for any $B \in \mathcal{B}$, there is a $B' \in \bigvee_{n=-\infty}^{\infty} f^{-n} \mathcal{B}(\alpha)$ or $B' \in \bigvee_{n=0}^{\infty} f^{-n} \mathcal{B}(\alpha)$, respectively, such that $\mu(B \triangle B') = 0$. If f is invertible and the stronger condition (B.19) holds, then α is called a strong or one-sided generator for f. Equivalent definitions of generators and one-sided generators by means of partition refinements converging to the point partition $\epsilon = \{\{x\} : x \in \Omega\}$ were given in Sect. 1.3.

Example 29 Since the sigma-algebra $\mathcal{B}_\Pi(S)$ of the one-sided and two-sided shift spaces are generated by the cylinder sets

$$C_{a_0,\dots,a_k} = \{\mathbf{s} = (s_n)_{n\in\mathbb{N}_0} : s_0 = a_0, \dots, s_k = a_k\} = \bigcap_{i=0}^{k} \Sigma^{-i} C_{a_i}$$

and

$$C_{a_{-k},\dots,a_0,\dots,a_k} = \{\mathbf{s} = (s_n)_{n\in\mathbb{Z}} : s_{-k} = a_{-k}, \dots, s_k = a_k\} = \bigcap_{i=-k}^{k} \Sigma^{-i} C_{a_i},$$

respectively, it follows that the partition

$$\gamma = \{C_a : a \in S\}$$

is a generator of both the one-sided and two-sided shifts.

Generating partitions can be found numerically; see, e.g., [40] for a general method based on relaxation algorithms. For higher dimensional maps, numerical techniques have been proposed for the dissipative Hénon map [87], the standard map [53], two-dimensional hyperbolic maps [26], etc. A method based on unstable period orbits was proposed in [63]. The construction of one-dimensional maps possessing generating partitions was studied in [99].

Theorem 24 (*Kolmogorov–Sinai Theorem*) *Let* $(\Omega, \mathcal{B}, \mu, f)$ *be a dynamical system.*

(a) *If* f *is an automorphism and* α *is a generator or a one-sided generator for* f, *then*
   $h(f) = H(f, \alpha)$.
(b) *If* f *is an endomorphism and* α *is a generator for* f, *then* $h(f) = H(f, \alpha)$.

The case of automorphisms with one-sided generators is uninteresting since then one can show that $h(f) = 0$ [202]. More interestingly, *Krieger's theorem* states that if f is an ergodic automorphism with $h(f) < \infty$, then f has a generator [67, 130, 169]. Although Krieger's proof is non-constructive, Smorodinsky [191] and Denker [65] provided methods to construct a two-sided generator for ergodic and aperiodic automorphisms. Denker's construction could even be extended by Grillenberger [66] to all aperiodic automorphisms. The existence of generators for endomorphisms was proved by Kowalski under different assumptions [128, 129]. At variance with the previous case, the construction of one-sided generators for endomorphisms remains an open problem till this very day; see [182] for some progress in this issue.

Example 30 Using the fact that the cylinder sets C_a are generators for the one-sided and two-sided $(\mathbf{p}, P)$-Markov shifts Σ on N symbols, one can prove

$$h_\mu(\Sigma) = -\sum_{i,j=1}^{N} p_i P_{ij} \log P_{ij}, \tag{B.20}$$

where μ is the Markov measure defined by $(\mathbf{p}, P)$ (see Example 25 (b)). Upon substituting $P_{ij} = p_j$ in (B.20), we get for $\mathbf{p}$-Bernoulli shifts

$$h_\mu(\Sigma) = -\sum_j^N p_j \log p_j,$$

where μ is the Bernoulli measure defined by $\mathbf{p}$ (see Example 25 (a)).

A second practical way of calculating (or, at least, estimating) the entropy is provided by the following theorem.

Theorem 25 [169, Ch. 5, Prop. 3.6] *Let $(\Omega, \mathcal{B}, \mu, f)$ be a measure-preserving dynamical system. If $\alpha_0 \leq \alpha_1 \leq \cdots$ is an increasing sequence of finite partitions of $(\Omega, \mathcal{B}, \mu)$ and $\vee_{n=0}^\infty \mathcal{A}(\alpha_n) = \mathcal{B}$ up to sets of measure 0, then*

$$\lim_{n \to \infty} h_\mu(f, \alpha_n) = h_\mu(f).$$

A third practical method calls for Pesin's theorem and Lyapunov exponents. Since this topic would take us too far away, we refer the interested reader to the specialized literature [142, 52, 72]. Due to the important role that the Lyapunov exponent(s) play in nonlinear dynamics, several numerical schemes have been developed to calculate them [193]. On the other hand, Pesin's theorem and its generalizations require the invariant measure to possess some properties—but invariant measures are in many interesting cases unknown. This fact limits the application of this method. For the calculation of the metric entropy in some one-dimensional systems, see [105].

B.2.2 Random Systems

Let $\mathbf{X} = \{X_n\}_{n\in\mathbb{N}_0}$ be a stationary stochastic process on a probability space $(\Omega, \mathcal{B}, \mu)$, taking on values in $S = \{0, \ldots, N-1\}$. In Sect. A.3 it is shown that $\mathbf{X}$ can be associated in a canonical way with a shift system $(S^{\mathbb{N}_0}, \mathcal{B}_\Pi(S), m, \Sigma)$, called its sequence space model, via $\Phi:\Omega \to S^{\mathbb{N}_0}$, $(\Phi(\omega))_n = X_n(\omega)$. The joint probability function $p(x_0, \ldots, x_{n-1})$ of the random process $\mathbf{X}$ is related to the measure of the cylinder sets $C_{x_0,\ldots,x_{n-1}}$, $x_0, \ldots, x_{n-1} \in S$, of the sequence space model in the following way:

$$
\begin{aligned}
p(x_0, \ldots, x_{n-1}) &= \mu\left\{\omega \in \Omega : X_0(\omega) = x_0, \ldots, X_{n-1}(\omega) = x_{n-1}\right\} \\
&= \mu\left\{\Phi^{-1}\left\{\mathbf{s} \in S^{\mathbb{N}_0} : s_0 = x_0, \ldots, s_{n-1} = x_{n-1}\right\}\right\} \\
&= m\left\{C_{x_0,\ldots,x_{n-1}}\right\} \\
&= m\{C_{x_0} \cap \cdots \cap \Sigma^{-(n-1)} C_{x_{n-1}}\}.
\end{aligned}
$$

Since the partition $\gamma = \{C_{x_0} : x_0 \in S\}$ is a generator of Σ (Example 29), we have

$$h_\mu(\mathbf{X}) = -\lim_{n\to\infty} \frac{1}{n} \sum_{x_0,\dots,x_{n-1}\in S} p(x_0,\dots,x_{n-1}) \log p(x_0,\dots,x_{n-1})$$

$$= -\lim_{n\to\infty} \frac{1}{n} H_m\left(\bigvee_{i=0}^{n-1} \Sigma^{-i}\gamma\right)$$

$$= h_m(\Sigma,\gamma)$$

$$= h_m(\Sigma)$$

by Theorem 24 (b). In words, the Shannon entropy rate of a stochastic process $\mathbf{X} = \{X_n\}_{n\in\mathbb{N}_0}$ coincides with the Kolmogorov–Sinai entropy rate of its sequence space model.

An important property of ergodic processes is the so-called *asymptotic equipartition property* or *Shannon–McMillan–Breiman theorem*.

Theorem 26 (*Shannon–McMillan–Breiman*) *If* $\mathbf{X} = \{X_n\}_{n\in\mathbb{N}_0}$ *is a finite-valued stationary ergodic process, then* $-\frac{1}{n}\log p(X_0,\dots,X_{n-1})$ *converges in probability to the entropy rate* $h(\mathbf{X})$.

Example 31 The sequence space model of a finite-state, time-homogeneous Markov chain $\mathbf{X} = \{X_n\}_{n\in\mathbb{N}_0}$ (Example 27) with transition matrix $P_{i,j}$, $0 \le i,j \le N-1$, and stationary probability vector $\mathbf{p}$ is the one-sided $(\mathbf{p}, P)$-Markov shift $\Sigma_{\mathbf{p},P}$. Therefore,

$$h(\mathbf{X}) = h(\Sigma_{\mathbf{p},P}) = -\sum_{i,j=0}^{N-1} p_i P_{ij} \log P_{ij}.$$

For the specific case

$$P = \begin{pmatrix} 1-p_{01} & p_{01} \\ p_{10} & 1-p_{10} \end{pmatrix} = \begin{pmatrix} 0.9 & 0.1 \\ 0.1 & 0.9 \end{pmatrix},$$

the stationary probability is

$$\mathbf{p} = \left(\frac{p_{10}}{p_{01}+p_{10}}, \frac{p_{01}}{p_{01}+p_{10}}\right) = \left(\frac{1}{2},\frac{1}{2}\right).$$

The upper curve in Fig. B.3 shows the entropy rates of order n, $h(X_0,\dots,X_{n-1})$, closing in on the true value $h(\mathbf{X}) = 0.469$ bits/symbol (horizontal line). The lower curve shows what happens in practice when $h(\mathbf{X})$ is estimated numerically in a naive way. Here the probabilities $p(x_0,\dots,x_{n-1})$ were estimated by the frequencies of the word $x_0,\dots,x_{n-1}$ in a sequence of 10,000 draws. In the left part of the experimental curve, we see the entropy rates of successive order $n = 1,2,\dots$ converging from above to the true value. For $n \approx 20$, the numerical values provide accurate estimates of the entropy. For greater lengths, the estimates tend toward zero along the parabola

$$h(n) = \frac{\log (N - n + 1)}{n}$$

due to undersampling.

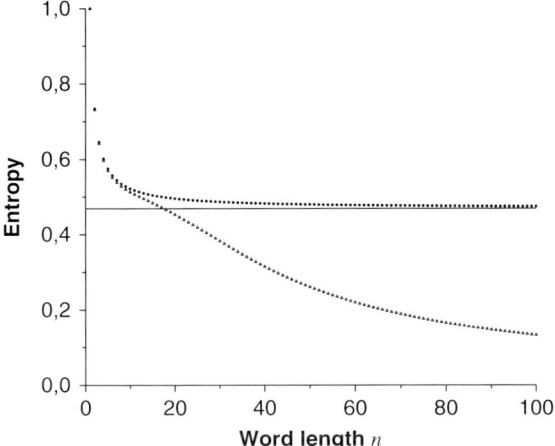

Fig. B.3 The *upper dotted line* shows the convergence of the entropy rate of order n to the true value, 0.469 bits/symbol (*horizontal line*), for an arbitrarily long sequence generated by a two-state Markov chain with transition probabilities $p_{01} = p_{10} = 0.1$. The *lower dotted line* shows what happens in practice due to undersampling

A particular case is of interest. Consider now not a general stationary stochastic process but the symbolic dynamics $X^\alpha = \{X_n^\alpha\}_{n \in \mathbb{N}_0}$ of the system $(\Omega, \mathcal{B}, \mu, f)$ with respect to a partition $\alpha = \{A_1, \ldots, A_{|\alpha|}\}$ (see (B.14)), and let $(S^{\mathbb{N}_0}, \mathcal{B}_\Pi(S), m, \Sigma)$ be the sequence space model of X^α; hence $S = \{1, \ldots, |\alpha|\}$ and

$$m(C_{a_0, a_1, \ldots, a_n}) = \mu(A_{a_0} \cap f^{-1} A_{a_1} \cap \cdots \cap f^{-n} A_{a_n})$$

for any cylinder set $C_{a_0, \ldots, a_n} = \{s \in S^{\mathbb{N}_0} : s_0 = a_0, \ldots, s_n = a_n\}$, with $a_0, \ldots, a_n \in S$. In this setting, the following question arises. When are the dynamical systems $(\Omega, \mathcal{B}, \mu, f)$ and $(S^{\mathbb{N}_0}, \mathcal{B}_\Pi(S), m, \Sigma)$ isomorphic (via $\Phi^\alpha : \Omega \to S^{\mathbb{N}_0}$, $(\Phi^\alpha(\omega))_n = X_n^\alpha(\omega)$)? Since $\{C_a : a \in S\}$ is a generator for Σ and $(\Phi^\alpha)^{-1} C_a = A_a$ for every $a \in S$, we need clearly that

$$\{(\Phi^\alpha)^{-1} C_a : 1 \le a \le |\alpha|\} = \{A_a : 1 \le a \le |\alpha|\} = \alpha$$

is also a generator for f. In other words, a generator for f gives a natural isomorphism between $(\Omega, \mathcal{B}, \mu, f)$ and the sequence space model associated with its symbolic dynamics. By Krieger's theorem we conclude that any ergodic, invertible dynamical system with finite entropy can be represented as a two-sided shift system. This result is useful in that it provides prototypes of ergodic, finite-entropy systems.

B.3 Topological Entropy

Topological entropy for continuous self-maps of compact topological spaces was introduced by Adler, Koheim, and McAndrews by means of open covers [3]. Later Dinaburg [70] and Bowen [36] found alternative approaches via separating and spanning sets in (not necessarily compact) metric spaces.

B.3.1 Generalities

Recall that a continuous or topological dynamical system is a pair (M,f), where M is a topological space and $f{:}M \rightarrow M$ is a continuous map. As compared to measure-theoretical dynamical systems, there is here no measurable structure involved (although M can be thought to be endowed with the Borel sigma-algebra); instead, continuity enters the scenario. Sometimes, continuity is weakened to piece-wise continuity, especially in conjunction with other properties like piecewise monotonicity.

Furthermore, in this section (M,d) denotes a metric space and $f{:}M \rightarrow M$ a uniformly continuous map. If, moreover, M is compact, then f needs only to be continuous (since every continuous self-map of a compact space is uniformly continuous).

Definition 19 Let K be a compact topological space, α an open cover of K, and $N(\alpha)$ the number of sets in a finite subcover of α with smallest cardinality. The entropy of the cover α is then defined as $H(\alpha) = \log N(\alpha)$.

If α is an open cover of K and $f{:}K \rightarrow K$ is continuous, then $f^{-1}\alpha$ is the open cover consisting of all sets $f^{-1}A$, $A \in \alpha$.

Definition 20 If α is an open cover of the compact space K and $f{:}K \rightarrow K$ is continuous, then the entropy of f relative to α is given by

$$h(f,\alpha) = \lim_{n\to\infty} \frac{1}{n} H\left(\bigvee_{i=0}^{n-1} f^{-i}\alpha\right) \tag{B.21}$$

and the topological entropy of f is given by

$$h(f) = \sup_{\alpha} h(f,\alpha). \tag{B.22}$$

It can be proved that the limit in (B.21) exists and the supremum in (B.22) can be taken over *finite* open covers of K.

In a metric space (M,d), the alternative definitions of topological entropy via spanning and separating sets may be more useful.

Definition 21 Let $n \in \mathbb{N}$, $\varepsilon > 0$, and $K \subset M$ compact. A subset $A \subset M$ is said to (n,ε)-span K with respect to $f{:}M \rightarrow M$ if for each $x \in K$ there exists $y \in A$ such that

$$\max_{0 \le i \le n-1} d(f^i(x), f^i(y)) \le \varepsilon.$$

Furthermore, let $r_n(\varepsilon, K)$ denote the smallest cardinality of any (n, ε)-spanning set for K with respect to f.

Definition 22 The topological entropy of $f{:}M \to M$ is

$$h_d(f) = \sup_K \lim_{\varepsilon \to 0} \limsup_{n \to \infty} \frac{1}{n} \log r_n(\varepsilon, K), \tag{B.23}$$

where the supremum is taken over all compact subsets of M.

The definition of topological entropy by means of separating sets is as follows.

Definition 23 Let $n \in \mathbb{N}$, $\varepsilon > 0$, and $K \subset M$ compact. A subset $A \subset K$ is said to be (n, ε)-separated with respect to $f{:}M \to M$ if $x, y \in A$, $x \ne y$, implies

$$\max_{0 \le i \le n-1} d(f^i(x), f^i(y)) > \varepsilon.$$

Furthermore, let $s_n(\varepsilon, K)$ denote the largest cardinality of any (n, ε)-separated subset of K with respect to f.

Thus, an (n, ε)-separated subset of Ω is a kind of microscope that allows us to distinguish orbits of length n up to a precision ε.

Definition 24 The topological entropy of $f{:}M \to M$ is

$$h_d(f) = \sup_K \lim_{\varepsilon \to 0} \limsup_{n \to \infty} \frac{1}{n} \log s_n(\varepsilon, K), \tag{B.24}$$

where the supremum is taken over all compact subsets of M.

If M is compact, then $h_d(f)$ can be shown [202] not to depend on the metric d (thus, it will be denoted by $h_{\text{top}}(f)$) and, moreover, definitions (B.23) and (B.24) can be simplified to

$$h_{\text{top}}(f) = \lim_{\varepsilon \to 0} \limsup_{n \to \infty} \frac{1}{n} \log r_n(\varepsilon, M) = \lim_{\varepsilon \to 0} \limsup_{n \to \infty} \frac{1}{n} \log s_n(\varepsilon, M). \tag{B.25}$$

Both $r_n(\varepsilon, M)$ and $s_n(\varepsilon, M)$ can be interpreted as the number of orbits of length n up to an error ε. For $\varepsilon \ll 1$,

$$e^{nh(f)} \sim r_n(\varepsilon, M) \quad \text{and} \quad e^{nh(f)} \sim s_n(\varepsilon, M),$$

where $\sim$ stands for "asymptotically as $n \to \infty$" (assuming the convergence of $\frac{1}{n} \log r_n(\varepsilon, M)$ and $\frac{1}{n} \log s_n(\varepsilon, M)$ in this limit), so the topological entropy measures the asymptotic exponential growth rate with n of the number of orbits of length n, up to error ε.

Definition 25 Let $f_1 : M_1 \rightarrow M_1$ and $f_2 : M_2 \rightarrow M_2$ be continuous maps of metric spaces and suppose that there exists a continuous surjective map $\phi : M_1 \rightarrow M_2$ such that $\phi \circ f_1 = f_2 \circ \phi$. Then we say that f_1 is topologically semiconjugate to f_2 or that f_2 is a factor of f_1 via the topological semi-conjugacy or factor map ϕ. In the case that ϕ is a homeomorphism, then f_1 and f_2 are said to be topologically conjugate and ϕ is said to be a topological conjugacy.

In particular, if two maps are metrically conjugate via a (measure-preserving) homeomorphism, then they are also topologically conjugate. Such is the case of the logistic and symmetric tent maps via the homeomorphism A.8 (Example 24). The qualifiers "topological" and "topologically" may be dropped if it is clear that they refer to a topological system.

Thus, conjugate maps are obtained from each other by a continuous change of coordinates. Therefore, properties that are independent of such changes of coordinates will be invariant under topological conjugacy, e.g., sensitivity to initial conditions, topological transitivity, number of periodic orbits of a given period.

Just as metric entropy is an invariant of metric conjugacy, so is topological entropy an invariant of topological conjugacy.

Theorem 27 *Let f_1 and f_2 be continuous self-maps of compact spaces. If f_1 and f_2 are topologically conjugate, then $h(f_1) = h(f_2)$. More generally, if f_2 is a factor of f_1, then $h(f_2) \leq h(f_1)$.*

Exercise 23 Show that the quadratic transformations $f_1(x) = vx(1 - x)$ on $[0, 1]$, $0 < v \leq 4$, and

$$f_2(y) = \tfrac{1}{2}(y^2 - v^2 + 2v)$$

on $[-v, v]$ are topologically conjugate via the homeomorphism

$$\phi(x) = v(1 - 2x) = f_1'(x).$$

In spite of not involving a measure-theoretical structure, topological entropy is tightly related to metric entropy through the following *variational principle*.

Theorem 28 *Let M be a compact metric space endowed with the Borel sigma-algebra $\mathcal{B}$, and $f : M \rightarrow M$ a continuous map. Then*

$$h_{top}(f) = \sup h_\mu(f), \tag{B.26}$$

where the supremum is taken over all f-invariant measures μ on the measurable space $(M, \mathcal{B})$.

Note that the set of f-invariant measures invoked in the variational principle (B.26) is non-empty by Theorem 20. Moreover, the supremum in (B.26) can be restricted to ergodic measures [202],

$$h_{\text{top}}(f) = \sup_{\mu \in E(M,f)} h_\mu(f), \tag{B.27}$$

where $E(M,f)$ is the set of f-invariant, ergodic measures on $(M,\mathcal{B})$. Measures μ such that $h_{\text{top}}(f) = h_\mu(f)$ are called *measures with maximal entropy* for obvious reasons.

In Sect. A.1 we defined the concept of generator of a measure-preserving transformation. In topological dynamics, there is also a concept of generator that plays a similar role with respect to the topological entropy. Given a compact metric space M and a map $f\colon M \to M$, a finite open cover $\alpha = \{A_1, \ldots, A_{|\alpha|}\}$ of M is said to be a *generator* for f if

(a) in case f is invertible, for any bisequence $(a_i)_{i\in\mathbb{Z}}$, $1 \leq a_i \leq |\alpha|$, the intersection

$$\bigcap_{i=-\infty}^{\infty} f^{-i} A_{a_i}$$

contains at most one point or

(b) in case f is non-invertible, for any sequence $(a_i)_{i\in\mathbb{N}_0}$, $1 \leq a_i \leq |\alpha|$, the intersection

$$\bigcap_{i=0}^{\infty} f^{-i} A_{a_i}$$

contains at most one point.

The topological dynamical systems that admit a generator have a simple characterization.

Definition 26 Let M be a compact metric space. A homeomorphism (correspondingly, a continuous map) $f\colon M \to M$ is said to be *expansive* if there exists $\delta > 0$, called an *expansivity constant* for f, such that

$$d(f^n(x), f^n(y)) \leq \delta$$

for all $n \in \mathbb{Z}$ (correspondingly, $n \in \mathbb{N}_0$) implies $x = y$. Expansive non-invertible maps and homeomorphisms for which the expansiveness condition holds already for non-negative iterates are collectively called *positively expansive maps*.

Alternatively, if $x \neq y$ and δ is an expansivity constant for f, then there exists $n \in \mathbb{Z}$ (correspondingly, $n \in \mathbb{N}_0$) with $d(f^n(x), f^n(y)) > \delta$. Notice that expansiveness differs from sensitive dependence in that *all* nearby points eventually separate by at least δ (for sensitive dependence it suffices this to occur for a single point in each neighborhood of the other). Intuitively, the orbits of an expansive map f can be resolved to any desired precision by taking n sufficiently large. Expansive maps

f have some nice properties like having a countable number of periodic points, and at least one invariant measure with maximal entropy [202]. Examples of expansive maps include the shift transformations and the hyperbolic toral automorphisms. On the other hand, there are no expansive maps of closed one-dimensional intervals [19, Thm. 2.2.31] nor expansive homeomorphisms of the circle [202]. Expansiveness and positively expansiveness are topological conjugacy invariants.

Theorem 29 *Let $f:M \to M$ be a map of the compact metric space (M,d). Then f is expansive if and only if f has a generator.*

Observe that the cylinder sets C_a are generators both in the measure-theoretical and in the topological senses because, among other considerations, they build a partition and an open cover at the same time. Therefore, shifts on sequence spaces are expansive transformations. Expansiveness is an invariant of topological conjugacy.

Theorem 30 *If $f:M \to M$ be an expansive map of the compact metric space (M,d) and α is a generator for f, then $h_{top}(f) = h(f,\alpha)$.*

Example 32 Let $S = \{0,\dots,k-1\}$ and Σ be the shift on the bisequence space $S^{\mathbb{Z}} = \{(s_n)_{n \in \mathbb{Z}}\}$. Then Σ has topological entropy $\log N$. Indeed, apply Theorem 30 with α comprising the cylinder sets $C_j = \{(x_n)_{n \in \mathbb{Z}}:x_0 = j\}$ to obtain

$$h_{top}(\Sigma) = \lim_{n\to\infty} \frac{1}{n} \log N \left(\bigvee_{i=0}^{n-1} \Sigma^{-i}\alpha \right) = \lim_{n\to\infty} \frac{1}{n} \log k^n = \log k.$$

Thus, if μ_0 is the Bernoulli measure on $(S^{\mathbb{Z}}, \mathcal{B}_{\Pi}(S))$ defined by the probability vector $\mathbf{p}_0 = (\frac{1}{k},\dots,\frac{1}{k})$, we have

$$h_{\mu_0}(\Sigma) = \log k = h_{top}(\Sigma).$$

This illustrates the existence of (in this case, unique) measures of maximal entropy. The result in the one-sided case is the same.

Example 33 Let $S = \{0,\dots,k-1\}$, $A = (a_{ij})_{i,j=0}^{k-1}$ be a $k \times k$ matrix whose entries a_{ij} are either 0's or 1's, and

$$\Omega_A = \{\omega \in S^{\mathbb{Z}}:a_{\omega_n\omega_{n+1}} = 1 \text{ for } \forall n \in \mathbb{Z}\}.$$

The space Ω_A is closed and shift invariant. The restriction

$$\Sigma_A := \Sigma|_{\Omega_A}$$

is called the two-sided *topological Markov chain* determined by the matrix A, a *Markov subshift*, or a *subshift of finite type* (see Sect. 1.1.2). One-sided topological Markov chains are defined analogously over $S^{\mathbb{N}_0}$. The matrix A is said to be irreducible if for any pair i,j there is $n > 0$ such that $a_{ij}^{(n)} > 0$, where $a_{ij}^{(n)}$ are the entries of A^n. If A is irreducible and Σ_A is a one-sided or two-sided topological Markov chain, then [202]

$$h_{top}(\Sigma_A) = \log \lambda, \tag{B.28}$$

where λ is the largest positive eigenvalue of A. A topological Markov chain Σ_A has a unique measure (called its Parry measure) of maximal topological entropy.

It can be proved [31, Sect. 4.3] that a C^2 piecewise expanding Markov map f is topologically conjugate (modulo 0) to the one-sided topological Markov chain Σ_A, where A is the transition matrix for f. Therefore, piecewise expanding Markov maps admit a symbolic description.

Example 34 Consider the *rooftop map f* defined by

$$f(x) = \begin{cases} ax + c & \text{if } 0 \le x \le c, \\ (1 - b)x & \text{if } c \le x \le 1, \end{cases}$$

$a > 1$, $b > 1$, and $c = \frac{1}{1+a}$; see Fig. B.4. Set $I_1 = [0, c)$ and $I_2 = [c, 1]$. Then f is C^∞ on I_1 and I_2 (lateral derivatives at the endpoints),

$$\left| f'(x) \right| = \begin{cases} a & \text{if } x \in I_1, \\ b & \text{if } x \in I_2, \end{cases}$$

and

$$f(\mathring{I}_1) = \mathring{I}_2, \quad f(\mathring{I}_2) \supset \mathring{I}_1 \cup \mathring{I}_2.$$

It follows that f is a smooth piecewise expanding Markov map with transition matrix

$$A = \begin{pmatrix} 0 & 1 \\ 1 & 1 \end{pmatrix},$$

see (B.2). Finally, from (B.28) we get

$$h_{top}(f) = h_{top}(\Sigma_A) = \log \frac{1+\sqrt{5}}{2}.$$

B.3.2 Topological Entropy of One-Dimensional Maps

Topological entropy, as metric entropy, is in general difficult to calculate and even to estimate. An exception worth mentioning because of its importance in applications is the case of one-dimensional interval maps.

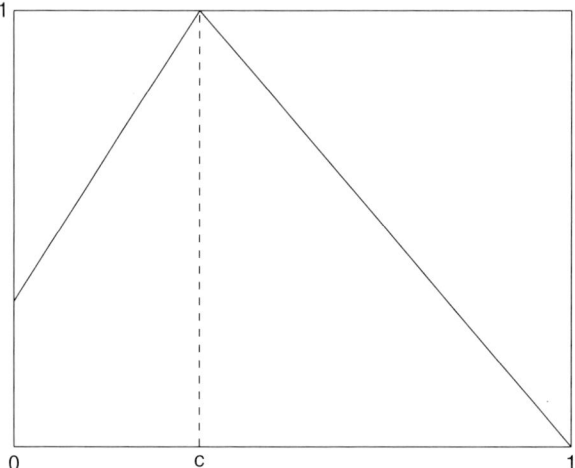

Fig. B.4 Rooftop map

Definition 27 Given an interval $I \subset \mathbb{R}$, a map $f{:}I \rightarrow I$ is said to be piecewise monotone if there is a finite partition of I into subintervals, such that f is continuous and monotone on each of those subintervals.

If $f{:}I \rightarrow I$ is piecewise monotone, there are different expressions for its topological entropy $h_{\text{top}}(f)$ that allow calculating it analytically in many cases. For instance [4, 155],

$$h_{\text{top}}(f) = \lim_{n \to \infty} \frac{1}{n} \log \text{lap}(f^n) \tag{B.29}$$

and

$$h_{\text{top}}(f) = \lim_{n \to \infty} \frac{1}{n} \log \left| \{x \in I{:}f^n(x) = x\} \right|, \tag{B.30}$$

where $\text{lap}(f^n)$ is the number of pieces of monotonicity of f^n (called laps of f^n) and $|\cdot|$ stands for the cardinality.

Other expressions of $h(f)$ are related to the notion of variation [4, 155]:

$$h_{\text{top}}(f) = \lim_{n \to \infty} \frac{1}{n} \log^+ \text{var}(f^n), \tag{B.31}$$

where, as usual, $\log^+ x = \max\{0, \log x\}$. Let us recall that the variation of a function $\varphi{:}I \rightarrow \mathbb{R}$ is given as

$$\text{var}(\varphi) = \sup \left\{ \sum_{i=1}^{s} |\varphi(x_i) - \varphi(x_{i-1})| \right\},$$

where the supremum is taken over all finite sequences $x_0 < x_1 < \cdots < x_s$ of elements of I. If φ is piecewise monotone, then (i) $\text{var}(\varphi) < \infty$, (ii) φ has finite derivative φ' almost everywhere on I, and (iii) φ' is integrable on I [95]. In this case,

$$\text{var}(\varphi) = \int_I \left|\varphi'(x)\right| dx. \tag{B.32}$$

Note that, for a piecewise monotone map φ, $\text{var}(\varphi)$ is closely related to the length of the graph of φ,

$$\text{len}(\varphi) = \int_I \sqrt{1 + |\varphi'(x)|^2} dx.$$

Indeed, since

$$\left|\varphi'(x)\right| < \sqrt{1 + |\varphi'(x)|^2} \leq \left|\varphi'(x)\right| + 1 \tag{B.33}$$

for all $x \in I$, we have

$$\text{var}(\varphi) < \text{len}(\varphi) \leq \text{var}(\varphi) + \text{len}(I), \tag{B.34}$$

upon integration of (B.33) over the interval I ($\text{len}(I)$ denotes the length of I). It follows

$$\lim_{n \to \infty} \frac{1}{n} \log^+ \text{len}(f^n) = \lim_{n \to \infty} \frac{1}{n} \log^+ \text{var}(f^n) = h(f), \tag{B.35}$$

since $\lim_{n \to \infty} \frac{1}{n} \log^+ \text{len}(I) = 0$.

Corollary 10 *If f is a continuous, piecewise monotone interval map of constant slopes $\pm s$, then*

$$h_{top}(f) = \log^+ s.$$

This result is very interesting for the following reason. If f is a continuous, piecewise monotone interval map and $h_{top}(f) = \log \beta > 0$, then f is semiconjugate to some continuous, piecewise monotone interval map of constant slopes $\pm\beta$ (via a non-decreasing map) [4]. If, moreover, f is topologically transitive, then "semiconjugate" can be replaced by "conjugate" in the previous statement (and the condition $h_{top}(f) > 0$ can be dropped because it is automatically satisfied).

Finally, let us mention that there are efficient algorithms for the numerical estimation of the topological entropy of piecewise monotone interval maps; see, for example, [27] for an algorithm that converges rapidly and provides both upper and lower bounds.

References

1. H.D.I. Abarbanel, *Analysis of Observed Chaotic Data*. Springer, New York, 1996.
2. M. Abraimowitz and I.A. Stegun (Eds.), *Handbook of Mathematical Functions*. Dover, New York, 1972.
3. R.L. Adler, A.G. Koheim, and M.H. McAndrews, Topological entropy, *Transactions of the American Mathematical Society* **114** (1965) 309–319.
4. L. Alsedà, J. Llibre, and M. Misiurewicz, *Combinatorial Dynamics and Entropy in Dimension One*. World Scientific, Singapore, 2000.
5. G. Alvarez, M. Romera, G. Pastor, and F. Montoya, Gray codes and 1D quadratic maps, *Electronic Letters* **34** (1998) 1304–1306.
6. J.M. Amigó, J. Szczepanski, E. Wajnryb, and M.V. Sanchez-Vives, Estimating the entropy of spike trains via Lempel-Ziv complexity, *Neural Computation* **16** (2004) 717–736.
7. J.M. Amigó, M.B. Kennel, and L. Kocarev, The permutation entropy rate equals the metric entropy rate for ergodic information sources and ergodic dynamical systems, *Physica D. Nonlinear Phenomena* **210** (2005) 77–95.
8. J.M. Amigó, L. Kocarev, and J. Szczepanski, Order patterns and chaos, *Physics Letters A* **355** (2006) 27–31.
9. J.M. Amigó and M.B. Kennel, Variance estimators for the Lempel-Ziv entropy rate estimators, *Chaos* **16** (2006) 043102.
10. J.M. Amigó, L. Kocarev, and I. Tomovski, Discrete entropy, *Physica D. Nonlinear Phenomena* **228** (2007) 77–85.
11. J.M. Amigó and M.B. Kennel, Topological permutation entropy, *Physica D. Nonlinear Phenomena* **231** (2007) 137–142.
12. J.M. Amigó, S. Zambrano, and M.A.F. Sanjuán, True and false forbidden patterns in deterministic and random dynamics, *Europhysics Letters* **79** (2007) 50001.
13. J.M. Amigó, L. Kocarev, and J. Szczepanski, Discrete Lyapunov exponent and resistance to differential cryptanalysis, *IEEE Transactions on Circuits and Systems II* **54** (2007) 882–886.
14. J.M. Amigó, S. Elizalde, and M.B. Kennel, Forbidden patterns and shift systems, *Journal of Combinatorial Theory, Series A* **115** (2008) 485–504.
15. J.M. Amigó, S. Zambrano, and M.A.F. Sanjuán, Combinatorial detection of determinism in noisy time series, *Europhysics Letters* **83** (2008) 60005.
16. J.M. Amigó and M.B. Kennel, Forbidden ordinal patterns in higher dimensional dynamics, *Physica D. Nonlinear Phenomena* **237** (2008) 2893–2899.
17. J.M. Amigó, The ordinal structure of the signed shift transformations, *International Journal of Bifurcation and Chaos* **19** (2009) 3311–3327.
18. C. Anteneodo, A.M. Batista, and R.L. Viana, Synchronization threshold in coupled logistic map lattices, *Physica D. Nonlinear Phenomena* **223** (2006) 270–275.
19. N. Aoki and K. Hiraide, *Topological theory of dynamical systems*. North Holland, Amsterdam, 1994.

20. D.K. Arrowsmith and C.M. Place, *Dynamical Systems*. Chapman and Hall, Boca Raton, 1996.
21. D. Arroyo, G. Alvarez, and J.M. Amigó, Estimation of the control parameter from symbolic sequences: Unimodal maps with variable critical point, *Chaos* **19** (2009) 023125.
22. R.B. Ash, *Information Theory*. Dover Publications, New York, 1990.
23. H. Atmanspacher and H. Scheingraber, Inherent global stabilization of unstable local behavior in coupled map lattices, *International Journal of Bifurcation and Chaos* **15** (2005) 1665–1676.
24. N. Ay and J.P. Crutchfield, Reductions of hidden information sources, *Journal of Statistical Physics* **120** (2005) 659–684.
25. E. Babson and E. Steingrímsson, Generalized permutation patterns and a classification of the Mahonian statistics, Séminaire Lotharingien de Combinatoire **44** (2000), Article B44b, 18.
26. A. Bäcker and N. Chernov, Generating partitions for two-dimensional hyperbolic maps, *Nonlinearity* **11** (1998) 79–87.
27. N.J. Balmforth, E.A. Spiegel, and C. Tresser, Topological entropy of one-dimensional maps: Approximations and bounds, *Physical Review Letters* **72** (1994) 80–83.
28. C. Bandt and B. Pompe, Permutation entropy: A natural complexity measure for time series, *Physical Review Letters* **88** (2002) 174102.
29. C. Bandt, G. Keller, and B. Pompe, Entropy of interval maps via permutations. *Nonlinearity* **15** (2002) 1595–1602.
30. C. Bandt and F. Shiha, Order patterns in time series, *Journal of Time Series Analysis* **28** (2007) 646–665.
31. A. Berger, *Chaos and Chance*, Walter de Gruyter, Berlin, 2001.
32. G.D. Birkhoff, Proof of a recurrence theorem for strongly transitive systems, *Proceedings of the National Academy of Science* **17** (1931) 650.
33. F. Blanchard, P. Kurka, and A. Maass, Topological and measure-theoretical properties of one-dimensional cellular automata, *Physica D. Nonlinear Phenomena* **103** (1997) 86–99.
34. S. Boccaletti and D.L. Valladares, Characterization of intermittent lag synchronization, *Physical Review E* **62** (2000) 7497–7500.
35. L. Boltzmann, Über the mechanischen Analogien des zweiten Hauptsatzes der Thermodynamik, *Journal für reine und angewandte Mathematik* **100** (1887) 201.
36. R.E. Bowen, Entropy for group endomorphisms and homogeneous spaces, *Transactions of the American Mathematical Society* **153** (1971) 401–414.
37. A. Boyarsky and P. Gora, *Laws of Chaos*. Birkhäuser, Boston, 1997.
38. W.A. Brock, W.D. Dechert, J.A. Scheinkman, and B. LeBaron, A test for independence based on the correlation dimension, *Econometrics Reviews* **15** (1996) 197–235.
39. A.A. Brudno, Entropy and complexity of trajectories of a dynamical system. *Transactions of the Moscow Mathematical Society* **44** (1983) 127–151.
40. M. Buhl and M.B. Kennel, Statistically relaxing to generating partitions for observed time-series data, *Physical Review E* **71** (2005) 046213: 1–14.
41. J. Bunge and M. Fitzpatrick, Estimating the Number of Species: A Review, *Journal of the American Statistical Association* **88** (1993) 364–373.
42. L.A. Bunimovich and Y.G. Sinai, Space-time chaos in coupled map lattices, *Nonlinearity* **1** (1988) 491–518.
43. L.A. Bunimovich and Y.G. Sinai, Statistical mechanics of coupled map lattices, In: K.Kaneko (Ed.), *Theory and Applications of Coupled Map Lattices*. Wiley, New York, 1993.
44. L.A. Bunimovich, Coupled map lattices: Some topological and ergodic properties, *Physica D. Nonlinear Phenomena* **103** (1997) 1–17.
45. Y. Cao, W. Tung, J.B. Gao, V.A. Protopopescu, and L.M. Hively, Detecting dynamical changes in time series using the permutation entropy, *Physical Review E* **70** (2004) 046217.
46. R. Carretero-González, Low dimensional travelling interfaces in coupled map lattices, *International Journal of Bifurcations and Chaos* **7** (1997) 2745–2754.

47. R. Carretero-González, D.K. Arrowsmith, and F. Vivaldi, One-dimensional dynamics for traveling fronts in coupled map lattices, *Physical Review E* **61** (2000) 1329–1336.

48. K. Cattell and J.C. Muzio, Synthesis of one-dimensional linear hybrid cellular automata, *IEEE Transactions on Computer-Aided Design of Integrated circuits and Systems* **15** (1996) 325–335.

49. A. Chao, Nonparametric estimation of the number of classes in a population, *Scandinavian Journal of Statistics, Theory and Applications* **9** (1984) 265–270.

50. H. Chaté and P. Manneville, Coupled map lattices as cellular automata, *Journal of Statistical Physics* **56** (1989) 357–370.

51. B.V. Chirikov and F. Vivaldi, An algorithmic view of pseudochaos, *Physica D. Nonlinear Phenomena* **129** (1999) 223–235.

52. G.H. Choe, *Computational Ergodic Theory*. Springer Verlag, Berlin, 2005.

53. F. Christiansen and A. Politi, Generating partition for the standard map, *Physical Review E* **51** (1995) R3811.

54. L.O. Chua, V.I. Sbitnev, and S. Yoon, A nonlinear dynamics perspective of Wolfram's New Kind of Science –Part II: Universal neuron, *International Journal of Bifurcation and Chaos* **13** (2003) 2377–2491.

55. L.O. Chua, V.I. Sbitnev, and S. Yoon, A nonlinear dynamics perspective of Wolfram's new kind of science –Part IV: From Bernoulli shift to $1/f$ spectrum, *International Journal of Bifurcation and Chaos* **15** (2005) 1045–1183.

56. R.W. Clarke, M.P. Freeman, and N.W. Watkins, Application of computational mechanics to the analysis of natural data: An example in geomagnetism, *Physical Review E* **67** (2003) 016203.

57. P. Collet and J.P. Eckmann, *Iterated Maps on the Interval as Dynamical Systems*, 5th printing. Birkhäuser, Boston, 1997.

58. M. Courbage, D. Mercier, and S. Yasmineh, Traveling waves and chaotic properties in cellular automata, *Chaos* **9** (1999) 893–901.

59. T.M. Cover and J.A. Thomas, *Elements of Information Theory*, 2nd edition. New York, John Wiley & Sons, 2006.

60. J.P. Crutchfield and K. Young, Inferring statistical complexity, *Physical Review Letters* **63** (1989) 105–108.

61. R. Dahlhaus, J. Kurths, P. Maass, and J. Timmer, *Mathematical Methods in Time Series Analysis and Digital Image Processing*. Springer Verlag, Berlin, 2008.

62. M. D'amico, G. Manzini, and L. Margara, On computing the entropy of cellular automata, *Theoretical Computer Science* **290** (2003) 1629–1646.

63. R. Davidchack, Y.C. Lai, E.M. Bollt, and M. Dhamala, Estimating generating partitions by unstable periodic orbits, *Physical Review E* **61** (2000) 1353–1356.

64. K. Denbigh, How subjective is entropy. In: H.S. Leff and A.F. Rex (Ed.), *Maxwell's Demon, Entropy, Information, Computing*, pp. 109–115. Princeton University Press, Princeton, 1990.

65. M. Denker, Finite generators for ergodic, measure-preserving transformations, *Zeitschrift für Wahrscheinlichkeitstheorie und verwandte Gebiete* **29** (1974) 45–55.

66. M. Denker, C. Grillenberger, and K. Sigmund, *Ergodic Theory on Compact Spaces*. Springer Lecture Notes in Math. **527**, Springer Verlag, Berlin, 1976.

67. M. Denker and W.A. Woyczynski, *Introductory Statistics and Random Phenomena*. Birkhäuser, Boston, 1998.

68. M. Denker, *Einführung in die Analysis Dynamischer Systeme*. Springer Verlag, Berlin, 2005.

69. R.L. Devaney, *Chaotic Dynamical Systems* (2nd edition). Westview Press, Boulder, 2003.

70. E.I. Dinaburg, The relation between topological entropy and metric entropy, *Soviet Mathematics* **11** (1970) 13–16.

71. Y. Dobyns and H. Atmanspacher, Characterizing spontaneous irregular behavior in coupled map lattices, *Chaos, Solitons & Fractals* **24** (2005) 313–327.

72. J.P. Eckmann and D. Ruelle, Ergodic theory of chaos and strange attractors, *Review of Modern Physics* **57** (1985) 617–656.

73. J.P. Eckmann, S.O. Kamphorst, and D. Ruelle, Recurrence plots of dynamical systems, *Europhysics Letters* **4** (1987) 973–977.

74. S. Elizalde and M. Noy, Consecutive patterns in permutations, *Advances in Applied Mathematics* **30** (2003) 110–125.

75. S. Elizalde, Asymptotic enumeration of permutations avoiding generalized patterns, *Advances in Applied Mathematics* **36** (2006) 138–155.

76. S. Elizalde, The number of permutations realized by a shift, *SIAM Journal of Discrete Mathematics* **23** (2009) 765–786.

77. R. Érdi, *Complexity Explained*. Springer Verlag, Berlin, 2007.

78. A. Fernández, J. Quintero, R. Hornero, P. Zuluaga, M. Navas, C. Gómez, J. Escudero, N. García-Campos, J. Biederman, and T. Ortiz, Complexity analysis of spontaneous brain activity in attention-deficit/hyperactivity disorder: Diagnosis implications, *Biological Psychiatry* **65** (2009) 571–577.

79. J. Ford, G. Mantica, and G.H. Ristow, The Arnold's cat: Failure of the correspondence principle, *Physica D. Nonlinear Phenomena* **50** (1991) 493–520.

80. A.M. Fraser and H.L. Swinney, Independent coordinates for strange attractors from mutual information, *Physical Review A* **33** (1986) 1134–1140.

81. J.B. Gao and H.Q. Cai, On the structures and quantification of recurrence plots, *Physics Letters A* **270** (2000) 75–87.

82. Y. Gao, I. Kontoyiannis, and E. Bienenstock, Estimating the entropy of binary time series: Methodology, some theory and a simulation study, *Entropy* **10** (2008) 71–99.

83. J. García-Ojalvo, J.M. Sancho, and L. Ramírez-Piscina, Generation of spatiotemporal colored noise, *Physical Review A* **46** (1992) 4670–4675.

84. M. Gardner, The fantastic combinations of John Conway's new solitaire game "life", *Scientific American* **223** (1970) 120–123.

85. A. Golestani, M.R. Jahed Motlagh, K. Ahmadian, A.H. Omidvarnia, and N. Mozayani, A new criterion to distinguish stochastic and deterministic time series with the Poincaré section and fractal dimension, *Chaos* **19** (2009) 013137.

86. S.W. Golomb, *Bulletin of the American Mathematical Society* **70** (1964) 747 (research problem 11).

87. P. Grassberger and H. Kantz, Generating partitions for the dissipative Hénon map, *Physics Letters A* **113** (1985) 235–238.

88. P. Grassberger, Finite sample corrections to entropy and dimension estimates, *Physics Letters A* **128** (1988) 369–373.

89. R.M. Gray, *Entropy and Information Theory*. Springer Verlag, New York, 1990.

90. F. Gu, X. Meng, E. Shen, and Z. Cai, Can we measure consciousness with EEG complexities?, *International Journal of Bifurcations and Chaos* **13** (2003) 733–742.

91. B. Hasselblatt and A. Katok, *A First Course in Dynamics*. Cambridge University Press, Cambridge, 2003.

92. G.A. Hedlund, Endomorphisms and automorphisms of the shift dynamical system, *Mathematical Systems Theory* **3** (1969) 320–375.

93. H. Herzel, Complexity of symbol sequences, *Systems, Analysis, Modelling, Simulations* **5** (1988) 435–444.

94. H. Herzel, A.O. Schmitt, and W. Ebeling, Finite sample effects in sequence analysis, *Chaos, Solitons & Fractals* **4** (1994) 97–113.

95. E. Hewitt and K. Stromberg, *Real and Abstract Analysis*. Springer Verlag, New York 1965.

96. F.C. Hoppensteadt, *Analysis and Simulation of Chaotic Systems* (2nd edition). Springer Verlag, New York, 2000.

97. K. Hiraide, Nonexistence of positively expansive maps on compact connected manifolds with boundary, *Proceedings of the American Mathematical Society* **110** (1990) 565–568.

98. M.W. Hirsch, S. Smale, and R.L. Devaney, *Differential Equations, Dynamical Systems, and an Introduction to Chaos*. Academic Press, San Diego, 2003.

99. C.S. Hsu and M.C. Kim, Construction of maps with generating partitions for entropy evaluation, *Physical Review A* **31** (1985) 3253–3265.
100. J. Hughes, J. Hellman, T.H. Rickets, and B.J.M. Bohannan, Counting the uncountable: Statistical approaches to estimating microbial diversity, *Applied and Environ. Microbiology* **67** (2001) 4399–4406.
101. L.P. Hurd, J. Kari, and K. Culik, The topological entropy of cellular automata is uncomputable, *Ergodic Theory and Dynamical Systems* **12** (1992) 255–265.
102. Y. Ishii and D. Sands, Monotonicity of the Lozi family near the tent-maps, *Communications in Mathematical Physics* **198** (1998) 397–406.
103. N. Israeli and N. Goldenfeld, Coarse-graining of cellular automata, emergence, and the predictability of complex systems, *Physical Review E* **73** (2006) 1–17.
104. S. Jalan, J. Jost, and F.M. Atay, Symbolic synchronization and the detection of global properties of coupled dynamics from local information, *Chaos* **16** (2006) 033124.
105. O. Jenkinson and M. Pollicott, Entropy, exponents and invariant densities for hyperbolic systems: Dependence and computation. In: M. Brin, B. Hasselblatt, and Y. Pesin (Eds.), *Modern Dynamical Systems and Applications*. pp. 365–384 Cambridge University Press, Cambridge, 2004.
106. K. Kaneko, Transition from torus to chaos accompanied by frequency lockings with symmetry breaking, *Progress in Theoretical Physics* **69** (1983) 1427–1442.
107. K. Kaneko, Period-doubling of kink-antikink patterns, quasiperiodicity in anti-ferro-like structures and spatial intermittency in coupled logistic lattice, *Progress in Theoretical Physics* **72** (1984) 480–486.
108. K. Kaneko, Pattern dynamics in spatiotemporal chaos, *Physica D. Nonlinear Phenomena* **34** (1989) 1–41.
109. K. Kaneko, Spatiotemporal chaos in one- and two-dimensional coupled map lattices, *Physica D. Nonlinear Phenomena* **37** (189) 60–82.
110. K. Kaneko, Chaotic traveling waves in a coupled map lattice, *Physica D. Nonlinear Phenomena* **68** (1993) 299–317.
111. H. Kantz, Quantifying the closeness of fractal measures, *Physical Review E* **49** (1994) 5091–5097.
112. H. Kantz and T. Schreiber, *Nonlinear Time Series Analysis*. Cambridge University Press, Cambridge, 1997.
113. N.J. Kasdin, Discrete simulation of colored noise and stochastic processes and $1/f^\alpha$ power law noise generation, *Proceedings of the IEEE* **83** (1995) 802–827.
114. A. Katok and B. Hasselbaltt, *Introduction to the Theory of Dynamical Systems*. Cambridge University Press, Cambridge, 1998.
115. S. Katok, *p-adic Analysis compared with real*. American Mathematical Society, Providence, 2007.
116. K. Keller and K. Wittfeld, Distances of time series components by means of symbolic dynamics, *International Journal of Bifurcation and Chaos* **14** (2004) 693–703.
117. K. Keller and M. Sinn, Ordinal analysis of time series, *Physica A* **356** (2005) 114–120.
118. K. Keller, H. Lauffer, and M. Sinn, Ordinal analysis of EEG time series, *Chaos and Complexity Letters* **2** (2007) 247–258.
119. M.B. Kennel and S. Isabelle, Method to distinguish possible chaos from colored noise and to determine embedding parameters, *Physical Review A* **46** (1992) 3111–3118.
120. M.B. Kennel, Statistical test for dynamical nonstationarity in observed time-series data, *Physical Review E* **56** (1997) 316–321.
121. M.B. Kennel and A.I. Mees, Context-tree modeling of observed symbolic dynamics, *Physical Review E* **66** (2002) 056209.
122. M.B. Kennel, J. Shlens, H.D.I. Abarbanel, and E.J. Chichilnisky, Estimating entropy rates with Bayesian confidence intervals, *Neural Computation* **17** (2005) 1531–1576.
123. B.P. Kitchens, *Symbolic Dynamics*. Springer Verlag, Berlin, 1998.

124. L. Kocarev and J. Szczepanski, Finite-space Lyapunov exponents and pseudo-chaos, *Physical Review Letters* **93** (2004) 234101.

125. L. Kocarev, J. Szczepanski, J.M. Amigó, and I. Tomovski, Discrete Chaos – Part I: Theory, *IEEE Transactions on Circuits and Systems I* **53** (2006) 1300–1309.

126. A.N. Kolmogorov, Entropy per unit time as a metric invariant of automorphism, *Doklady of Russian Academy of Sciences* **124** (1959) 754–755.

127. I. Kontoyiannis, P.H. Algoet, Y.M. Suhov, and A.J. Wyner, Nonparametric entropy estimation for stationary processes and random fields, with applications to English text. *IEEE Transactions on Information Theory* **44** (1998) 1319–1327.

128. Z.S. Kowalski, Finite generators of ergodic endomorphisms, *Colloquium Mathematicum* **49** (1984) 87–89.

129. Z.S. Kowalski, Minimal generators for aperiodic endomorphisms, *Commentationes Mathematicae Universitatis Carolinae* **36** (1995) 721–725.

130. W. Krieger, On entropy and generators of measure-preserving transformations, *Transactions of the American Mathematical Society* **149** (1970) 453–464.

131. A.P. Kurian and S. Puttusserypady, Self-synchronizing chaotic stream ciphers, *Signal Processing* **88** (2008) 2442–2452.

132. J. Kurths, D. Maraun, C.S. Zhou, G. Zamora-López, and Y. Zou, Dynamics in complex systems, *European Review* **17** (2009), 357–370.

133. J.C. Lagarias, Pseudorandom numbers, *Statistical Science* **8** (1993) 31–39.

134. A. Lasota and J.A. Yorke, On the existence of invariant measures for piecewise monotonic transformations, *Transactions of the American Mathematical Society* **186** (1973), 481–488.

135. A.M. Law and W.D. Kelton, *Simulation, Modeling, and Analysis*, 3rd edition. McGraw-Hill, Boston, 2000.

136. B. LeBaron, A fast algorithm for the BDS statistics, *Studies in Nonlinear Dynamics & Econometrics* **2** (1997) 53–59.

137. A. Lempel and J. Ziv, On the complexity of an individual sequence, *IEEE Transactions on Information Theory* **IT-22** (1976) 75–78.

138. M. Li and P. Vitányi, *An Introduction to Kolmogorov Complexity and Its Applications*. Springer Verlag, New York, 1997.

139. D. Lind and B. Marcus, *Symbolic Dynamics and Coding*. Cambridge University Press, Cambridge, 2003.

140. T. Liu, C.W.J. Granger, and W.P. Heller, Using the correlation exponent to decide whether an economic series is chaotic. *Journal of Applied Econometrics,* Supplement: Special Issue on Nonlinear Dynamics and Econometrics (Dec., 1992) S25–S39.

141. G. Manzini and L. Margara, A complete and efficiently computable topological classification of linear cellular automata over $\mathbb{Z}_m$, *Theoretical Computer Science* **221** (1999) 157–177.

142. R. Mañé, *Ergodic Theory and Differentiable Dynamics*. Springer Verlag, Berlin, 1987.

143. M.T. Martin, A. Plastino, and O.A. Rosso, Generalized statistical complexity measures: Geometrical and analytical properties, *Physica A* **369** (2006) 439–462.

144. N. Marwan, M.C. Romano, M. Thiel, and J. Kurths, Recurrence plots for the analysis of complex systems, *Physics Reports* **438** (2007) 237–329.

145. C. Masoller and A.C. Martí, Random delays and the synchronization of chaotic maps, *Physical Review Letters* **94** (2005) 134102.

146. M. Matilla-García, A non-parametric test for independence based on symbolic dynamics, *Journal of Economic Dynamic & Control* **31** (2007) 3889–3903.

147. M. Matilla-García and M. Ruiz Marín, A non-parametric independence test using permutation entropy, *Journal of Econometrics* **144** (2008) 139–155.

148. M. Matsumoto and T. Nishimura, Mersenne Twister: A 623-dimensionally equidistributed uniform pseudo-random number generator, *ACM Trans. on Modeling and Computer Simulation* **8** (1998) 3–30.

149. W. Meier and O. Staffelbach, The self-shrinking generator. In: Proc. of Eurocrypt'94, Lecture Notes in Computer Science. vol. 950, pp. 205–214. Springer Verlag, Berlin, 1994.

150. W. de Melo and S. van Strien, *One-Dimensional Dynamics*. Springer Verlag, Berlin, 1993.
151. A.J. Menezes, P.C. van Oorschoot, and S.A. Vanstone, *Handbook of Applied Cryptography*. CRC Press, Boca Raton, 1997.
152. M.E. Mera and M. Morán, Geometric noise reduction for multivariate time series, *Chaos* **16** (2006) 013116.
153. N. Metropolis, M. Stein, and P. Stein, On finite limit sets for transformations on the unit interval, *Journal of Combinatorial Theory, Series A* **15**, 25–44 (1973).
154. J. Milnor, Non-expansive Hénon Maps, *Advances in Mathematics* **69** (1988) 109–114.
155. M. Misiurewicz and W. Szlenk, Entropy of piecewise monotone mappings, *Studia Mathematica* **67** (1980) 45–63.
156. M. Misiurewicz, Strange attractors for the Lozi mappings. In: R.G. Helleman (Ed.), *Nonlinear Dynamics*, Vol. **357**, pp. 348–358 The New York Academy of Science, New York 1980.
157. M. Misiurewicz, Permutations and topological entropy for interval maps, *Nonlinearity* **16** (2003) 971–976.
158. M. Mitchell, *Complexity —A Guided Tour*. Oxford University Press, New York, 2009.
159. R. Monetti, W. Bunk, T. Aschenbrenner, and F. Jamitzky, Characterizing synchronization in time series using information measures extracted from symbolic representations, *Physical Review E* **79** (2009) 046207.
160. M. Morse and G.A. Hedlund, Symbolic Dynamics, *American Journal of Mathematics* **60** (1938) 815–866.
161. J. von Neumann, The general and logical theory of automata. In: L.A. Jeffress (Ed.), *Cerebral Mechanisms in Behavior*. Wiley, New York, 1951.
162. M. Newman, A.L. Barabási, and D.J. Watts, *The Structure and Dynamics of Networks*. Princeton University Press, Princeton, 2006.
163. E. Olbrich, N. Bertschinger, N. Ay, and J. Jost, How should complexity scale with system size?, *The European Physical Journal B* **63** (2008) 407–415.
164. G.J. Ortega and E. Louis, Smoothness implies determinism in time series: A measure based approach, *Physical Review Letters* **81** (1998) 4345–4348.
165. E. Ott, *Chaos in Dynamical Systems*. Cambridge University Press, Cambridge, 2002.
166. N.H. Packard, J.P. Crutchfield, J.D. Farmer, and R.S. Shaw, Geometry from a time series, *Physical Review Letters* **45** (1980) 712–716.
167. L. Paninski, Estimation of entropy and mutual information, *Neural Computation* **15** (2003) 1191–1253.
168. H.O. Peitgen, H. Jürgens, and D. Saupe, *Chaos and Fractals*. Springer Verlag, New York, 2004.
169. K. Petersen, *Ergodic Theory*. Cambridge University Press, Cambridge, 1983.
170. S.D. Pethel, N.J. Corron, and E. Bollt, Symbolic dynamics of coupled map lattices, *Physical Review Letters* **96** (2006) 034105.
171. S.D. Pethel, N.J. Corron, and E. Bollt, Deconstructing spatiotemporal chaos using local symbolic dynamics, *Physical Review Letters* **99** (2007) 214101.
172. J. Pieprzyk, T. Hardjorno, and J. Seberry, *Fundamentals of Computer Security*. Springer Verlag, Berlin, 2003.
173. W.H. Press, S.A. Teukolsky, W.T. Vetterling, and B.P. Flannery, *Numerical Recipes: The Art of Scientific Computing*. Cambridge University Press, Cambridge, 2007.
174. R.C. Robinson, *An Introduction to Dynamical Systems*. Pearson Prentice Hall, Upper Saddle River NJ, 2004.
175. M.G. Rosenblum, A.S. Pikovsky, and J. Kurths, Phase synchronization of chaotic oscillators, *Physical Review Letters* **76** (1997) 1804–1807.
176. O.A. Rosso, H.A. Larrondo, M.T. Martin, A. Platino, and M.A. Fuentes, Distinguishing noise from chaos, *Physical Review Letters* **99** (2007) 154102.
177. D.J. Rudolph, *Fundamentals of Measurable Dynamics*. Oxford University Press, Oxford, 1990.
178. http://topo.math.u-psud.fr/ sands/Programs/Lozi/index.html.

179. A.N. Sarkovskii, Coexistence of cycles of a continuous map of a line into itself, *Ukrainian Mathematical Journal* **16** (1964) 61–71.

180. P.R. Scalassara, M.E. Dajer, C. Dias Maciel, C. Capobianco Guido, and J.C. Pereira, Relative entropy measures applied to healthy and pathological voice characterization, *Applied Mathematics and Computation* **207** (2009) 95–108.

181. A.O. Schmitt, H. Herzel, and W. Ebeling, A new method to calculate higher-order entropies from finite samples, *Europhysics Letters* **23** (1993) 303–309.

182. O. Schmitt, *Remarks on the Generator-Problem* (Thesis). University of Göttingen, 2001.

183. T. Schreiber, Detecting and analyzing nonstationarity in a time series using nonlinear cross predictions, *Physical Review Letters* **78** (1997) 843–846.

184. R. Sexl and J. Blackmore (Eds.), *Ludwig Boltzmann - Ausgewahlte Abhandlungen* (Ludwig Boltzmann Gesamtausgabe, Band 8). Vieweg, Braunschweig, 1982.

185. C.R. Shalizi and J.P. Crutchfield, Computational mechanics: Pattern and prediction, structure and simplicity, *Journal of Statistical Physics* **104** (2001) 817–879.

186. C.E. Shannon, A mathematical theory of communication, *Bell System Technical Journal* **27** (1948) 379–423, 623–653.

187. L.A. Shepp and S.P. Lloyd, Ordered cycle length in a random permutation, *Transactions of the American Mathematical Society* **121** (1966) 340–357.

188. M.A. Shereshevsky, Expansiveness, entropy and polynomial growth for groups acting on subshifts by automorphisms. *Indagationes Mathematicae* **4** (1993) 203–210.

189. Y.G. Sinai, On the Notion of Entropy of a Dynamical System, *Doklady of Russian Academy of Sciences* **124** (1959) 768–771.

190. M. Sinn and K. Keller, Estimation of ordinal pattern probabilities in fractional Brownian motion, arXiv:0801.1598.

191. M. Smorodinsky, *Ergodic Theory, Entropy* (Lectures Notes in Mathematics) Vol. **214**. Springer Verlag, Berlin, 1971.

192. D. Sotelo Herrera and J. San Martín, Analytical solutions of weakly coupled map lattices using recurrence relations, *Physics Letters A* **373** (2009) 2704–2709.

193. J.C. Sprott, *Chaos and Time-Series Analysis*. Oxford University Press, Oxford, 2003.

194. J.C. Sprott, High-dimensional dynamics in the delayed Hénon map. *Electronic Journal of Theoretical Physics* **3** (2006) 19–35.

195. S.P. Strong, R. Koberle, R.R. de Ruyter van Steveninck, and W. Bialek, Entropy and information in neural spike trains. *Physical Review Letters* **80** (1998) 197–200.

196. J. Szczepanski, J.M. Amigó, E. Wajnryb, and M.V. Sanchez-Vives. Application of Lempel-Ziv complexity to the analysis of neural discharges, *Network: Computation in Neural Systems* **14** (2003) 335–350.

197. F. Takens, Detecting strange attractors in turbulence, In: D. Rand and L.S. Young (Eds.), *Dynamical Systems and Turbulence*, Lecture Notes in Mathematics, vol. 898. Springer, Berlin, 1981, pp. 366–381.

198. T. Toffoli and N. Margolus, *Cellular Automata Machines*. The MIT Press, Cambridge MA, 1987.

199. S. Ulam, Random process and transformations, Proceedings of the International Congress of Mathematicians 2 (1952), 264–275.

200. D.B. Vasconcelos, S.R. Lopes, R.L. Viana, and J. Kurths, Spatial recurrence plots, *Physical Review E* **73** (2006) 056207.

201. S.B. Volchan, What is a Random Sequence, *The American Mathematical Monthly* **109** (2002) 46–63.

202. P. Walters, *An Introduction to Ergodic Theory*. Springer Verlag, New York, 2000.

203. L. Wang and N.D. Kazarinoff, On the universal sequence generated by a class of unimodal functions, *Journal of Combinatorial Theory, Series A* **46** (1987) 39–49.

204. A. Wolf, J.B. Swift, H.L. Swinney, and J.A. Vastano, Determining Lyapunov exponents from a time series, *Physica D. Nonlinear Phenomena* **16** (1985) 285–317.

205. S. Wolfram, Computation theory of cellular automata, *Communications in Mathematical Physics* **96** (1984) 15–57.
206. S. Wolfram, Universality and complexity in cellular automata, *Physica* **10D** (1984) 1–35.
207. S. Wolfram, *A New Kind of Science*. Wolfram Media, Champaign, 2002.
208. X-S. Zhang, R.J. Roy, and E.W. Jensen, EEG complexity as a measure of depth anesthesia for patients, *IEEE Transactions on Biomedical Engineering* **48** (2001) 1424–1433.
209. J. Zhang and M. Small, Complex networks from pseudoperiodic time series: Topology versus dynamics. *Physical Review Letters* **96** (2006) 238701.
210. G.C. Zhuang, J. Wang, Y. Shi, and W. Wang, Phase synchronization and its cluster feature in two-dimensional coupled map lattices, *Physical Review E* **66** (2002) 046201.
211. J. Ziv and A. Lempel, Compression of individual sequences via variable-rate coding *IEEE Transactions on Information Theory* **IT-24** (1978) 530–536.
212. L. Zunino, D.G. Pérez, M.T. Martín, M. Garavaglia, A. Plastino, and O.A. Rosso, Permutation entropy of fractional Brownian motion and fractional Gaussian noise, *Physics Letters A* **372** (2008) 4768–4774.
213. L. Zunino, D.G. Pérez, M.T. Martín, M. Garavaglia, A. Plastino, and O.A. Rosso, Fractional Brownian motion, fractional Gaussian noise, and Tsallis permutation entropy, *Physica A* **387** (2008) 6057–6068.

Index